Space Microelectronics

Volume 1

Modern Spacecraft Classification, Failure, and Electrical Component Requirements

Artech House
Space Technology and Applications Series

Bruce R. Elbert, Series Editor

For further information on these and other Artech House titles, including out-of-print books available
through our In-Print-Forever® (IPF®) program, contact:

Artech House	Artech House
685 Canton Street	16 Sussex Street
Norwood, MA 02062	London SW1V 4RW U.K.
Phone: 781-769-9750	Phone: +44 (0)171-973-8077
Fax: 781-769-6334	Fax: +44 (0)171-630-0166
e-mail: artech@artechhouse.com	e-mail: artech-uk@artechhouse.com

Find us on the World Wide Web at: www.artechhouse.com

Space Microelectronics

Volume 1

Modern Spacecraft Classification, Failure, and Electrical Component Requirements

Anatoly Belous
Vitali Saladukha
Siarhei Shvedau

ARTECH HOUSE

BOSTON | LONDON
artechhouse.com

Library of Congress Cataloging-in-Publication Data
A catalog record for this book is available from the U.S. Library of Congress

British Library Cataloguing in Publication Data
A catalog record for this book is available from the British Library.

ISBN-13: 978-1-63081-257-7

Cover design by John Gomes

© 2017 Artech House
685 Canton Street
Norwood, MA

10 9 8 7 6 5 4 3 2 1

Contents

Preface

This book was published in Moscow and appeared in Russian bookstores at the beginning of 2015. The structure of the book, the contents of its chapters, and the form of material presented in the Russian edition is based on two areas of the authors' expertise.

On the one hand, the book reflects the practice of the authors' scientific research and the results of numerous lectures and seminars held by them over the years for specialists in space instrument engineering and students and professors of the leading technical universities of Russia and Belarus, institutes and research laboratories of China, India, Germany, and Israel, as well as for technical managers of laboratories, institutes, and enterprises of Roscosmos State Corporation (the Russian equivalent of NASA) that research, develop, and manufacture spacecraft and microcircuits for rocket and space industry.

On the other hand, the book integrates the authors' extensive practice in the semiconductor industry—their more than 40 years of experience in managing large teams of developers and manufacturers of microelectronic devices for space applications. The authors have made valuable personal contributions to the implementation of the majority of the Russian space programs, having shared both sorrow at failures and joy at victories with developers of rockets and satellites in their struggle for space exploration. Apart from the authors' original materials that haven't been published before, the book features over 1,200 references, including over 600 to Russian editions (with 150 publications of the authors) and over 500 to publications in English, as well as almost 100 to online sources.

Reader responses following the book's release proved that the right choice of approach had been taken and at least partially justified the authors in the eyes of their wives who blamed them for almost a three-year withdrawal from parenting and grandparenting duties while the authors spent all their "leisure" time working at the writing desk. First, in 2015 the book was honored as the best-selling technical book in the stores of Russia's largest cities: Moscow, Saint-Petersburg, and Novosibirsk. Second, its audience turned out to be unexpectedly wide in terms of readers' occupations: apart from specialists in the field of space electronics, for whom the book was primarily intended, and developers of military radioelectronics, it attracted the interest of specialists in high-performance computing devices and energy-efficient technologies, as well as developers of unmanned aircraft, navigation devices, automotive electronics, and many others. The leading Russian technical universities that are in some way or another connected with the study of radioelectronics and its applications provided their libraries with copies of the book and recommended professors to use it for study courses in a wide range of fields related to the design of microelectronic devices and spacecraft; heads of the leading enterprises of the rocket and space industry advised their managers and design engineers to study the book and make use of it in their daily practice.

When Artech House Publishers offered to release the English edition of the book, the authors thought they could rest on their laurels and bask in the spotlight. After the offer was gladly accepted, the hard work with the publisher's experts began and it was soon that the authors were brought back to Earth. It wasn't that the experts considered the book to be poor or the data unreliable. However, they did have a number of questions, critical remarks, and specific proposals on how to improve the quality of the material presented and application of a more systematic and coherent presentation from simple to complicated.

It must be admitted that the majority of proposals were accepted by the authors.

As a result, unlike the Russian two-volume edition, the book now consists of three volumes, each of them dwelling on separate major aspects of space microelectronics.

The first volume is focused on the study of modern spacecraft: their classification, design versions, causes and consequences of failures and accidents, and also the main requirements of the electronic components.

The second volume deals with the most challenging practical aspects of designing microcircuits for space applications.

The third volume provides a detailed examination of issues arising from the impact of streams of high-velocity microparticles (or *cosmic dust*, as specialists put it) on spacecraft, mechanisms of their influence on equipment and microcircuits, as well as methods of protection against them.

The preface to the Russian edition mainly describes the background and history of writing the book, rather than a summary of the encyclopedic material of the book. Therefore, the authors decided to prepare a more specific preface to the English edition.

Although the development of modern microcircuits and semiconductor devices is complicated itself, the designs of products and electronic control systems for space and military electronics (rockets and satellites) have to meet very special requirements due to the particular conditions of their operation and manufacturing. Accordingly, the developers have to be much more qualified and skilled than the developers of commercial electronics.

Roughly speaking, in the case of a latent error in the design of a microcircuit for household appliances (a TV set, phone, or microwave oven), in the worst case (microcircuit failure) a replacement of either the integrated circuit or the device itself will be needed. It is a very different matter if such a latent error manifests itself during operation of a spacecraft at any stage of its functioning, either during the launch, in orbit, or during a mission on some other planet where any repairs are out of the question.

Sometimes technical, environmental, and economic consequences of such failures; for instance, in strategic missile systems, are hard to imagine, but they are certain to be of a disastrous nature without mentioning the blow to the prestige of a spacefaring nation.

Therefore, the main part of the book is focused on features of designing the so-called element base for control systems of rockets and satellites (i.e., microcircuits and semiconductor devices), as well as on the methods of their protection against specific factors of outer space, primarily including radiation, electromagnetic interference, high-energy microparticles (cosmic dust), extreme temperatures, and mechanical overloads, on peculiarities of choosing proper protecting and absorbing materials, methods of required additional and special testing, and also on designing onboard control systems.

It should be pointed out that since many of the issues examined in the book haven't been considered in popular and standard study courses before, development engineers often treat effects that emerge in space microcircuits during their operation onboard a spacecraft as something mystical and even call them black magic effects. That's why the authors have attempted to present the basics of a white magic and debunk the popular myth that something extraordinary and unexplainable happens to microcircuits in outer space—it only requires study of the principles of physics underlying such effects and the application of them properly on Earth during the design of microcircuits, boards, and electronic units of on-board systems. The corresponding chapters explain the mechanisms of the impact of physical principles and the rules of their practical application.

This book allows its readers to reach the following main objectives:

1. To study and think through the statistical data on causes of accidents and failures of launch vehicles and spacecraft that have been summarized and systematized by the authors following their collection from open sources as well as from personal numerous meetings and negotiations with engineers and managers of the rocket and space industries of Russia, Europe, China, and India, in order to take all possible measures to eliminate such accidents and failures (or reduce their probability) during the design of rocket and space equipment.

2. To study the main physical mechanisms of the impact of destabilizing factors of outer space (various kinds of radiation, high-energy galactic ions, and particles of cosmic dust) on radioelectronic equipment, structures, and elements of spacecraft design, microcircuits, and semiconductor devices, and based on this knowledge to apply the ways, technology, materials, methods, and technical solutions described in this book to protect radioelectronic products for civil, special, and space applications against negative impact.

3. This book allows students and engineers who specialize in designing control systems to study the structure and features of the construction of optimal onboard electronic systems for data processing to study and apply in daily practice such methods of system and device design that will reduce the risk of failures and accidents of electronic devices onboard a spacecraft.

4. Students and design engineers of space microcircuits will find it useful to study the chapters focused on designing advanced space-grade microelectronic products (i.e., memory microcircuits, microprocessors, interface and logic microcircuits, and power control microcircuits).

5. This book can help managers, heads of space enterprises, those in the semiconductor industry, and officials responsible for creation and implementation of programs and projects for development of new space equipment. It will be also helpful to engineers in avoiding problems connected with the use of counterfeit microcircuits, realizing the importance and features of

using such microelectronic devices as system-in-package and system-on-a-chip for spacecraft.

Finally, this book can be interesting both for specialists in this field and common readers, since it includes a range of facts not widely known and characterizes the hard profession of cosmonauts (astronauts, taikonauts), and provides previously publicly unavailable facts and features of the space race among the leading three countries of the space club—the United States, Russia, and China.

The authors express their sincere gratitude to the experts of Artech House, whose critical remarks and recommendations have undoubtedly contributed to the quality of material presented, and in particular to Aileen Storry of Artech House for her truly angelic patience during the preparation of this edition for publication.

CHAPTER 1

Modern Spacecraft

1.1   Space Industry Development

Space technologies today have become such an important part of modern life that if rejected the development of civilization would be thrown backwards. Many of the problems of modern industrial society in general and the problems of sovereign, independent, and industrialized nations can be solved only with the help of space technologies.

The ultimate aim of any modern state's policy in the field of space industry is entering a new level of science, technology, and industry. Protection of state, scientific, and economic interests in the field of space activities; state defense, and security improvement; enhancement of workforce capacity; development of modern and future national space systems, devices, strategic materials, communication and control means; accumulation and development of scientific knowledge about the Earth, the universe, and outer space; extension of international space cooperation in the field of joint scientific research and space exploration; and a vast range of other issues are of extreme importance.

Speaking pragmatically, the status of a spacefaring nation should bring profit and contribute to the country's social and economic development.

Apart from obvious tasks of national defense and security (space reconnaissance, space communication and control, ballistic missile early-warning systems) and technological independence, space technologies ensure solutions to various economic and engineering problems in the following spheres:

- Global multichannel communication;
- Multiprogram radio and television broadcasting;
- High precision online navigation of transport and other objects;
- Reliable meteorological support;
- Emergency alerting;
- Environmental monitoring;
- Global emergency rescue;
- Study of natural recourses—remote search for mineral deposits;
- Cutting-edge technology and materials;
- Expansion of knowledge about the universe;
- Medical and biological research.

The needs of countries in these spheres are currently fulfilled in the framework of national space programs, aimed at creating future space systems and technologies.

At the same time space systems and technologies from different countries tend to integrate to ensure the most efficient use of national budget allocated to space programs' development, facilitating the development and joint use of space capabilities to solve socioeconomic, defense, and scientific problems that countries face. The necessity and significance of such integrations are supported by the countries' needs for:

- Inexpensive quality Earth remote sensing data, including at space service centers;
- Development of international cooperation in the field of joint use of space capabilities;
- Establishment of stable cooperation of companies and organizations on development of future space systems and technologies, including micro- and nanotechnologies.

Successful implementation of the joint space programs of Russia and Belarus [1–3] can serve as an example of scientific, technological, and industrial integration in the post-Soviet area.

In 1998 the joint program Development and Application of Space Systems and Technologies for Space Information Receiving, Processing and Distribution (Cosmos-BR) was launched and one of its objectives was reactivation of scientific and industrial connections between space industry enterprises and institutions of the two countries after the dissolution of the USSR.

Cosmos-BR was followed by the second joint program Development and Research of Future Space Systems and Technologies in the interest of Social and Economic Development of the Russia and Belarus (Cosmos-SG). It resulted in the creation of a modern technical complex for receiving space data from Earth remote sensing system, a wide range of modern hardware and software for processing and interpretation of space images, a mobile sample of control and correction station for high precision positioning, and many others.

In 2011 the third joint Russian-Belarusian program *Development of Base Elements, Technologies of Creation and Applications of Orbital and Land Means of Multipurpose Space System for 2008–2011 (Cosmos-NT)* was implemented. In the framework of this program, more than 25 experimental examples of space systems and 18 experimental high technologies in the interest of space industry in Belarus and Russia were created.

The joint program *Development of Nanotechnologies for the Production of Materials, Devices, and Space Engineering Systems and Their Adaptation to other Sectors of Technology and Mass Production for 2009–2012 (Nanotechnology-SG)* was also carried out.

The results of these programs are significant due to the need for improving stability of space systems to the outer space environment, reducing their weight-size parameters, and making the production of components for space and other industries less expensive.

It is important to note two major directions of further Russian-Belarusian cooperation in the sphere of space technologies and their applications.

First, it is the development of ground-based infrastructure for providing Russia and Belarus with Earth remote sensing data and creation of relevant space service centers. This is the most important objective of space systems and technology integration of Russia and Belarus. It will ensure access to quality space data for a wide range of users, creation of systems, technologies, and software for improvement of reliability, working efficiency, and working life of space systems of remote sensing of the Earth.

The second direction is joint technical development of materials, equipment, and key elements of space systems, which aims at considerable reduction of spacecraft (SC) weight and size, improvement of reliability and operational life of elements, devices and systems of rocket and space equipment, and reduction of development and operation costs.

In particular, this will allow creating and utilizing standardized technology for a wide range of rocket and space equipment elements with enhanced performance characteristics, including:

- Temperature-control and electric power supply systems;
- Spacecraft onboard control systems;
- A wide range of unified element and component base voltage comparators;
- Special units and parts of small spacecraft;
- Highly reliable high-temperature units of propulsion system;
- Nanoelectronic low-energy devices;
- Functional nanoscale sensors for various purposes and many other parts, which will allow to enhance functional characteristics and reliability of space systems during their launch and operation.

The production of space and rocket equipment in Russia has undergone considerable changes in the past decade. Output has dramatically decreased, and a number of stock-produced items have been taken out of production, while at the same time new high-tech products are developed. However, the rocket and space industry has retained its organizational and technological structure, plant and equipment, and principles of preproduction engineering.

State policy on development and technological modernization of Russian rocket and space industry is implemented in accordance with *Development Strategy for the Rocket and Space Industry Through to 2015* and further targeted programs [4–6]:

- Federal Space Program (FSP) of the Russian Federation for 2006–2015 (FSP-2015);
- Federal targeted program (FTP) *Global Navigation System* for 2002–2011 (FTP *GLONASS*);
- *FTP Development of the Military-Industrial Complex of the Russian Federation in 2007–2011 and through to 2015* (FTP Development of MIC-2015);
- FTP *National Technology Base Development in 2007–2011*, as well as in the following projects:
- FTP *Development of Nano Industry Infrastructure in the Russian Federation* for 2007–2010;

• FTP *Development of Electronic Component Base and Radio Electronics* for 2008–2015.

It should be noted that success of any space or military program is directly dependent on standardization, functional capabilities, and quality of the component base, which serves as the foundation for any modern equipment either general or special purpose.

Space microelectronics is an independent field of scientific and technical process that includes a whole system of interdependent fields ranging from research of new materials and physical mechanisms of microchip space application to the development of new technologies and methods of microchip design, enhancement of their noise tolerance, reliability, and tolerance to ionizing effects, reduction of static and dynamic energy consumption and increase of both chip density and operation (clock) frequencies.

Priority fields of the Russian Federation state policy in this sphere are as follows:

1. *Creation of space complexes and systems* of new generation, technical characteristics of which ensure competitive advantage on the world market, including:
   - Development of modern launch vehicles and modernization of existing rocket carriers, development of new rocket carriers and upper-stage rockets;
   - Creation of medium capacity launchers for manned spacecraft of new generation space satellites with longer operation life;
   - Preparation for implementation of cutting-edge projects in the sphere of space technology and space exploration.
2. *Completion and further development of GLONASS system*, in particular:
   - Deployment of a satellite constellation on the basis of new generation spacecraft with long operation life (at least 12 years) and enhanced technical characteristics;
   - Creation of ground control stations and equipment for end consumers, its promotion on the world market, ensuring GLONASS and GPS compatibility.
3. *Development of the satellite constellation*, including creation of a communications satellite constellation that contributes to the growth of fixed-line telephony, mobile and personal communication (in the entire territory of the Russian Federation); creation of a meteorological satellite constellation that can transmit information real time. In the long term, competitive performance in the data transmission market will require a quantum leap in operational life of communications satellites. It can be achieved through the creation of recoverable satellites that can be maintained, refilled with rocket propellant, repaired and upgraded in orbit. Such technological development is planned to result in creation of massive orbital platforms by 2025 that will support developed and other equipment, including power generation systems that can be maintained or replaced.
4. *Enhancement of Russia's space market representation, including*:
   - Retaining leadership in the traditional space service markets (commercial launches—up to 30%);

- Enhancement of representation in commercial spacecraft market;
- Enhancement of representation in the foreign markets of space-and-rocket equipment components and relevant technologies;
- Entering high-tech world market sectors (production of satellite communications and navigation ground equipment, Earth remote sensing);
- System creation and modernization of the Russian segment of the International Space Station (ISS).

5. *Modernization of ground-based space infrastructure and technological level of rocket-and-space industry, in particular*:
   - Technical and technological refurbishment of the industry, adoption of new technologies, optimization of the field technological structure;
   - Development of a system of launch sites, re-equiping ground control systems, communication systems, and experimental and production bases of the space-and-rocket industry.

The state program on Development of Space and Rocket Complex of the Russian Federation through to 2025 includes the development of the following high-priority technological fields:

- Spaceborne navigation systems;
- Space data relay systems;
- Spaceborne hydrometeorological systems;
- Spaceborne systems of Earth remote sensing;
- Satellite communications and broadcasting;
- Spaceborne geodetic and cartographic support;
- Spaceborne communications and battle management;
- Missile warning via spacborne systems;
- Spaceborne electronic reconnaissance;
- Spaceborne electro-optical surveillance;
- Space systems for multipurpose radiolocation surveillance;
- Spaceborne offshore surveillance systems.

Providing space and rocket equipment with a modern radiation-resistant electronic component base is one of the major problems in implementation of a national program for space and rocket equipment development, which is going to be solved within the framework of the federally targeted program for development of radio electronics industry (under the coordination of The Ministry of Industry and Trade of the Russian Federation).

Independent access to space will be ensured by means of development and use of the Plesetsk Cosmodrome, renting the Baikonur Cosmodrome and creation of a new national Vostochny Cosmodrome in compliance with the November 6, 2007 Presidential Decree *On Vostochny Cosmodrome*.

The results of this program are to be as follows:

- Creation and commissioning of the *Angara-A5* space-launch vehicle; modernization and efficient operation of Plesetsk and Baikonur Cosmodromes.
- Creation of the first (2015) and the second (2018) phases of the Vostochny Cosmodrome facilities.

- Deployment of orbital spacecraft in the interests of state needs in the following amounts: 95 and 113 spacecraft are to be deployed in 2015 and 2020, respectively, including the Russian segment of International Space Station consisting of six and seven modules (2015 and 2018, respectively) is to be deployed.
- Providing necessary components for the orbit fleet of GLONASS system, equipping it with GLONASS-K spacecraft with extended functional capabilities. By 2015 the system will ensure positioning accuracy of 1.4 meters and of approximately 0.6 meters by 2020.
- Creation of scientific-technical capacity for future prototypes of rocket and space equipment, including preparation of transport and power modules for a future propulsion system with flight development tests in 2018.
- Development of competitive production technologies, satellite communication and Earth remote sensing technologies, navigation, search and rescue technologies, as well as technologies for emergency monitoring, tracking and monitoring of mobile objects with the use of space-based automatic identification system and personal radio beacons.

In the field of fundamental space research there will be projects, implementation of which will allow Russia to catch up with leading space powers to reach top positions the in main space science fields and become one of the world's leaders in cosmic research in the long term. More specifically, it is planned:

- To create three space observatories—*Spektr-UV*, *Spektr-M* (*Millimetron*), and *Gamma-400* to study astrophysical objects in different ranges of electromagnetic spectrum and gamma radiation in high-energy ranges;
- To deploy a program for in-depth study of the Moon—to launch missions of *Luna-Glob* orbiter and *Luna-Resurs* landing probe (phases 1 and 2), and a mission to return Moon soil samples for thorough examination;
- To develop a whole range of new technologies for interplanetary flights and on-planet manned activities;
- To create a future manned transport system that can deliver humans to the Moon.

The results of the state program for expansion of service scope by 2020 include:

- In terms of space communication, broadcasting, and relay systems:
  - Creation of a full-scale multifunctional space data relay system that will enhance efficiency of the national low-orbit Earth remote sensing spacecraft, as well as efficiency of launch vehicles and the Russian segment of the International Space Station;
  - Increase of the number of satellites (up to 39 vehicles) in the orbital fleet of fixed-line telephony, mobile presidential communication, and radio and TV broadcasting systems, which will allow providing communication service almost on the entire territory of the Russian Federation, including the Arctic Region.
- In terms of Earth remote sensing and hydrometeorological survey:
  - Increase of the number of satellites in the orbital fleet up to 24 by deploying conceptually new space systems designed for solution of cartographic

problems, environment monitoring, operational monitoring of emergencies, natural resources inventory, support for efficient agriculture and water usage, and Arctic Region monitoring.

Successful implementation of the state program will contribute to the development and utilization of national space hardware in the interest of social and economic spheres, enhancement of Russia's representation on the world space market, achievement of ambitious goals to research and explore outer space, ensuring guaranteed space access from the Russian territory, and preserving the leading positions in manned flights [7].

1.2   Classification of Modern Spacecraft

Spacecraft (SC) is a generic term for hardware used for a variety of purposes in space, including research and other kinds of work close to or on the surface of various celestial bodies. Rocket carriers or launch vehicles (LV) deliver spacecraft to the orbit.

Spacecraft are also called spaceships (SS) or space vehicles (SV) if their main aim is transportation of people and equipment in the upper atmosphere or the so-called near space, or further.

According to fields of their application spacecraft are classified into:

- Suborbital;
- Low-Earth orbit spacecraft in the geocentric orbit of Earth artificial satellites (EAS);
- Interplanetary (expeditionary);
- Planetary (Moon and Mars rovers, and the like.)

There's a distinction between automated satellites (AES) and manned spacecraft. Manned spacecraft include all kinds of manned spaceships (SSs) and orbital space stations (OS). Although modern orbital space stations orbit the Earth in near space and can be called space vehicles, they are traditionally referred to as spacecraft.

The term *space vehicle* is often used to name active (maneuvering) AESs to differentiate them from passive satellites. However, in most cases the terms space vehicle and spacecraft are synonymous and interchangeable.

Currently in projects for creation of orbital hypersonic vehicles as a part of aerospace systems (ASS), other terms (e.g., *earth-to-orbit vehicles*, or ETOV), the definitions ASS *space plane* or *aerospace vehicle* are used. Such vehicles are designed for manned flights both in space vacuum environment and dense Earth atmosphere.

There are dozens of countries that have AESs. However, only a few of them (the USSR/the Russian Federation, the United States, China, Japan, India, and countries of Europe/ESA) utilize complex technologies for creation of automatic recoverable and interplanetary SC. Only the first three of these countries have manned SSs. (In addition, Japan and Europe have SC that are seen in orbit as ISS modules and space delivery vehicles).

These three countries also have technology for AES interception in orbit (although Japan and Europe are also quite close to it since they perform docking).

SC can be also classified according to their operating modes, functions, ability to return to the Earth, weight, types of control, types of propulsion system, and suitability for near or outer space (see Figure 1.1).

According to *operating mode* spacecraft are divided into:

- Artificial Earth satellites—a generic term for all spacecraft in geocentric orbit (orbiting the Earth);
- Automatic interplanetary stations (space probes)—spacecraft designed for flights to celestial bodies to explore them by either orbiting or flying by, and some of the probes then leave the solar system;

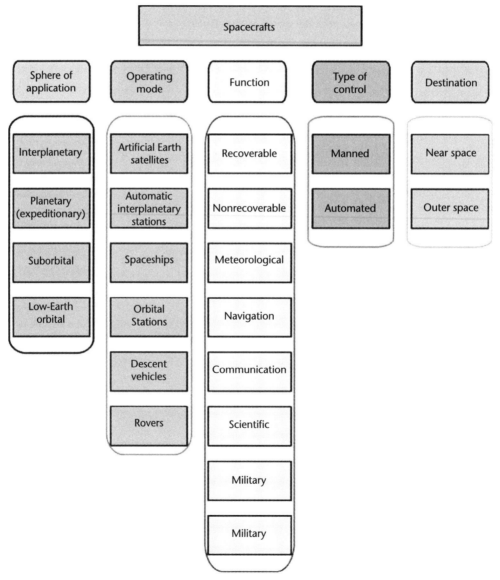

Figure 1.1   Classification of spacecraft.

- Automated and manned spaceships designed to deliver cargo and people to the Earth's orbit; there are also projects of flights to other planets' orbits;
- Orbital stations—spacecraft designed for a long-term crew's stay and work in Earth's orbit;
- Descent vehicles used to carry people or cargo from orbit or interplanetary trajectory to the surface of a planet;
- Rovers—automatic laboratory facilities or vehicles designed to move across the surface of a planet or other celestial body.

According to *ability to return* to the Earth SC are divided into two groups:

- Recoverable spacecraft designed to transport people and cargo to the Earth performing soft or hard landing;
- Nonrecoverable spacecraft usually deorbit and burn up in the atmosphere when their service life has expired.

According to their *functions* spacecraft are classified into:

- Meteorological;
- Navigation;
- Communications, television broadcasting, and telecommunications satellites;
- Scientific;
- Geophysical;
- Geodetic;
- Astronomical;
- Earth remote sensing;
- Reconnaissance and military satellites;
- Other.

It should be noted that many modern spacecraft perform several functions.
According to their *weight* SC are divided into:

- Femto: up to 100 g;
- Pico: up to 1 kg;
- Nano: 1–10 kg;
- Micro: 10–100 kg;
- Mini: 100–500 kg;
- Medium-sized: 500–1000 kg;
- Large: over 1000 kg.

Therefore, any SC with weight over 10 kg but less than 100 kg should be called a microsatellite, and a SC that weighs over 500 kg but less than 1,000 kg is called a medium-sized spacecraft.

There are manned and automated SC according to the type of control. According to propulsion system SC are divided into: high-thrust engine-equipped SC, nuclear-powered propulsion system SC, and chemical-propellant (liquid-propellant or solid-propellant) SC.

There are SC designed for near and outer space.

SC for near space include low-Earth spacecraft (EAS, observatories, manned orbital vehicles) SC for outer space include interplanetary SC (planetary satellites, descent and flyby spacecraft).

1.3   Spacecraft Designs and Structures

The world's first spacecraft was launched in the USSR on October 4, 1957. The first manned spacecraft *Vostok 1* with Soviet cosmonaut Y. A. Gagarin was launched on April 12, 1961. Over the past seven decades after the first launch (not taking into account 20 years of research and tests before it) SC designs have been constantly improved. A considerable contribution to the development of SC designs was made by so-called test spacecraft that were specifically designed for testing and adjusting elements, systems, items, components and units of design under real-life conditions, as well as for finding ways of their efficient application and unification.

The USSR used mainly various modifications of *Kosmos* series as test SC, whereas the United States had a wide range of test SC: *ATS, GGTS, OV, DODGE, TTS, SERT, RW*, and the like.

Despite a large variety of SC designs, they commonly consist of a main body with a set of different structure members (the so-called sustaining equipment) and special-purpose electronic equipment [7].

The SC body is a structure where all the spacecraft elements and equipment are installed. For example, an automated SC typically includes the following subsystems: orientation and stabilization subsystem, thermal-control subsystem, electric power subsystem, telemetry and tracking, navigation and control subsystem, command and data handling subsystems, various actuators, and other.

Manned SC also have life-sustaining and emergency recovery subsystems.

Spacecraft special-purpose equipment includes optic (electrooptic), photographic, television, infrared, radar, radiotechnical, spectrometric, x-ray, radio and relay, radiometric, calorimetric, and other equipment. A detailed description of these systems is provided in the next section, but it should be noted that all these systems (their structure, functions, and configuration) have most up-to-date electronic component base (ECB).

Since there are different applications of SC, there are different SC configurations: launch vehicles (LV) that place spacecraft into orbits and trajectories, upper-stage and retrorocket units that have main propulsion unit and vernier thruster, integral fuel cells, supporting system units (that ensure SC transfer from low orbit to higher or interplanetary orbit and vice versa as well as correction of trajectory parameters, and the like.).

The term *arrangement* is closely connected with SC configuration and means a most efficient and compact layout of spacecraft systems. There are internal and exterior (aerodynamic) SC arrangements [7].

Development of SC design is a complex process because a number of factors (often contradictory ones) are to be taken into consideration. For example, it is necessary to ensure minimum contact with ground support systems (especially for LV), crew safety and comfort (for manned SC), safe operation and maintenance during launch and flight, proper parameters of stability, control, thermal control, as well as aerodynamic characteristics and many others.

Developers' tasks are all the more difficult due not only to spacecraft weight, but also because cost and production time are to be minimized while ensuring that the spacecraft is reliable, multifunctional, and so forth.

Evolution of SC designs is presented in the following figures. Figure 1.2 shows *Vostok 1*, the world's first spacecraft to take a human to low-Earth orbit.

The spaceship launched from Baikonur made a single (but the first in the history of humankind) orbit around the Earth. It flew in fully automatic mode, so that the first cosmonaut was a passenger who could overtake control at any time. Though according to our classification, this flight should not be considered not a *manned* one, but a flight in fully automatic mode, this is the case when classification doesn't reflect the essence of a process (phenomenon or event).

Figure 1.3 shows a general view of one of the first (1977) SC of the *Voyager* series (most famous ones are *Voyager-1* and *Voyager-2*) designed for reaching far

Figure 1.2   Manned spaceship *Vostok 1*.

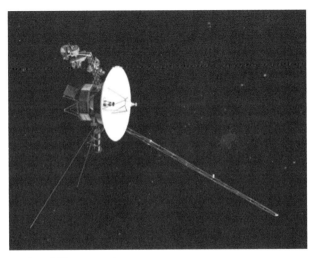

Figure 1.3   Space probe of *Voyager* series.

destinations (space probes). According to some literary sources, the space probe weighing 723 kg launched on September 5, 1977, and was designed for exploration of the Solar System and its vicinity, is still in working state (to the surprise of its developers), and now has a new (additional) mission—defining the boundaries of the Solar System, including the *Kuiper belt* (asteroid belt) though originally its mission was to explore two planets, Jupiter and Saturn.

Such long active operation of the SC was achieved due to efficient engineering solutions for spaceborne equipment and a good choice of relevant ECB for its onboard systems.

Figure 1.4 shows general view of a NASA automatic interplanetary satellite designed for exploration of Pluto and its natural satellite Charon. The probe reached Pluto in 2015 and will now explore the asteroid belt (*Kuiperbelt*).

A comparison of Figure 1.3 and Figure 1.4 shows that the general views of the two spacecraft constructions are quite similar; however, there are differences between their elements and structure members due to their different functions. Although unification of structural solutions has its limits, the unification of technical solutions for spaceborne systems is crucial and can be achieved by the means of standardization of electronic component and technical specifications. Unfortunately, developers of national rocket and space equipment haven't completely achieved this aim yet.

Figure 1.5 shows, against the background of Saturn rings general view of *Cassini-Huygens*, another interplanetary probe designed for exploration of two objects of the Solar System—Saturn (its rings and satellites) and Titan.

This spacecraft presents the result of globalization in space industry since it was developed by joint efforts of several space powers and organizations—the United States (NASA), the European Space Agency, and the Italian Space Agency.

Figure 1.4   General view of NASA's *New Horizons* automatic interplanetary satellite.

Figure 1.5   Interplanetary probe Cassini-Huygens.

The spacecraft consists of two self-sustaining parts—*Cassini* orbiter and *Huygens* descent module (probe) designed for autolanding and autonomous operation on Titan.

This example demonstrates high efficiency of joint efforts of developers from different countries and schools for implementing ambitious research programs. The combined interplanetary spacecraft was launched in 1997. In 2004 it successfully reached Saturn, performed the whole research cycle, and continued its mission. On December 25, 2004, *Huygens* probe separated normally from the main spacecraft and reached the vicinity of Titan. Having performed all the planned maneuvers, it landed on the surface under difficult atmospheric conditions.

As a result, the *Cassini* orbiter became the first artificial satellite of Saturn.

Figure 1.6 shows the general view of the widely known recoverable space shuttle spacecraft.

As we can see, it has a hybrid design—rocket carrier plus spacecraft itself which is fundamentally different from the previously mentioned spacecraft shown in Figures

Figure 1.6   American space shuttle.

1.2–1.5. The main purpose of space shuttles was to deliver cargo to orbital space stations and back. Altogether five space shuttles were built. Two of them crashed. There was also one prototype (test spacecraft—according to our classification). The program existed for 20 years (from 1981 to 2011) but was cancelled as it was found inefficient.

Each shuttle was planned to perform two launches a month (24 launches a year) and to perform a total of a hundred flights. In reality they were launched far less often (135 launches altogether) due to a number of reasons. The biggest number of launches (39) was performed by the most proven shuttle, *Discovery*.

Finally, we should mention one of the most complex interplanetary space-craft—manned interplanetary spacecraft with landing recoverable modules, the development of which requires analysis and the use of all the previous knowledge, experience, and recent advances in microelectronics, space equipment building, materials science, space psychology, and other fields.

Figure 1.7 shows a photograph of one of *Apollo* interplanetary spacecraft (*Apollo 11*) made in 1969 right after its landing on the Moon. Its design and inter-nal arrangement is different from all the previously mentioned spacecraft due to its special purpose and functions.

Here are some facts that have already become part of history: legendary crew commander Neil Armstrong and pilot Edwin Aldrin spent 21 hours, 36 minutes, and 21 seconds in a special Lunar Module on the surface of the Moon. Michael Collins, the Command Module pilot, remained in near-lunar orbit, keeping constant radio and video communication with both the two astronauts and ground support

Figure 1.7 Interplanetary spacecraft *Apollo 11* on the Moon.

center. Astronauts (not cosmonauts, to our regret) stepped onto the lunar surface and spent 2 hours, 31 minutes, and 40 seconds there, after which they returned to the Lunar Module successfully (or almost successfully), lifted off the Moon, then docked with the Command Module just as successfully (in accordance with the plan), and came back to the Earth.

In order to preclude failures of the onboard electronic systems of space vehicles such as command and lunar modules, it is vital to make the right choice and to apply to the optimum the integrated circuits of the element-component base that will serves as the basis for design of similar electronic systems. The appropriate regulations and techniques for selection of such a component element base are detailed in the book.

Although this book deals only with specific technical aspects, in order to understand the role of microelectronics in space device engineering, it is important to indicate an unpleasant episode that is known only to a small group of specialists and that could have turned the great triumph of the mission into a catastrophe.

It is known that the size of manned spacecraft, both the Command and Lunar Modules, had strict limits. Coming back from the historical mission to the Moon in massive space suits and closing the access door, one of the astronauts accidentally damaged (broke) a plastic switch by the door. The switch had only two positions: up and down. But its important function was to control the ascent engine of the Lunar Module: down position stood for "sleep mode" and up position meant "start" (turning on automatic system for lifting off the Moon).

Since the switch broke in down position, astronauts and flight support specialists had to spend an extremely unpleasant half an hour getting the mechanical metallization contacts in the up position.

This seemingly comical situation is marked by the following fact: After the crew commander reported the situation to the astronaut in the Command Module and the latter reported it to the NASA ground support center, there were five minutes of confusion that was followed by the first valuable piece of advice from a number of experts that are always present at Mission Control Centers: "Take a ballpoint pen refill and mechanically move the slide lever up with the help of the refill's sharp (writing) point." It's quite a reasonable recommendation on the Earth, but where do you get a ballpoint pen on the Moon? However, nonstandard thinking is one of the psychological features of both our cosmonauts and American astronauts. (Cosmonaut/astronaut candidates are tested for nonstandard thinking and as a result about 10%–15% of candidates are rejected, although they meet all other requirements).

The "stranded" Lunar Module astronauts found and utilized a sharp and thin equivalent of a ballpoint pen refill; afterward, the ascent engine worked normally, and docking with the Command Module and triumphant return to the Earth also went normally.

Thus, a conclusion can be made that although design of a spacecraft depends on its special purpose, the foundation of a spacecraft's onboard electronic system is a modern electronic element base that meets all the specific requirements, which we will discuss in the following chapters.

Figure 1.8 shows the main application spheres of electronic component base in spacecraft. On the right, the main functions of a SC onboard system are provided: data collection and storage, data processing and generation of relevant commands,

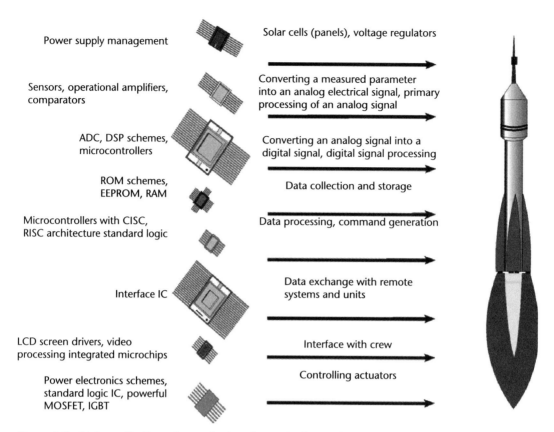

Figure 1.8 Main application spheres of microelectronic element base in SC.

ground support interface with crew, controlling various actuators, power supply management, converting measuring parameters into analog electrical signals and primary processing of analog signals, and so forth.

On the left, specific kinds (classes) of electronic component base that perform previously mentioned functions are presented: sensors, operational amplifiers, interface microchips, drivers, power electronics schemes, standard logic ICs, digital and analog microchips, voltage comparators, solar cells (panels), and so forth.

1.4 Spacecraft Onboard Systems

1.4.1 Classification of Spacecraft Onboard Systems

In order to ensure viability of SC both under operation and design as well as implementation of special-purpose programs, it is necessary to complete quite a wide range of tasks that are common to all kinds of SC. These tasks include [8]:

- Ensuring data exchange with ground control center (GCC);
- Ensuring power supply;
- Distribution of electrical power on SC among loads;
- Collection, storage, processing, and transmission of telemetry information;

- Controlling SC systems and equipment in accordance with SC flight program and its actual state;
- Ensuring SC design thermal control;
- Detection and control of SC attitude sensing;
- Ensuring SC space motion (center of mass motion);
- Ensuring SC angular motion (motion around center of mass);
- Determining and predicting SC location in orbit;
- Controlling rotating solar panels (if available).

When the first SC were designed, each task was solved independently with the help of independent specialized systems that have their own sensing equipment, actuators, and preset automatic controls. When SC interiors became more complex and the number of applications grew, there arose a need for centralization of control and monitoring devices to manage SC onboard systems, primarily in terms of efficient use and replenishment of energy, priorities, and time of flight and routine operations, autonomous response to contingency situations on the base of online diagnostics, and onboard equipment testing. This problem was solved with development and application of computing means with advanced software (SW) on SC. Similar to the *ground* control complex (GCC), the term *onboard* control complex (OCC) has appeared and it encompasses the main SC onboard systems, including onboard computer system, guidance, navigation and control system, onboard equipment control system, onboard command channel equipment, onboard status measuring system, as well as relevant OCC software, which is the key integration element of OCC.

In order to examine the principles of OCC design concepts for automated special application SC (communications satellites, SC for stellar space observation, Earth remote sensing spacecraft) we shall list the main onboard systems (Figure 1.9) [8]:

- Onboard computer system that includes computing means and mediation devices (communications interfaces and adapters), which ensure information interaction with onboard components and provide computational power for controlling and monitoring SC systems;
- Guidance, navigation, and control system (attitude and motion control) designed for both controlling SC motion as a point mass (center of mass motion) and angular motion (motion around mass center);
- Onboard equipment control system that ensures power source switching, amplification, and conversion of electric signals as well as command issue for SC systems and devices in accordance with time and logic conditions;
- Onboard status measuring system that ensures collection, processing and transmission of telemetry information (measurement results that describe status of SC systems and current processes) to GCC;
- Onboard equipment of command channel or command radio link which is a radio technical complex that ensures timely service information transfer from GCC to OCC;
- Combined propulsion unit consisting of a set of engines that ensures SC motion relative to orbit and SC angular motion;
- System that ensures thermal control inside the SC;

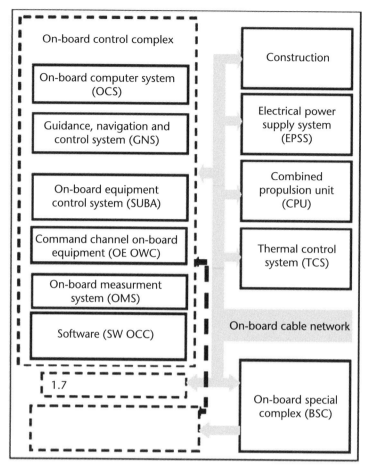

Figure 1.9   Spacecraft onboard systems.

- Electric power supply system (EPSS) that converts primary energy (solar) into electric power.

The onboard cable network is considered a structure element in some classes of spacecraft.

Application of computing and new design and technological solutions, modern element base and SW integration technology have created basis for creation of integrated OCCs. Online system control and smart execution of SC flight programs with regard to external conditions, status of onboard systems, and available resources have allowed transferring a number of control and monitoring functions to OCS, or more specifically, its SW. These functions tend to concentrate more and more in onboard computer system (SW OCC) in the course of software and hardware development. Software has developed as an independent (and one of the most important) component of OCC. OCC software has its hierarchy [8]:

- The first or lower level is represented by drivers for exchange with equipment and programs for computing process organization.

- The second level consists of programs for control and monitoring of onboard devices and equipment.
- The third level includes programs for onboard systems' flight mode and calculation programs.
- The fourth or the superordinate level includes programs for planning and organization of OCC operation modes and monitoring of SC systems' state.

According to SW OCC architecture (see Figure 1.10) there are service (dispatcher, exchange, control of OCS configuration, timing, and the like), and functional programs (turning on/off devices, calculation programs for preparatory and accompanying information, programs for control actions generation for specific devices, and the like.) Each program (program module) has setting parameters and logical-information links with other programs. SW OCC design requires determined circulation of information between programs of all levels with control data coming from downward (from superordinate level programs to lower level programs) and monitoring and diagnostic information coming upward. In order to ensure real-time

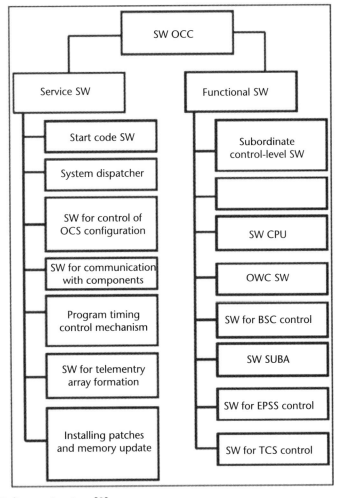

Figure 1.10   Software structure [8].

operation of SW OCC, each program has a sequence and specific time of access to onboard computer and computing power.

The integrative nature of SW OCC allows performing not only control and monitoring functions but also such important functions and tasks as:

- SC flight (orbit-by-orbit) planning;
- Optimization of onboard resource consumption;
- Autonomous SC functions;
- Online response to contingency situations.

Such functions as acquiring digital information from onboard systems (directly or via OMS), information exchange with GCC (via OE OWC), processing and usage of collected data in SW computing tasks have made OCS and SW OCC an integral part of onboard control complex. Command channel and OSM are an indispensable part of OCC as a source of circulating information and elements that support logic and information interfaces on SC. It should be noted that links between OCS, command channel and OMS are both physical (wired, via onboard cable network and interface mediation devices) and virtual (informational, via data exchange channels).

Onboard equipment control system (SUBA) is an integral part of OCC. Two most important functions of SUBA have an integrative nature and present a prerogative of OCC [8]:

- Ensuring physical (wired) interface with SC systems and equipment and controlling them by means of relevant command and signal generation;
- Providing onboard loads with power.

All the rest of the previously mentioned systems perform their specific tasks that are essential to SC, but they aren't integrative in terms of OCC structure. It should be noted that guidance and navigation subsystem (GNS) is often classified as a part of OCS by SC developers due to the following aspects:

- GNS tasks (orientation, stabilization, and guiding SC to implement its mission) are of primary importance.
- This system (together with SUBA) is one of the first onboard systems that was designed and developed for the first SC.
- Control programs for GNS are closely connected with control programs for other systems and superordinate level control programs of SW OCC and other.

GNS includes optical spectral sensors and angular rate sensors, converter units and control signal formers as well as actuators in the form of powered gyro (e.g., flywheels or gyrodynes). Propulsion system thrusters also serve as actuators for the GNS. The GNS can also be supplemented with various navigation devices and satellite navigation equipment.

For example, Rocket and Space Corporation *Energia* creates GNS on the basis of an adjustable strapdown inertial navigation system that allows for determining the current position of SC body axes relative to inertial frame of reference by means of integrating components of absolute angular velocity (Figure 1.11) [8].

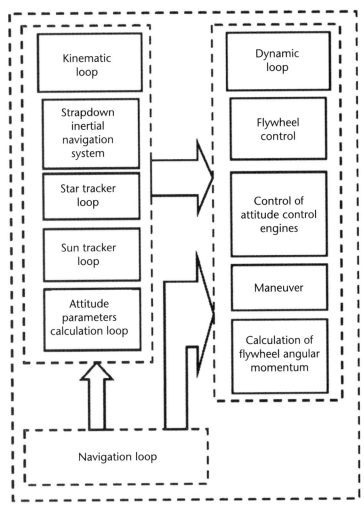

Figure 1.11   Structure of GNS based on a double-loop strapdown inertial navigation system [8].

Operation of GNS includes operation of three control loops:

- Navigation loop determines (predicts) real-time SC location in orbit with given accuracy using the initial conditions, determined by GCC or by onboard navigation equipment with the help of prediction models of Earth gravity and magnetism and information from SC onboard navigation sensors.
- Kinematic attitude control loop measures error angle between GNS instrument base and some reference base that is set in attitude mode, in addition to this, kinematic loop adjusts (corrects) the given reference base.
- Dynamic attitude control loop aligns SC base and reference base with given accuracy, moreover, it ensures SC stabilization during orbit adjustment and other dynamic operations.

Support systems of automated SC of the studied class have similar requirements that are determined by payload (or onboard special complex, BSC) of support systems. Flagship companies of space industry now tend to develop universal space platforms (USP) suitable for different kinds of automated SC. Integration of an existing USP with a new BSC by means of continuity of USP devices, minimization of time required for design and on-ground training can considerably reduce the time and cost of SC creation, while preserving its quality and reliability.

The continuity level of an existing space platform for application on SC with new special equipment is determined with regard to parameters of equipment and requirements to the USP. The main parameters of BSC are power consumption, mass, number of supply feeders, interface types, and so forth. The tasks, defined by operation conditions of BSC, include:

- Ensuring accuracy of spacecraft orientation and stabilization during operation of BSC;
- Ensuring and estimation of guidance accuracy;
- Keeping proper parameters of working orbit and spacecraft service life;
- Execution of set operation modes of BSC and so forth.

In such a scheme OCC is viewed as the USP core [8]. In other words, OCC includes main support systems and onboard SW that ensures integration of USP systems while executing SC functional tasks on program-logic and physical levels by means of relevant interfaces.

The main tasks of OCC include:

- Collection, primary processing, storage, and telemetering of online monitoring information and its application in control tasks of USP;
- Organization of information and command interaction with BSC;
- Coordination of USP and SC operation during on-the-ground training and normal flight execution in automatic mode and according to GCC information;
- Diagnostics of USP and its systems status; detection, localization, and addressing contingency situations in automatic mode.

While solving these tasks onboard control complex proves to have the following quality characteristics:

- High level of integration (i.e., organization of integrated work of USP systems and interface support while executing all the flight modes and operations in real time);
- Strict hierarchy of OCC multilevel structure and determinacy of information flows (command and control information moving downwards from the core of OCC to each element; monitoring and diagnostic information moving upward, from peripheral elements to the core);
- Advanced control flexibility, implemented in control algorithms of SW OCC, which ensures efficient resource consumption and application as well as adaptability to failures and addressing contingency situations on the basis of GCC commands and analyzed diagnostic information from sensing equipment.

A proven platform suitable for a whole class of spacecraft is especially attractive to customers of advanced space products. OCC is planned to be developed and improved in the following fields:

- Enhancement of reliability and term of guaranteed operation life of devices and OCC in general by means of application of modern ECB and functional backup on hardware and software levels;

- Reduction of OCC total mass and economy of interface resources in the interest of payload by means of interface unification, optimization of onboard cable network location, and transition to doubling instead of tripling of certain hardware elements of OCC while preserving its special functions and parameters.

1.4.2 Peculiarities of Design of Onboard Information Control Complexes with the Use of Programmable Logic Microchips

Modern onboard equipment requires high level of structural integration of its elements. As previously mentioned, the main element that ensures control of this complex equipment is onboard information control complex (OICC). Such tasks require a lot of resources and are performed by onboard computer complexes. Both minimal and maximum permissible computational power might be needed for efficient performance depending on special purpose and parameters of a spacecraft. For example, most modern high-orbit communication satellite systems require operation of relay satellites in the mode of direct relay and frequency translation that can be ensured by switching subsets of transmit/receive equipment and doesn't require complex computing tasks in the framework of spacecraft mission. Another example is low-orbit electro-optical satellites for observation and Earth remote sensing, that are going to be examined in detail in the following section; operation of such satellites requires constant receiving and complex computer processing of ground images together with monitoring location and orientation of the spacecraft to ensure accuracy of received data. Constructing onboard equipment based on chips developed individually for each particular case and purpose is quite an expensive and time-consuming stage of development. Moreover, due to rapid development of microelectronics, technologies soon become outdated. As previously mentioned, one of the main tendencies of modern onboard equipment development is unification of its elements. That is why proven standardized technologies of design are often used for construction. They are implemented according to the following principles [9]:

1. Multimodule construction of OICC based on standard microcontrollers and microprocessors;
2. Hierarchical structure of OICC with the central element (onboard microcomputer) having the main computing power and using additional elements for performing subtasks;
3. Constructing according to the *system-on-a-chip* principle (i.e., integration of all the functions of OICC on a single very-large-scale integration chip).

The first alternative is based on the use of standardized modules and microchips. The PC 104 standard can serve as an example, and it provides stackable construction of onboard equipment, where each module has a standardized size (form factor) and switchable elements for interaction. The central module that ensures control is the central processor module.

The second alternative of constructing OICC is based on a modular principle with the use of several microcontrollers that have a simplified underlying structure, for example, based on PIC-controllers.

Hierarchical structure of OICC requires a high-performance onboard computer. However, such computers as a rule have excessive size, weight, and power consumption for onboard apparatus. That is why this alternative is applicable only for medium-sized and large spacecraft.

System-on-a-chip concept, which will be thoroughly examined in Chapter 8 of Volume 2, is the most suitable for onboard equipment of advanced spacecraft since it ensures maximum integration of equipment elements together with small size and power consumption. Implementation of this concept may require special integrated chips developed specifically for a certain spacecraft. However, development and single-item production of such elements consumes time and funds so is efficient only in case of serial production of spacecraft and platforms. Moreover, modernization and modification of such elements becomes more complicated.

Generally, *system-on-a-chip* includes different kinds of blocks: programmable processor cores, blocks of specialized integrated chips, programmable logic blocks, memory blocks, peripheral devices, analog components, and various interface circuits. Not all of the blocks must be physically located on one chip: processors and memory blocks can be used as separate components.

Efficiency of high-integrity OICC based on *system-on-a-chip* technology can be increased with the use of programmable logic devices (PLD). PLD is a widely known and widespread technology; however, increased integration and speed of element base has allowed active application of PLD for spacecraft onboard equipment.

The use of PLD enhances technological efficiency of digital devices design and allows transferring the whole process to a PC. Figuratively speaking, any digital device can be implemented on the basis of PLD right at the developer's desk with the help of a PC and a relevant programming unit.

The class of programmable logic devices includes integrated chips based on the following technologies:

- Field programmable gate array (FPGA);
- (E)EPROM technology based complex programmable logic devices (EPLD);
- CMOS fast flash complex programmable logic devices (CPLD);
- Mask programmable logic devices (MPLD).

EPLD, CPLD, and MPLD technologies are a combination of fully programmable gate arrays and/or macrocell banks. Macrocells form functional blocks that perform various combinatory or sequential logic functions. Field-programmable gate array (FPGA) is a more sophisticated technology since it allows implementation of underlying fabric by means of programming of underlying array.

Quick Logic, Actel, Xilinx, and Altera are major developers and manufacturers of the programmable logic component base. Actel company products stand out in the use of PLD in space systems. The feature of Actel PLD is creation of a metal jumper during programming. This technology ensures high reliability and flexible resources for tracking and doesn't require configuration ROM. There are commercial, industrial, and radiation-resistant modifications of company's products that meet all the reliability and stability requirements for severe conditions. Designing chips on the base of PLD includes the following operations [9]:

- Input of description of designed logic circuit in several high-level languages (description of element's functional behavior, graphical input of its circuit on the basis of standardized medium scale integration chips or with the help of libraries of standard components);
- Expert choice of PLD model for implementation of the described circuit with regard to the necessary number of programmable cells and conditions of component's functioning;
- Translation of logic circuit description into the PLD model and its optimization (translation, optimization, location);
- Functional and timing simulation of the designed component to check whether it meets requirements;
- Verification of the designed component and modification of its model;
- Uploading well-functioning model to PLD with the help of PC via appropriate port or by means of programming ROM in case with FPGA technology, which loads configuration data when powered.

CPLD chips with FAST FLASH memory keep configuration data after power is turned off. PLD components are most often described in VHDL—language designed to describe logic structure of digital components. Moreover, component libraries designed for the majority of applications are distributed by PLD manufacturers free of charge.

Figure 1.12 shows an alternate design of OICC of an advanced spacecraft [9]. In this case PLD serves as the central element that collects and processes special information as well as controls and monitors status of all the onboard equipment elements. The structure of PLD-based OICC requires internal buses for information exchange between different components. Approximate number of logic gates is about 1–1.5 million which is implemented in a programmable logic integrated chip of the ProAsic series. Figure 1.13 [9] shows a board made by PLD manufacturer together with ProAsic Plus integrated chip and elements that ensure interaction with external devices.

Due to well-established software models and libraries of standard components, further modernization of onboard equipment won't cause additional expenses and in case of mass application of this technology it will allow serial production of advanced spacecraft.

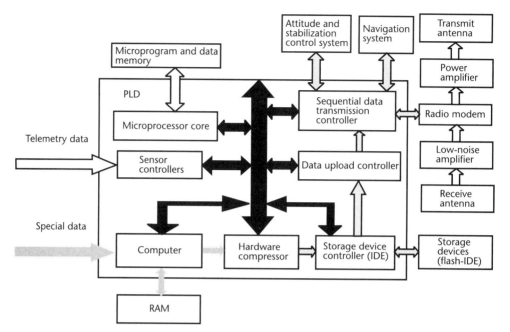

Figure 1.12   An example of PLD-based spacecraft OICC [9].

Figure 1.13   Evaluation board for programmable logic microcircuits.

1.5   Earth Remote Sensing Spacecraft

The development of technologies at the beginning of the twenty-first century has led to remarkable growth in the sphere of Earth remote sensing (ERS) [10–15]: new generation of spacecraft imaging systems has emerged that allows producing images of ultra-high spatial resolution (*GeoEYE-1* has spatial resolution of up to

41 cm). Imaging systems have hyperspectral and multiband multispectral modes (*WorldView-2* satellite has up to 8 bands). Recent tendencies include emergence of a new generation of satellites of ultra-high resolution with advanced characteristics (French system *Pleiades*), development of small satellite constellations for high-resolution, operational and global sensing of Earth surface (German satellite constellation *RapidEye*, new high-resolution satellite in *DMC* satellite constellation, advanced satellites *SkySat*, *NovaSAR*, and other). Apart from traditional fields of ERS technologies (improvement of spatial resolution, adding new spectral bands, automation of image processing and operational data delivery), new technologies for operational video imaging of Earth objects from space emerge (e.g., projects of SkyBox Imaging company, United States).

In this section we provide comparative characteristics of the most interesting ERS spacecraft of high and ultrahigh resolution that either have already been or are planned to be launched in the next 3–4 years [16].

Space activities are universally considered to be any kind of activities directly relating to the straightforward execution of work associated with exploration and making use of outer space. At present, more than 60 countries are actively involved in space activities. Nearly all the developed states of the world successfully use space technologies in communications and broadcasting, remote probing the Earth's surface (meteorological survey, cartography, geodesy, etc.), in navigation, and in scientific research. Space in the twenty-first century has become a sphere of ambition-driven attainments and clashes of interests of the countries in pursuit of near-Earth space environment exploration.

More than 50 years have elapsed since the Soviet Union launched the first spacecraft in 1957. During this time the world has seen the successful launches of more than 6800 satellites, manned spacecraft, long mission manned orbit stations, and automated interplanetary stations. The United States launched their first satellite Explorer-1 on January 31, 1958. The first Chinese spacecraft Dongfanghong-1 was launched on April 24, 1970.

Generalized data about the space activities of the United States, China, India, and Russia (USSR) are indicated in Figure 1.14 and in Table 1.1. Figure 1.14 shows the timeline of successful satellite launches made by the United States, Russia, and

Figure 1.14   Number of satellites successfully put into orbit from 1957 to 2010.

China from 1957 to 2010. Table 1.1 provides data on the total number of launched satellites, including the spacecrafts that were successfully deployed and operational in orbit, both for the last decade and the entire period of space activity.

It can be seen from Figure 1.14 that in the USSR the peak of space activity occurred from 1970 until 1991. During these years the USSR launched about a hundred space craft annually. Then, Russian space activity plummeted sharply, and in the first decade of the twenty-first century Russia successfully launched into orbit a mere 214 spacecraft or, on average, just over 21 spacecraft annually.

For space activity up to 2010 inclusive, Russia launched 3,479 spacecraft, of which 3,250 were successfully put into orbit. Thus, the sustainability ratio of a successful launch of an individual satellite in Russia is equal to 93.4%. The actual orbital strength of the civil, dual-purpose, and military satellites of Russia as per December 31, 2010, comprised 74 spacecraft. From the start of the first launches, the average period of the active existence of Russian spacecraft was elevated and in the 1990s reached its historical maximum. However, at present there is a steady drop in the average value of the satellites' active existence, despite the assurances of the producers about their constant advancement [2, 3]. The latter determined the substantial cuts in the orbital strength of Russia during the recent decade.

It follows from the schedule, that in the 1960s the United States launched, on average, about 70 satellites per year. During this time the American scientists and engineers mastered the construction technologies for the satellites with the active existence period incremented from 10 to 15 years and, as a consequence, in the 1970s the United States more than twice cut the number of launches, on average, to 30 spacecraft per year. Specifically, this very yearly average level of launches is maintained by the United States during the past 30 years. The peak of satellite launches by the United States at the end of the 1990s is determined by creation of the low orbit satellite systems of communication (Iridium, Globalstar, and Orbcomm).

For the entire period of the space activities from 1958 until 2010, the United States launched 2,402 satellites for their own requirements; from them 2,147 were successfully deployed into orbit. Thus, reliability of the successful launch of an

Table 1.1 Spacecraft Created and Launched into Orbit by Russia (USSR), United States, China, and India

Country	All World	Russia (USSR)		U.S.		China		India	
	1957–2010	1957–2010	2001–2010	1957–2010	2001–2010	1970–2010	2001–2010	1975–2010	2001–2010
Number of SC produced and launched into orbit/total successful	6,853/ 6,264	3,479/ 3,250	222/ 214	2,402/ 2,147	372/ 344	147/ 138	87/ 87	58/ 53	31/ 27
Number of SC, operational as per the status of December 31, 2010	958	74		440		69		29	

*SC produced and launched in the same country.

individual satellite in the United States is equal to 89.4%. The actual orbital United States satellite grouping of the civil, dual and military purpose as per the status of 31.12.2010 comprised 440 SC with the specific assigned missions. It is worthwhile noting that the United States enterprises built more than 300 satellites for other states.

In the 1970s, the Chinese People's Republic (China) joined the club of the space states of the world. From the schedule, represented in Figure 1.1, it ensues that in China the peak of the space activities was evident in the first decade of the twenty-first century. During these years China successfully launched into orbit 87 satellites from 87, including 20 satellites in 2010. Thus, reliability of the successful launch of an individual satellite in China in the recent ten years was equal to 100%, such a success for a decade was unattainable for both the United States and Russia.

During the period from 1970 until 2010 China built 147 satellites for its own requirements and 6 SC for other countries. From 147 satellites, 138 were success-fully deployed. Reliability of the satellite successful launch in the CPR for the entire period is equal to 93.9%, well surpassing the reliability of launchers in Russia and United States. Besides, China purchased nine satellites from the European and American producers. The Chinese orbital grouping as per the status of 31.12.2010 was 69 SC strong.

China created its own space system of the direct television broadcasting and the satellites for transmission of educational programs. China created several systems of the Earth surface remote probing, including the network of the geostationary and low orbital meteorological satellites. China successfully realizes the plans for the Moon explorations, having launched to it a re-entry space craft and delivered to the Earth the lunar probe samples. China entered the international space market of the launch services about 20 years ago. However, the satellite launches in the period from 1992 until 1996 were accident ridden, and China quit the market of the launch services for 12 years. At the end of the first decade of the twenty-first century, China successfully came back to the international space market, with a complex service, including the design development of satellites, their deployment into orbit, and their financing. This empowered China to build satellites for close to ten countries.

China leadership undertakes the measures to ensure the steady development of the space industry. Among them are perfection of the legislation base and methods of production management, ensuring activities in space with observance of the defi-nite standards. The state supports innovations in the sphere of space technologies and promotes creation in the space industry of the encouragement system, making it possible to strengthen the potential of its technological renewal. The state support of the industry is combined with the market principles, applied for its development. In 1999 the rocket-space industry was restructured by decision of the State Council of China. The only in the country corporation of the rocket-space industry (China Aerospace Corporation) was transformed into two independent state commercial corporations, being competitors: China Aerospace Science and Industry Corp. (CASIC), earlier known as China Aerospace Machinery and Electronics Corpora-tion (CAMEC) and China Aerospace Science and Technology Corp (CASQ) [4, 5].

Thus, at present in China the entire scientific research and production activi-ties on the rocket-space (military and civil grade) machinery is concentrated in two major state space corporations. Both corporations are the state commercial enterprises and have the structure, making it possible to realize in full volume the scientific researches, design developments and manufacture of the military and civil space products. At present the Chinese rocket-space industry is one of the major ones not only in Asia, but in the entire world regarding both the personnel strength and the volumes of sales.

During the same period in the major aerospace corporations of the United States, Lockheed Martin and Boeing, 140,000 and 160,000 people were employed. Now production of satellites and carrier rockets in the United States is concentrated in four companies: Lockheed Martin, Boeing, Space Systems/Loral, and Orbital Sciences Corporation.

In Russia the rocket-space industry comprises 100 enterprises with 320,000 employees [7]. Satellites are made at no less than 10 enterprises.

In 1975 the Republic of India emerged in the list of the space powers of the world. It follows from the schedule, represented in Figure 1.1, that in India the peak of the space activities occurred in the first decade of the twenty-first century. In these years India successfully launched into orbit 27 satellites from a total of 31.

During the period of the space activities from 1975 to 2010 inclusive India launched into orbit 62 satellites (58 from them were of its own production); from them, 57 were successfully put into orbit. Thus, reliability of the successful launch of an individual satellite in India for the entire period constituted 91.96%, which exceeds the launch reliability in the United States but was below the satellite launch reliability in China and Russia. The actual orbital force of the satellites of the civil, dual, and military purpose in India as per the status of January 2011 comprises 29 SC.

India has created the space system of the direct television broadcasting, which is being quite dynamically developed. The entire country receives the educational program via the special satellite EdUSt. In India a network of the geostationary and low orbital meteorological satellites has been created.

1.5.1 ERS Spacecraft of the Russian Federation and the Republic of Belarus

In accordance with the Federal Space Program in 2012 small SC *Kanopus-V* was launched (see Table 1.2 for the technical characteristics). It was designed for providing operational information to relevant departments of Roscosmos, Ministry of Emergency Situations of Russia, Ministry of Natural Resources and Environment of Russia, Federal Service for Hydrometeorology and Environmental Monitoring of Russia (Roshydromet), the Russian Academy of Sciences, and other interested institutions. The tasks of the satellite include:

- Detection of forest fire sources, large environmental pollutant emissions;
- Monitoring of manmade and natural emergencies including natural disasters and hydrometeorological phenomena;
- Monitoring of agriculture and natural (including water and coastal) resources;
- Land use;
- Operational observation of specified regions of the Earth's surface.

Apart from the *Kanopus-V* satellite, the Russian ERS constellation consists of satellites *Resurs-DK1* (launched in 2006) and *Monitor-E* (launched in 2005). The features of *Resurs-DK1* are improved operational and precision characteristics of received images (1m resolution in panchromatic mode and 2–3m resolution in multispectral mode). Satellite data is actively used for topographic and thematic mapping, data support for efficient use of natural resources and economic activities, inventory of forests and agricultural lands, and other purposes.

Electro-optical SC *Resurs-P* launched on June 25, 2013, continues the mission of the Russian high-resolution environmental satellites. Technical solutions for SC *Resurs-DK1* served as the basis for development of SC *Resurs-P* (see Table 1.3). Placing the spacecraft in circular sun-synchronous orbit at 475 km altitude has allowed improvement of observing conditions. Revisit time has been improved and now is 3 instead of 6 days.

The satellite has panchromatic and 5-band multispectral modes. In addition to electro-optical high-resolution equipment, the satellite has hyperspectral spectrometer (GSA) and high and medium resolution wide capture multispectral optical sensors (SHMS-VR and SHMS-SR).

It is planned to enlarge the Russian ERS constellation and to launch satellites of *Obzor* series. The constellation *Obzor-O* consisting of four electro-optical spacecraft is designed for operational multispectral imaging of Russia, adjacent territories of neighboring countries, and specific Earth regions. Two spacecraft are to be launched at the first phase (2017) and another two spacecraft are to be launched at the second phase (2018–2019). *Obzor-O* system will provide space imagery for the Ministry of Emergency Situations of Russia, the Ministry of Agriculture, the Russian Academy of Sciences, the Federal Service for State Registration, Cadastre, Cartography, and other ministries, institutions, and regions of Russia. It is planned to install hyperspectral equipment prototypes on SC *Obzor-O No. 1* and *No. 2*.

According to technical specification of *Obzor-O* system (see Table 1.4), it is to consist of four satellites capable of imaging in eight spectral bands, including visible and infrared bands. The visible band has a resolution of 5 meters and the infrared band has a resolution of at least 20 meters. At the first stage of project implementation when two spacecraft are in orbit, *Obzor-O* will cover the whole territory of Russia in no more than 30 days. When all four satellites are in orbit, they will cover it in no more than seven days.

Table 1.2 Main Technical Characteristics of SC *Kanopus-V* Imaging System

Imaging Mode	Panchromatic	Multispectral
Spectral range, μm	0.52–0.85	0.54–0.60 (green)
		0.63–0.69: 0.6–0.72 (red)
		0.75–0.86 (near IR)
Spatial resolution (at nadir), m	2.1	10.5
Swath width, km	Over 20 (at 510 km altitude)	
Acquisition capacity, million km²/day	Over 2	
Revisit time, days	5	
Data-transmission rate (X-band), Mbps	2 × 122.8	

Radar SC *Obzor-R* is designed (see Table 1.5) for X-band imaging at any time of the day (under any weather conditions) in the interest of social and economic development of the Russian Federation. *Obzor-R* will provide the Ministry of Emergency Situations of Russia, the Ministry of Agriculture, the Federal Service for State Registration, Cadastre and Cartography and other ministries, institutions and regions of Russia with radar imagery.

SC *Belarusian Space Vehicle* (*BKA*) launched together with Russian SC *Kanopus-V* provides coverage of the whole country's territory. According to international classification of spacecraft, *BKA* belongs to the class of minisatellites (it is completely identical to SC *Kanopus-V*). Payload of *BKA* includes panchromatic and multispectral cameras with 20 km swath width. Panchromatic and multispectral images of objects on Earth surface have 2.1m and 10.5m resolution, respectively. Such resolution is sufficient for various purposes connected with monitoring, for example, location of fire sources and the like. However, in the future the country might need a satellite with higher resolution. That is why Belarusian scientists have already started designing a spacecraft with 0.5m resolution which is expected to be launched after 2017.

Table 1.3 Main Characteristics of *Resurs-P* Imaging System

| Imaging Mode | Electro-Optical Equipment of High Resolution | | SHMS | | GSA |
	Panchromatic	Multispectral	SHMS-VR	SHMS-SR	
Spectral range, μm	0.58–0.80	0.45–0.52 (blue) 0.52–0.60 (green) 0.61–0.68 (red) 0.72–0.80; 0.67–0.70; 0.70–0.73 (red + near IR)	Panchromatic mode 0.43–0.70 Multispectral mode 0.43–0.51 (blue) 0.51–0.58 (green) 0.60–0.70 (red) 0.70–0.90 (near IR-1) 0.80–0.90 (near IR-2)		0.4–1.1 (96–255 spectral bands)
Spatial resolution (at nadir), m	1	3–4	12 (panchromatic mode) 24 (multispectral mode)	60 (panchromatic mode) 120 (multispectral mode)	25
Geolocation accuracy, m	CE90 mono = 3.1–21				
Swath width, km	38		96	480	25
Field of view, km	950		1300		950
Acquisition capacity, million km²/day	1				
Revisit time, days	3				

Table 1.4 Main Technical Characteristics of the *Obzor-O* Imaging System

Imaging Mode	Multispectral	
	First Stage	*Second Stage*
Spectral range, μm	7 simultaneously working spectral bands: 0.50–0.85 0.44–0.51 0.52–0.59 0.63–0.68 0.69–0.73 0.76–0.85 0.85–1.00	8 simultaneously working spectral bands: 0.50–0.85 0.44–0.51 0.52–0.59 0.63–0.68 0.69–0.73 0.76–0.85 0.85–1.00 1.55–1.70
Spatial resolution (at nadir), m	max. 7 (for 0.50–0.85 band) max. 14 (for other bands)	Max. 5 (for 0.50–0.85 band) max. 20 (for o.55–1.70 band) max. 14 (for other bands)
Radiometric resolution, bpp*	12	
Geolocation accuracy, m	30–45	20–40
Swath width, km	min. 85	min. 120
Acquisition capacity of each spacecraft, million km²/day	6	8
Revisit time, days	30	7
Data-transmission rate, Mbps	600	

*bpp—bits per pixel

Table 1.5 Main Technical Characteristics of *Obzor-R* Imaging System

Spectral Range	*X-band (3.1 cm)*			
Revisit time, days	*2 (in the latitude of 35–60°N)*			
Mode	*Nominal Spatial Resolution, m*	*Field of View, km*	*Swath Width, km*	*Polarization*
Ultra-fine	1	2 × 470	10	Single (optional—H/H, V/V, H/V, V/H)
Fine	3	2 × 600	50	Single (optional—H/H, V/V, H/V, V/H); Dual (V/ (V+H) and H/ (V+H))
Stripmap narrow	5	2 × 600	30	
	3	2 × 470		
Stripmap	20	2 × 600	130	
	40		230	
Stripmap wide	200	2 × 600	400	
	300		600	
	500	2 × 750	750	

In July 2013 LV *Soyuz-FT* with *Fregat* upper stage was used to launch a whole range of spacecraft into orbit from Baikonur Cosmodrome: *Kanopus-V, MKA-FKI (Zond-PP)*, Belarusian *BKA*, German *TET-1*, and Canadian *ADS-1B (ExactView-1)*.

Kanopus-V

Spacecraft *Kanopus-V No.1* is a part of space complex for operational monitoring of man-caused and natural emergencies, developed by the All-Russian Scientific and Research Institute of Electromechanics with Plant named after A.G. Iosifian (Federal State Unitary Enterprise NPP VNIIEM) at the request of Roscosmos.

Kanopus-V is designed for providing operational hydrometeorological data to address the following main tasks (see Table 1.6):

- Monitoring man-caused and natural emergencies, including natural hydro-meteorological phenomena;

Table 1.6 Main Technical Characteristics of SC *Kanopus-V*

Application	Earth Remote Sensing
SC Mass	400 kg
Payload mass	110 kg
SC dimensions	0.9×0.75m
Active lifespan	5 years
Working orbit	Sun-synchronous
Altitude	~510 km
Inclination	~98°
Orbital period	94.75 min
Orientation accuracy	5 ang. min
Stabilization accuracy	0.001 °/s
Retargeting rate capability (± 40°)	2 min
Daily average power consumption	300W
Imaging System Characteristics	
Panchromatic camera	
Swath width	23 km
Resolution (pixel projection)	2.1m
Spectral range	0.52–0.85 μm
Multispectral Camera	
Swath width	20 km
Resolution (pixel projection)	10.5m
Spectral ranges	0.54–0.6; 0.63–0.69; 0.69–0.72; 0.75–0.86 μm
Characteristics of Targeted Radio Link	
Memory capacity	24 Gbyte
Range of operating frequencies	8084–8381.5 MHz
Number of transmission bands	2
Data transmission rates	61.4 Mbps

- Mapping;
- Detection of forest fire sources, large environmental pollutant emissions;
- Monitoring abnormal physical phenomena to predict earthquakes;
- Monitoring of agriculture, water, and coastal resources;
- Land use;
- Highly operational observation of specific areas of Earth surface.

MKA-FKI (Zond-PP)

MKA-FKI No. 1 (see Table 1.7) is a part of a group of small spacecraft designed for fundamental space research (MKA-FKI). The scientific spacecraft complex is designed by the Federal State Unitary Enterprise Lavochkin Research and Production Association (FSUP NPO Lavochkin) at the request of Roscosmos.

MKA-FKI allows performing the following scientific tasks:

- Researching temperature and moisture state of forest and marsh systems;
- Studying biometric parameters of vegetation;
- Studying water salinity;
- Researching glacial and cryogenic areas;
- Studying energy exchange in ocean-land-atmosphere system;
- Researching geothermal activities;
- Mapping of soil moisture.

Table 1.7   Main Technical Characteristics of *MKA-FKI No. 1*

Application	Scientific Earth Remote Sensing
SC mass	110 kg
Active lifespan	3 years
Working orbit	Sun-synchronous
Altitude	~817km
Inclination	~97.4°
Orientation accuracy	6 ang.min
Stabilization accuracy	0.0015 °/s
Daily average power consumption	220W
Operating frequencies	
Transmitting	2268.946–2271.054 MHz
Receiving	2289.238–2091.346 MHz
Memory capacity	8 Gb
Data transmission rate	max. 3 Mbit/s
Payload	A small-size dual-beam L-band panoramic UHF radiometer Zond-PP
Mass	13 kg
Field of view	800 km
Width of the main antenna beam	22° × 15°

BKA

Belarusian Space Vehicle (BKA) is a part of Belarusian Space System for Earth Remote Sensing designed in accordance with the agreement between the National Academy of Sciences of Belarus and FSUP NPP VNIIEM.

Tasks of the Belarusian Space System:

- Providing the Republic of Belarus with regular and operational Earth remote sensing data of high spatial resolution;
- Ensuring operational receiving of space data at ground stations;
- Controlling *BKA* at all stages of its active lifespan;
- Creating basis for future development of multisatellite system for operational Earth observation.

The data received from the Belarusian Space System is applied in the following spheres:

- Monitoring land use and agriculture;
- Monitoring natural and renewable resources;
- Emergency monitoring;
- Updating topographic maps;
- Environmental monitoring.

ExactView 1 (ADS-1B)

ExactView 1 (EV 1) (see Table 1.8); also known as *ADS-1B*, is a spacecraft for ship tracking created by a leading developer of small satellites Surrey Satellite Technology Ltd. (SSTL) for a Canadian space equipment manufacturer COM DEV Canada.

The spacecraft, built on the *SSTL-100* platform, is part of the *exactEarth AIS* constellation of COM DEV company and will serve government agencies that monitor maritime traffic through busy shipping bands and harbors, as well as provide information on global maritime traffic.

The spacecraft payload features a VHF band receiver, onboard data handling subsystem (OBDH), and a C-band transmitter to relay data to Earth.

Tasks of OBDH include not only data processing but also data storage before the satellite passes it over to the ground station. In addition to this, the spacecraft has two reserve powerful S-band transmitters. The spacecraft is controlled with the help of S-band radio link.

TET-1

TET-1 is a technological microsatellite created by a German company Kayser-Threde GmbH for German Aerospace Center (DLR) (see Table 1.9).

TET is the core element of DLR's On-Orbit Verification Program (OOV), which provides industry and research institutes with means for demonstration of space technology in orbit.

SC *TET-1* is based on the platform of the Bi-Spectral Infra-Red Detection (BIRD) satellite that was launched in 2001. BIRD's design concept was adapted with regard

Table 1.8   Characteristics of SC *EV 1*

Application	Communication, Navigation
Working orbit	Sun-synchronous
Altitude	817 km
Inclination	98.88°
Ordering party	COM DEV Canada
Contractor	SSTL
Platform	SSTL-100
SC mass	95 kg
Active lifespan	5 years
Orientation accuracy	
Roll and pitch angles	< 3°
Yaw angle	< 5°
Daily average power consumption	66 W
Payload	
C-band transmitter	1
Operating frequency	5183 MHz
Data transmission rate	20 Mbps
VHF-band receiver	1
Operating frequency	162 MHz
Command S-band radio link	
Memory capacity	16 GB
Number of transmission bands	3
Data transmission	
Low speed	38.4 Kbps
High speed	Up to 8 Mbps
Transmission operating frequencies	
Low speed	2230 MHz
High speed	2233.333 MHz
Relay	2275.11 MHz

to newly available components to improve spacecraft characteristics, payload size and mass as well as system reliability.

The new satellite platform has a different principle of onboard data control.

Moreover, satellite sensors and actuators were modified and a new payload was added.

TET-1 satellite has 11 payloads. They include a lithium polymer battery, flexible solar cells, picosatellite and nanosatellite propulsion systems, infrared and visible light cameras for Earth imaging, and other tested electronic and structure members. Nine payloads are located in the payload section and three payloads are installed on the middle solar panel and PL panel.

Table 1.9   Characteristics of *TET-1*

Application	Demonstration of Technology
Working orbit	Sun-syncronous
Altitude	539–560 km
Inclination	97.6°
Ordering party	German Aerospace Center (DLR)
General contractor	Kayser-Threde GmbH
SC mass	117.3 kg
Active lifespan	1 year
Payload power consumption	
continuous power	20W
peak power	160W
Receiving frequency	
GPS signals	1575.42 MHz
Control command frequency	2032.5 MHz
Transmission frequency	2203.707 MHz

Soyuz-FG Launch Vehicle

Launch vehicle (LV) *Soyuz-FG* (see Table 1.10) is a modification of *Soyuz* series rocket. Energetic efficiency of its central block and lateral block engines was improved as compared with the serial LV *Soyuz*.

LV *Soyuz-FG* was designed and constructed by the Progress State Research and Production Space Centre (TsSKB-Progress) in Samara at the request of the Federal Space Agency.

LV *Soyuz-FG* implements parallel staging of lateral blocks at the end of the first stage burning and tandem staging of the second stage after it finishes burning. At the first stage engines of four lateral blocks and central block are burning and at the second stage only central block engine is burning.

In comparison to LV *Soyuz*, *Soyuz-FG* main propulsion system of first and second stages has an improved energetic efficiency due to the monopropellant injectors provided in the injector assemblies for better mixing.

Launch vehicle *Soyuz-FG* ensures orbital injection of the whole range of spacecraft launched by LV *Soyuz*.

The control system of LV *Soyuz-FG*, borrowed from basic LV *Soyuz* and slightly modified, ensures high injection accuracy. The first and second stages of LV *Soyuz-FG* as well as the third stage are equipped with radiotelemetric monitoring systems borrowed from *Soyuz* rocket to control LV's systems, assemblies and structures at launch phase.

The first stage of the launch vehicle consists of four lateral blocks of conical shape attached to the central block by means of spherical joints.

The second stage (central block) consists of a tail section with a single-start engine RD-108A (comprised of four cruise thrust chambers and four steering

Table 1.10   The Main Characteristics of Stages of LV *Soyuz-FG*

	First Stage (*Lateral Block*)	*Second Stage* (*Central Block*)	*Third Stage* (*Block I*)
Number of blocks	4	1	1
Length, m	19.6	27.1	6.7
Diameter, m	2.68	2.95	2.66
Gross mass, t	43.4	99.5	25.3
Empty mass, t	3.80	6.55	2.41
Engine	RD-107A	RD-108A	RD-0110
Number	1	1	1
Propellant:			
Oxidizer/fuel	Liquid oxygen/ kerosene	Liquid oxygen/ kerosene	Liquid oxygen/ kerosene
Thrust, kN:			
Sea level/Vacuum	838.5/1021.3	792.48/990.18	–/297.93
Burn time, s	118	280	230

nozzles), hydrogen peroxide tank with an inside-mounted liquid nitrogen tank, fuel tank, intertank bay, oxidizer tank, and equipment bay.

The third stage (block I) consists of an adapter module, fuel tank, oxidizer tank, tail section, and engine; it is mounted on the central block and is connected with it by a truss structure.

A picture is worth a thousand words—even the first images of *Kanopus-V* and *BKA* satellites prove that national industry has crossed an important milestone. The Russia and Belarus have its own powerful and efficient system for monitoring Earth from space.

NPP VNIIEM has demonstrated new space images (including color/multispectral images) produced by *Kanopus-V* and *BKA* (the Belarusian Space Vehicle, BelKa) spacecraft during flight tests and calibration of mission payload. At the moment images of objects in Greece, the United Arab Emirates, and Bahrain are released to public [15].

Even the first test images have a high geometric and radiometric quality, and it will be improved when the satellites are placed in normal operation. The spatial resolution is about 2m per pixel, which meets performance requirements. One can easily identify a typical outline of a national transport aircraft at Jebel Ali Airport. However, it's not the main point.

Image spatial resolution is high, though not record-setting, but the main point is that ERS satellites are conceptually new spacecraft in national industry that are conceptually different from their predecessors. We shall mention some of the differences.

The orbital arrangement of the satellites and sun-synchronous orbit ensure guaranteed efficiency of the system and imaging of any region of the Earth.

The guaranteed access to and use of standard spectral ranges provide solutions to a number of economic objectives, for example, classification of vegetation. One cannot underestimate its importance—not so long ago imaging in a shortwave blue range was an insoluble problem for a national high resolution satellite.

The satellites feature a pressurized platform and have a much lower mass than previous national precursors. This has reduced commissioning expenses and will reduce maintenance expenses in the future.

Joint creation of satellites with Belarus has introduced Belarussia's powerful industry of precision and optical equipment to space production cooperation. Belarusian industry has retained both its material resources and professional continuity.

The following figures show several innovative microelectronic devices developed in Belarus and used as ECB for electro-optical system of ERS SC.

These are charge-coupled devices (CCD) the physics of which is based on charge exchange between MOS capacitors that are closely located on a semiconductor chip. By manipulating biases of MOS capacitors, charges can be not only accumulated but also transferred, divided, and combined (i.e., basically information can be processed on such discrete elements). Some 45 years ago Bell Laboratories produced the first CCD structures with eight discrete elements whereas modern devices have more than 100 million elements (pixels). There was an active development of CCD as multielement photoreceivers and photodetectors for various electro-optical devices and cameras, including high resolution ERS SC.

In 1981–1984 SC shuttle (STS-7 and STS-11) had two CCD modifications with a line array of 1728 pixels (with the total length of 6912 pixels).

The parameters were as follows:

- Photo area format 2048×128;
- Cell size 8×8 μm;
- Scanning frequency 2×20 Mhz;
- Sectionalization 8, 16, 32, 64, 128;
- Four-phase control of photo area;
- Dynamic range—2000;
- Charge capacity ≥ 150 ke;
- Direct and mirror topology of integrated assembly.

Figure 1.15 shows first integrated microassemblies of PCCD for Earth remote sensing spacecraft that have six interdependent CCD arrays located on a silicon board. Figure 1.16 and Figure 1.17 show their further development.

Earth remote sensing satellites *Kanopus-V* and *BKA* are oriented toward actual needs of local consumers and accumulate precise factual and contextual information on a global scale, which is a primary objective at this point of history (Figure 1.18).

For the first time in Russian practice optical systems will have an open rational polynomial coefficients (RPC) model, a mathematical model that will facilitate data management.

The system was designed with regard to both technological and general scientific changes connected with the shift of geographical paradigm. The principles of neogeography and situational awareness are embedded in the system, and it will ensure high specific performance, compliance with requirements of customers, and high capacity for development and modernization. In particular, flight missions are defined with the help of neogeographic geointerface Neoglobus developed by

Figure 1.15   First photosensitive CCD arrays produced by Integral JSC.

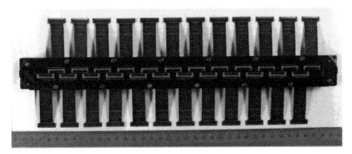

Figure 1.16   CCD microassemblies BAI 2093M6.

Figure 1.17   CCD microassemblies BAI 2093M4.

NPP VNIIEM that allows optimizing flight missions with regard to a number of factors—presence or absence of retrospective imaging, terrain relief, SC location, meteorological conditions, general context, and the like.

It is crucial to create an optimally balanced, efficient, and user-friendly system of remote sensing because a nonsystematic remote sensing that produces so much

Figure 1.18 Artificial Palm Island in Dubai (the United Arab Emirates). Panchromatic image. Image: VNIIEM.

output that it can't be processed can disorganize state control. The United States has already faced this problem, and Russia should learn from the mistakes of the United States.

What value does constellation *Kanopus-V-BKA* present to the Union State? In order to answer this question, it is important to consider the recent past situation.

Half a century after Yuri Gagarin's flight, Russia was inferior in terms of Earth remote sensing not only to India or Canada, but also to Nigeria. Space images weren't exported but were imported. Even such spacecraft as *Resurs-DK* that could sometimes produce submeter resolution images were only an exception. The fact that the world's largest country lacks information about itself and has to acquire it on the foreign markets opened prospects for manipulation. Fabrication of space images and monopolization of key orbital resources began.

Launches of *Kanopus* and *BKA* changed the situation—Russia and Belarus now have heir own well-operating ERS system. The Russia and Belarus accumulate experience examines opportunities and restores industry with precision instrument-making in the first place.

In conclusion it's important to note an obvious fact—the more the country finances the development of space industry, the more powerful infrastructure is developed as a result of national space program implementation.

1.5.2 ERS Spacecraft of Ukraine

SC *Sich-2* (see Table 1.11) was launched within the framework of the national program of Ukraine to develop systems for space monitoring and providing national

Table 1.11 Main Technical Characteristics of *Sich-2* Imaging System

Imaging Mode	Panchromatic	Multispectral
Spectral range, μm	0.51–0.90	0.51–0.59 (green)
		0.61–0.68 (red)
		0.80–0.89 (near IR)
Spatial resolution (at nadir), m	8.2	
Swath width, km	48.8	

economy with geo-information. The satellite is equipped with an electro-optical sensor with three spectral and one panchromatic band as well as with a middle infrared scanner and a set of scientific equipment named *Potential*. The main tasks of SC *Sich-2* include monitoring of agriculture and land resources, water bodies, forest vegetation and monitoring of emergency regions. The satellite was taken out of operation in May 2013. The State Space Agency of Ukraine plans to launch SC *Sich-3-O* with the resolution higher than 1 meter. The satellite is designed at DO Yuzhnoye.

1.5.3 ERS Spacecraft of the United States

In the United States the ERS industry is developing mainly in the field of ultra-high resolution. On February 1, 2013, Digital Globe and Geo Eye, two leading American companies that provide ultra-high resolution data, merged.

The new company retains the name Digital Globe.

As a result of the merger Digital Globe has now unique opportunities for providing a wide range of space images and geo-information services. Despite the monopolistic position of the joint company in a highly profitable market segment, the bulk of its income (75–80%) is gained from a defense order within a 10-year program *EnhancedView* (EV) worth a total of $7.35 billion USD. The program foresees government procurement of commercial satellite services of interest to the National Geospatial Intelligence Agency (NGA).

At the moment Digital Globe operates the following ERS satellites of ultra-high resolution: *WorldView-1* (50 cm resolution), *WorldView-2* (46 cm), *QuickBird* (61 cm), *GeoEye-1* (41 cm), and *IKONOS* (1m). The total daily acquisition capacity of the system is over 3 billion square kilometers.

In 2010 Digital Globe concluded a contract with Ball Aerospace for development and construction of the *WorldView-3* satellite with contract value of $180.6 million USD. Exelis VIS received a contract for construction of satellite onboard imaging of total $120.5 million USD. *WorldView-3* imaging system of will be similar to that of *WorldView-2* (see Table 1.12). Moreover, imaging modes will include SWIR mode (8 bands, 3.7 meter resolution) and CAVIS mode (12 bands, 30 meter resolution).

The development of advanced SC *GeoEye-2* started in 2007. The spacecraft is planned to have 0.25–0.3 m resolution in panchromatic mode and improved spectral characteristics. Exelis VIS is the manufacturer of the sensor. Originally it was planned to launch the spacecraft in 2013. However, after the merger of Digital Globe and Geo Eye, it was decided to put the spacecraft into storage until one of the satellites in orbit needs to be replaced or demand makes the launch profitable for the company.

Table 1.12 Main Technical Characteristics of SC *WorldView-3* Imaging System

Imaging Mode	Panchromatic	Multispectral
Spectral range, µm	0.50–0.90	0.40–0.45 (coastal)
		0.45–0.51 (blue)
		0.51–0.58 (green)
		0.585–0.625 (yellow)
		0.63–0.69 (red)
		0.63–0.69 (red edge)
		0.77–0.895 (near IR-1)
		0.86–1.04 (near IR-2)
Spatial resolution (at nadir), m	0.31	1.24
Maximum off-nadir angle, degrees	40	
Radiometric resolution, bpp	11	
Geolocation accuracy, m	CE90 mono = 3.5	
Swath width, km	13.1	
Revisit time, days	1	
Stereo imaging	Yes	
File formats	GeoTIFF, NITF	

SC *Landsat-8* (Table 1.13) (project *LDCM—Landsat Data Continuity Mission*) was launched on February 11, 2013. The satellite will continue to fill the *Landsat* image bank that has existed for 40 years and covers the whole surface of the Earth. SC *Landsat-8* is equipped with two sensors: electro-optical (Operational Land Imager, OLI) and thermal (Thermal Infrared Sensor, TIRS).

1.5.4 ERS Spacecraft of France

Astrium GEO-Information Services, geo-information division of multinational company Astrium Services, is the major commercial operator of ERS satellites in France. The company was created in 2008 as a result of a merger of SpotImage (France) and Infoterra Group of Companies. Astrium Services-GEO-Information is the operator of *SPOT* and *Pleiades* optical satellites of high and ultra-high resolution, new generation radar satellites *TerraSAR-X* and *TanDEM-X*. The headquarters

Table 1.13 Twelve Main Technical Characteristics of SC *Landsat-8* Imaging System

Imaging Mode	VNIR	SWIR	PAN	TIR
Spectral range, µm	0.43–0.45 (coastal)	1.36–1.39 (Cirrus)	0.50–0.68	10.40–12.50
	0.45–0.52 (blue)	1.56–1.66 (SWIR-1)		
	0.53–0.60 (green)	2.10–2.30 (SWIR-2)		
	0.63–0.68 (red)			
	0.85–0.89 (near infrared)			
Spatial resolution (at nadir), m	30	30	15	100
Radiometric resolution, bpp	12			

of Astrium Services-GEO-Information is in Toulouse and there are 20 offices and 100 distributors around the world.

Astrium Services is a part of the European Aeronautic Defense and Space Company (EADS).

Earth observation satellite system Satellite Pour L'Observation de la Terre (SPOT) was designed by the National Center for Space Studies (CNES) together with Belgium and Sweden. SPOT includes a range of space and ground systems. At the moment *SPOT-5* (launched in 2002), *SPOT-6* (launched in 2012), and *SPOT-7* (see Table 1.14) (launched in 2014) are operating in orbit.

SPOT-4 was taken out of operation in January 2013. *SPOT-6* and *SPOT-7* have identical characteristics.

Launch of SC *Pleiades-1A* and *Pleiades-1B* in 2011–2012 marked the beginning of French program for ultra-high resolution Earth imaging and competitive struggle with American commercial ERS systems.

Pleiades High Resolution program is a part of the European ERS satellite system and has been operated by CNES since 2001.

Pleiades-1A and *Pleiades-1B* satellites (see Table 1.15) are synchronized in one orbit so that they provide daily imagery of one and the same surface area. New-generation space technologies, such as fiber-optic gyrometers, provide satellites with unprecedented maneuvering capability. The satellites can image any area on

Table 1.14 Main Technical Characteristics of *SPOT-6* and *SPOT-7* SC Imaging Systems

Imaging Mode	Panchromatic	Multispectral
Spectral range, µm	0.48–0.71	0.50–0.59 (green)
		0.61–0.68 (red)
		0.78–0.89 (near IR)
Spatial resolution (at nadir), m	2	8
Geolocation accuracy, m	CE90 = 10	
Swath width, km	60	
Stereo imaging	Yes	
Acquisition capacity, million km²/day	3	

Table 1.15 Basic Technical Characteristics of the Surveying Equipment KA Pleiades

Shooting Mode	Panchromatic	Multichromatic
Spectral range, µm	0.48–0.83	0.43–0.55 (blue)
		0.49–0.61 (green)
		0.6–0.72 (red)
		0.79–0.95 (close IR)
Spatial resolution (nadir), m	0.7	2.8
Geopositioning accuracy, m	CE90 = 4.5	
Shooting band width, km.	20	
Stereo-frame option	Yes	
Shooting capacity, mln sq. km / nautical day	1.5	

an 800-km swath in less than 25 seconds with geolocation accuracy of less than 3 meters (CE90) without ground reference site and less than 1 meter with the use of ground reference site. The daily acquisition capacity in panchromatic and multi-spectral mode is over 1 million square kilometers per day.

1.5.5 ERS Spacecraft of Japan

ALOS is the most renowned ERS satellite of Japan (electro-optical imaging of 2.5m resolution in panchromatic mode and 10m resolution in multispectral mode as well as radar L-band imaging of 12.5m resolution). SC *ALOS* was created within the framework of Japan Space Program and is financed by Japan Aerospace Exploration Agency (JAXA).

ALOS was launched in 2006. On April 22, 2011, problems occurred with satellite control. On May 12, 2011, after three weeks of attempts to recover the satellite, its power was commanded off. At the moment only archived images are available.

ALOS will be replaced with two spacecraft: an electro-optical spacecraft and a radar spacecraft. Thus, JAXA specialists diverged from combining optical and radar systems on one platform, which was the case with *ALOS* satellite that had two optical cameras (PRISM and AVNIR) and a radar (PALSAR).

ALOS-2 radar satellite was launched in 2014 (see Table 1.16). Electro-optical SC *ALOS-3* is planned to launch in 2016 (see Table 1.17). The satellite will be capable of imaging in panchromatic, multispectral and hyperspectral modes.

It is important to note the Japanese project *Advanced Satellite with New System Architecture for Observation (ASNARO)* (see Table 1.18) that was initiated by Institute for Unmanned Space Experiment Free Flyer (USEF) in 2008. Innovative technologies for creation of minisatellite platforms (100–500 kg mass) and imaging systems are the basis of the project. One of *ASNARO* objectives is to create a new-generation ultrahigh resolution minisatellite that can compete with similar

Table 1.16 Main Technical Characteristics of SC *ALOS-2* Imaging System

Spectral Range, μm		L-band
Revisit time, days		14
Data-transmission rate, Mbit/s		800
Mode	*Nominal space resolution, m*	*Swath width, km*
Spotlight	1–3	25
StripMap	3–10	50–70
ScanSAR	100	350

Table 1.17 Main Technical Characteristics of SC *ALOS-3* Imaging System

Imaging Mode	*Panchromatic*	*Multispectral*	*Hyperspectral*
Spatial resolution (at nadir), m	0.8	5	30
Swath width, km	50	90	30

Table 1.18   Main Technical Characteristics of SC *ASNARO* Imaging System

Imaging Mode	Panchromatic	Multispectral
Number of spectral bands	1	6
Spatial resolution (at nadir), m	0.5	2
Radiometric resolution, bpp	12	
Swath width, km	10	

satellites from other countries due to data cost-cutting and reduced terms of satellite construction. *ASNARO* satellite is designed for Earth surface imaging for interested Japanese government institutions.

1.5.6   ERS Spacecraft of India

The country has created one of the most efficient ERS programs on the base of the state system for space industry financing. India successfully operates a group of spacecraft of various applications, including those of *RESOURCESAT* and *CARTOSAT* series.

In April 2011 *RESOURCESAT-2* was designed for prevention of natural disasters, and water and land resource control joined the satellite constellation in orbit (see Table 1.19).

SC *RISAT-1* (see Table 1.20), equipped with a multifunctional C-band frequency range (5.35 GHz) radar was launched on April 26, 2012. The satellite is designed for 24-hour all-weather Earth imaging in various modes. The satellite produces images in C-band wavelength range with variable polarization (HH, VH, HV, VV).

A constellation of cartographic electro-optical satellites of the *CARTOSAT* series is operating in orbit. *CARTOSAT-2c* was launched in 2016. The satellite is equipped with electro-optical equipment of unprecedented 25 cm spatial resolution.

1.5.7   ERS Spacecraft of China

In the past eight years, China has created a multipurpose constellation of ERS satellites that consists of several space systems: imagery intelligence satellites and satellites designed for oceanography, cartography, monitoring of natural resources, and emergency situations.

In 2011 China launched more ERS satellites than any other country: two imagery intelligence satellites *Yaogan (YG)-12* (equipped with electro-optical system of submeter resolution) and *Yaogan (YG)-13* (equipped with synthetic aperture radar); SC *HaiYang(HY)-2A* equipped with microwave radiometer for oceanographic purposes; multipurpose satellite *ZiYuan(ZY)-1-02C* designed for monitoring of natural resources of interest to the Ministry of Land and Resources (2.3m resolution in panchromatic mode and 0.5m resolution in multispectral mode, 54 and 60 km swath width); optical microsatellite (35 kg) *TianXun* (TX) of 30m resolution.

In 2012 China again became the leader in terms of number of launches—the national ERS satellites constellation (not including meteorological satellites) gained another five satellites: *Yaogan(YG)-14* and *Yaogan(YG)-15* (imagery intelligence), *ZiYuan(ZY)-3* and *TianHui(TH)-2* (cartographic satellites), and radar SC *HuanJing(HJ)-1C*.

TH-1 and TH-2 are the first Chinese satellites that capture triplet stereo images for geodetic surveying and mapping (see Table 1.21). The satellites are identical in terms of technical characteristics and their mission. Each satellite is equipped with

Table 1.19 Main Technical Characteristics of SC *RESOURCESAT-2* Imaging System

Imaging Mode	LISS-4		LISS-3 (Multispectral)	A WiFS (Multispectral)
	Monospectral	Multispectral		
Spectral range, μm	0.62–0.68	0.52–0.59 (green) 0.62–0.68 (red) 0.77—0.86 (near infrared)	0.52–0.59 (green) 0.62–0.68 (red) 0.77–0.86 (near IR) 1.55–1.70 (middle IR)	0.52–0.59 (green) 0.62–0.68 (red) 0.77–0.86 (near IR) 1.55–1.70 (middle IR)
Spatial resolution (at nadir), m	5.8		23.5	56
Radiometric resolution, bpp	10		10	12
Swath width, km	70		141	740

Table 1.20 Main technical characteristics of SC *RISAT-1* imaging system

Spectral Range	C-band			
Mode	Nominal Spatial Range, m	Swath Width, km	Incidence Angle, Degree	Polarization
High resolution spotlight (HRS)	< 2	10	20–49	Single
Fine resolution stripmap-1 (FRS-1)	3	30	20–49	
Fine resolution stripmap-2 (FRS-2)	6	30	20–49	Quadruple
Medium resolution scanSAR (MRS)/ coarse resolution scanSAR (CRS)	25/50	120/240	20–49	Single

Table 1.21 Main Technical Characteristics of SC *TH-1* and SC *TH-2* imaging Systems

Imaging Mode	Panchromatic	Multispectral	Stereo Triplet
Spectral range, μm	0.51–0.69	0.43–0.52 (blue) 0.52–0.61 (green) 0.61–0.69 (red) 0.76–0.90 (near IR)	0.51–0.69
Planned operating life, years	2	10	5
Geolocation accuracy, m	CE90 = 25		
Swath width, km	60	90	60
Temporal resolution, days	9		
Stereo imaging	Yes		

three cameras: a stereo camera for triplet stereo images, a high resolution panchromatic camera, and a multispectral camera. The cameras provide imagery of the Earth's surface for scientific research and monitoring of land resources, geodetic surveying, and cartography.

The satellites are designed for various applications:

- Creation and upgrading of topographic maps;
- Creation of digital terrain models;
- Creation of 3D models;
- Landscape changes monitoring;
- Land use monitoring;
- Monitoring of agricultural crops and crop-yield prediction;
- Forest use and forest health monitoring;
- Monitoring of irrigation facilities;
- Water quality monitoring.

1.5.8   ERS Spacecraft of the European Space Agency

In order to ensure global environmental monitoring, in 1998 the institutions of the European Union decided to establish the Global Monitoring for Environment and Security (GMES) program under the auspices of the European Commission in cooperation with the European Space Agency (ESA) and the European Environment Agency (EEA). Being today's major system for Earth observation, GMES provides the European Union's users with accurate, timely, and easily accessible information to improve the management of the environment, understand and mitigate the effects of climate change, ensure civil security, and for other purposes.

GMES in its final operational stage will be comprised of a complex of observation systems: ERS satellites, ground stations, marine vessels, atmospheric sounders, and the like [16].

GMES space segment will build on two types of ERS systems: *Sentinel* satellites designed especially for the GMES program (operated by ESA) and national (or international) ERS satellite systems that are part of the GMES contributing missions (GCMs).

The first Sentinel satellites were launched in 2014. It is prudent that they will capture images with the help of various technologies (e.g., radars and electro-optical multispectral sensors).

In order to implement the *GMES* program, five *Sentinel* ERS satellites are developed under general supervision of ESA and each of the satellites will carry out a specified mission connected with Earth monitoring.

Thus, two satellites will be assigned to each *Sentinel* mission so that the fullest coverage and acceleration of imaging revisit are ensured and *GMES* reliability and data integrity are improved.

The *Sentinel-1* mission will be a constellation of two radar satellites in polar orbit equipped with synthetic aperture radar (SAR) for C-band imaging capable of producing images regardless of weather and time of the day. The first satellite was launched in 2014 and the second one in 2015. The *Sentinel-1* mission, designed especially for *GMES* program, will continue C-band imaging carried out by *ERS-1, ERS-2, Envisat* (operated by ESA) and *RADARSAT-1, 2* (operated by MDA, Canada).

The *Sentinel-1* constellation will provide imagery (see Table 1.22) of the whole Europe, Canada and the main waterways each 1–3 days regardless of weather conditions. Radar data will be delivered within an hour after imaging, which is a big step forward as compared with existing radar satellite systems.

A pair of *Sentinel-2* satellites will regularly provide high resolution Earth imagery ensuring continuous acquisition of data with characteristics analogous to that of *SPOT* and *Landsat* programs (see Table 1.23).

Sentinel-2 will be equipped with a electro-optical multispectral sensor for capturing images of 10–60m resolution in visible, near infrared (VNIR), and shortwave infrared (SWIR) spectral bands, including 13 spectral bands that will ensure imaging of vegetation changes (also temporal changes) and minimize atmospheric effect on quality of images.

Average orbit altitude of 785 km and presence of two satellites will provide revisit time of 5 days at the equator and 2–3 days in mid-latitudes. The first satellite was launched in 2015. The second is planned for 2017.

Expanded swath width and fast revisit time allow monitoring of rapidly changing processes (e.g., flora characteristics during vegetation period).

The *Sentinel-2* mission has a unique combination of large territory coverage and frequent re-imaging, which results in systematic acquisition of multispectral full-coverage high resolution Earth imagery.

The main objective of the *Sentinel-3* mission is monitoring ocean topography, sea and land surface temperature, and sea and land color with high level of accuracy

Table 1.22   Main Technical Characteristics of SC *SENTINEL-1* Imaging Systems

Spectral Range	C-band		
Revisit time, days	1–3		
Mode	Nominal spatial resolution, m	Swath width, km	Polarization
Interferometric Wide Swath	5 × 20	250	Dual (optional—HH/HV or VV/VH)
Extra Wide Swath	20 × 40	400	
Stripmap	5 × 5	80	
Wave	20 × 5	20 × 20	Single (optional—VV or HH)

Table 1.23   Main Technical Characteristics of SC *SENTINEL-2* Imaging Systems

Imaging Mode	VNIR										SWIR		
Spectral bands	1	2	3	4	5	6	7	8	8a	9	10	11	12
Spectral range, μm	0.44	0.49	0.56	0.66	0.70	0.74	0.78	0.84	0.86	0.94	1.38	1.61	2.19
Spatial resolution (at nadir), m	60	10	10	10	20	20	20	10	20	60	60	20	20
Swath width, km	290												
Revisit time, days	from 5 (at the equator) to 2–3 (in mid-latitudes)												

and reliability to support ocean system forecasting as well as monitoring environment and climate.

Sentinel-3 is the successor of proven *ERS-2* and *Envisat* satellites. Two *Sentinel-3* satellites allow fast revisit time. The satellite orbit (815 km) will ensure delivery of a full data package every 27 days. The first satellite of *Sentinel-3* pair was launched in 2016, right after the launch of *Sentinel-2*. *Sentinel-3B* satellite is planned to launch in 2018. *Sentinel-4* and *Sentinel-5* missions are designed for providing GMES services with data on atmospheric composition. Both missions will be implemented on the platforms of meteorological satellites operated by the European Organization for the Exploration of Meteorological Satellites (EUMETSAT). It is planned to launch the satellites in 2017–2019 [16].

1.5.9   Earth Remote Sensing Spacecraft of Other Countries

Target-specific Earth remote sensing is actively developed not only in the previously mentioned economies but also in a number of other countries [16].

Canada plans to further develop the *RADARSAT* satellite series, strengthening its leading position on the market of radar imaging. *RADARSAT-1* and *RADAR-SAT-2* satellites are currently in orbit.

On January 9, 2013, MDA announced concluding a contract of $706 million value with the Canadian Space Agency to create and launch a constellation of three radar satellites, the *RADARSAT Constellation Mission* (RCM). Duration of the contract is seven years.

RCM will ensure 24-hour radar coverage of the whole country. The data may include revisit images of the same regions at different times of the day, which will significantly improve monitoring of coastal areas, the north, arctic waterways, and other regions of strategic and defense importance. *RCM* system will also include an automatic image interpretation system that together with operational data delivery will instantaneously detect and identify marine vessels across the world's oceans. Substantial acceleration of data processing speed is anticipated.

The Republic of Korea created a national Earth remote sensing system in 1992 within the framework of space program.

Korea Aerospace Research Institute (KARI) developed a series of Earth observation satellites, *Korean MultiPurpose Satellite* (*KOMPSAT*). SC *KOMPSAT-1* was used for military purposes until 2007. *KOMPSAT-2* satellite was placed in orbit in 2006.

KOMPSAT-3 launched in 2012 continues *KOMPSAT* mission and is designed for producing Earth surface digital images of 0.7m spatial resolution in panchromatic mode and 2.8m in multispectral mode.

KOMPSAT-5 project is a part of the Korean national development plan of its Ministry for Education, Science, and Technology (MEST), which started in 2005. The project is developed by the Korea Aerospace Research Institute (KARI). The primary aim is to develop a radar satellite system for the monitoring purposes. The system will be capable of producing images in C-band with variable polarization (HH, VH, HV, VV).

In the *United Kingdom* DMC International Imaging Ltd.(DMCii) operates the *Disaster Monitoring Constellation* (DMC) working both in the interest of countries-owners of satellites and providing space imagery for commercial use. The DMC provides state agencies and commercial customers with operational imagery of disaster areas. The satellites provide data for the purposes of agriculture, forestry, and other. The satellites were developed by Surrey Satellite Technology Ltd. (SSTL) from the UK. All the satellites are placed in sun-synchronous orbit to ensure daily global coverage. The first generation of DMC spacecraft was launched starting in 2002 and now comprises six satellites.

SC *UK-DMC-2* was launched in 2009 as a part of DMC. It carries out imaging in multispectral mode with the resolution of 22m on a swath of 660m. *DMC-3a,b,c* satellites of improved characteristics were launched in 2015. They produce images of 1m resolution in panchromatic mode and 4m resolution in a 4-band multispectral mode (including infrared band) on a swath of 23 km.

At the moment SSTL is finishing the development of a new low-budget radar satellite: 400-kilogram SC *NovaSAR-S* will have a *SSTL-300* platform and an innovative radar for S-band imaging. SSTL's approach to engineering and design allows full-scale deployment of *NovaSAR-S* mission within 24 months of receipt of order.

NovaSAR-S will also carry out radar imaging in four modes with the resolution of 6–30m and various polarization combinations. Technical parameters of the satellite are optimized for a wide range of purposes, including flood monitoring, crops assessment, forestry monitoring and classification of vegetation cover, disaster management, monitoring of water areas—in particular maritime traffic tracking, and detection of oil spills.

Spain is developing a national ERS satellite constellation. The *Deimos-1* satellite was placed in orbit in 2009 as a part of DMC international constellation. It conducts imaging in multispectral mode with the resolution of 22m on a swath of 660m. The satellite is operated by Deimos Imaging, the company that originated as a result of cooperation of Spanish aerospace engineering company Deimos Space and ERS Laboratory of the University of Valladolid (LATUV). The primary objective of the new company is to develop, launch, and operate commercial ERS systems. The company is located in Valladolid, Spain.

Deimos Imaging developed the *Deimos-2* high resolution satellite launched in 2014. SC *Deimos-2* is designed for acquiring inexpensive multispectral Earth remote sensing data of high quality. Together, *Deimos-1* and *Deimos-2* comprise satellite system Deimos Imaging.

In the next two years the implementation of the national program for Earth observation *Programa Nacional de Observación de la Tierra por Satélite* (*PNOTS*) will start. A satellite named *Paz* (which means *peace* in Spanish, the satellite is also known under the name *Satélite Español de Observación SAR*, *SEOSAR*) is the first Spanish radar double-purpose satellite and one of the program's parts. The satellite is capable of imaging under any weather conditions by day and night and its primary purpose is fulfilling orders of Spain government that are connected with security and defense. SC *Paz* will be equipped with a synthetic aperture radar developed by Astrium GmbH on the platform of *TerraSAR-X* radar satellite.

SC *Amazonia-1* will be also equipped with British electro-optical system *RAL-Cam-3* capable of producing images of 10m resolution on a swath of 88 km. Radar minisatellite *MAPSAR* (*multiapplication purpose*) is a joint project of INPE and the German Aerospace Center (DLR). The satellite is designed to work in three modes (3, 10, and 20m resolution). It was launched in 2015.

In this section we didn't aim to analyze all the new and advanced national ERS systems of high and ultra-high resolution. More than 20 countries now have their own Earth observation satellites. Apart from the previously mentioned ones, the following countries have such systems: *Germany* (constellation of electro-optical satellites *RapidEye*, radar satellites *TerraSAR-X* and *TanDEM-X*), *Israel* (SC *EROS-A,B*), *Italy* (radar SC *COSMO-SkyMed-1-4*), and others. This kind of space club gains new countries and ERS systems every year. In 2011–2012 satellites were launched by *Nigeria* (*Nigeriasat-X* and *Nigeriasat-2*), *Argentina* (SAC-D), *Chili* (*SSOT*), *Venezuela* (*VRSS-1*), and others. *Gokturk-2* satellite (2.5m resolution in panchromatic mode, 10m resolution in multispectral mode) was launched in December 2012 and it continued with the Turkish ERS program (the third satellite of *Gorturk* series launched in 2015. The *United Arab Emirates* are planning to launch their own ultra-high resolution satellite *Dubaisat-2* (1m resolution in panchromatic mode, 4m resolution in multispectral mode).

Conceptually new systems for space monitoring are under development. Terra Bella (formerly Skybox Imaging) located in Silicon Valley, is working on *Skysat*, the world's most powerful innovative ERS satellite constellation. It will provide high resolution space imagery of any Earth area several times a day. Data will be used for operational response to emergency situations, environment monitoring, and the like. The imaging will be made in panchromatic and multispectral modes. *SkySat-1*, the first satellite of the constellation, was launched in 2013. When the whole constellation is in orbit (it is planned to launch 20 satellites altogether), it will be possible to observe any part of the Earth in real time. It is also planned to perform video imaging from space.

1.6   Earth Remote Sensing Radar Stations

Today's synthetic aperture radars (SAR) find wide application among Earth remote sensing radar stations (RS) [11, 12, 17]. SAR produces radar images of ultra-high resolution since its resolution is considerably higher compared with systems that have real size antenna aperture.

This section provides a brief but quality review of ERS satellites that have standard view RS with SAR [17].

Analysis shows that there is quite a number of space-based systems with remote sensing equipment that have SAR and can produce radar images of quality comparable with images of Earth's surface produced by optical and electro-optical systems.

Modern synthetic aperture radars will have large data capability provided that technologies for Earth remote sensing, algorithms, and hardware for data processing and interpretation are improved and there's advance in the development of element base designed for the implementation of global aerospace system of ongoing tracking of natural and man-caused dynamic processes, ecosystem monitoring, and armed conflict monitoring.

Equipment for space-based SAR has already been developed and it generally meets the requirements of civil and military data consumers within the limits of radio regulations on the use of frequency bands for Earth remote sensing.

Use of measuring technologies is an essential factor for acquiring thematic data from Earth radar sensing, as stated in [17].

Apart from amplitude (energy) images that measure radar cross-section (RCS) of objects and specific RCS of land areas with different polarization properties, interferometric processing of complex images is increasingly used (terrain mapping, differential interferometry, measuring the speed of compact and distributed objects, polarimetric interferometry) [17].

SAR on spacecraft is actively used for sensing in different wave bands. It provides high resolution radar images (0.5–1.5m) in 3-cm wavelength for the purposes of imagery intelligence, detection of small objects in open terrain, as well as images of 3m resolution in 3- to 5.6-cm wavelength for vegetation monitoring and other purposes of Earth remote sensing (ERS). As compared with S-band sensing, L-band sensing in 23-cm band with 3m resolution has its advantages: it detects vehicles under leaves and there's considerable improvement of signal phase stability during interferometric processing (terrain mapping and detection of changes in operational environment using differential interferometry). Also estimation of biomass amount and identification of vegetation are improved.

An important development direction of SAR caused by advances in spatial and radiometric resolution is a wide application of radio-wave imaging and multipositional sensing to produce 3D radar images of objects and details of their shapes as well as application of automated methods for classification of objects by their radar images and image texture.

Multipositional sensing and miniaturization of receive sensors of inexpensive minisatellites and unmanned aerial vehicles allows permanent monitoring of local regions for military operations support and emergency situation monitoring.

Advanced Earth remote sensing spacecraft under development are integrated in Geo-Information System (GIS) with the help of multimode, multifunction SAR, which makes it possible to conduct more accurate calculations of geoid shape and consequently to create high precision digital terrain maps, climate research, and ocean-atmospheric interaction research.

A fundamental problem of Earth remote sensing is the inverse problem of identifying and describing properties of the studied surface, ground, or underground (underwater) objects with the use of radar information. Only the first steps have been made to address this problem.

A wide range of possibilities for Earth remote sensing emerged at the end of 1960s with practical application of civil research on radar and scanning methods that were developed for military reconnaissance and allowed imaging in bands of electromagnetic waves that hadn't been used before and acquiring qualitatively different information, including images of Earth surface parts of different area size. There emerged such imaging methods as multizone and infrared (IR) imaging, and thermal and passive UHF mapping. Systems that are based on these methods allow acquiring data on Earth surface by analyzing UHF and IR radiation together with reflected electromagnetic waves in visible and near IR spectrum ranges. In fact, all sensors in such systems are passive ones (i.e., they only detect the energy emitted by Earth surface or energy of Sun radiation re-emitted from Earth surface).

Radar stations that operate in microwave range of electromagnetic spectrum provide conceptually new possibilities. Side-looking radar (SLR) installed on a ERS SC transmits a narrow high frequency electromagnetic impulse signal focused toward the Earth's surface by an antenna. The signal is reflected from the scanned surface and again received by antenna and then registered by onboard receiving equipment. Since such systems use their own radiation energy and operate in relatively long waves, imaging can be carried out at any time of the day under any weather and cloud conditions because electromagnetic radiation of a certain wave range transmitted by antenna, easily goes through clouds, rain, and fog. Moreover, such systems provide unique opportunities for monitoring dynamic processes on the Earth's surface.

Due to the previously mentioned reasons radar methods for Earth remote sensing have a wide range of applications. SLR with real aperture (also known as noncoherent SLR) and synthetic aperture radars (so-called coherent SLR) [11] are used. Resolution of noncoherent SLR is defined by the size of real antenna aperture and its advantage is relative simplicity of radar and data processing system as well as a wider swath width. The disadvantage is low resolution. Coherent SLR provides higher resolution but requires quite a complex data processing system.

Side-looking radars present the most universally applicable and informative sensors for Earth remote sensing in the microwave range. Spatial resolution of ERS radar equipment (10–100m for SAR and 1–2 km for noncoherent SLR) is comparable to that of optical systems [10].

At the moment RS with synthetic antenna aperture have the widest range of applications. Their operation is based on motion of a RS onboard antenna that successively creates a large antenna array along the flight track. Onboard antenna has a small size and a wide array pattern. At each point of fight trajectory SAR records amplitude and phase information of viewed territory and objects that corresponds to their instantaneous coordinates relative to the SC. This information (radio hologram) is formed by combination of received radio signals.

Despite complexity and high cost of space-based SAR (SB SAR), its development and operation allow addressing a number of important objectives that can't be managed by other remote sensing systems. SB SAR is used for navigational purposes. In order to geo-reference radar images (RI) accurately, RI must have high spatial resolution which can be ensured only by synthetic aperture.

That is why we shall examine only those ERS SC that are equipped with SAR. Apart from orbital parameters of motion, mission payload of each ERS SC has a number of technical characteristics and operating parameters (radiation power, frequency range, signal polarization, scanning swath width, and the like.) However the key parameter of remote sensing systems is resolution. Resolution is specified in terms of spatial resolution and radiometric (intensity) resolution. Radiometric resolution is determined by the number of discrete quantization levels into which intensity is digitized and depends on dynamic range width.

Spatial resolution depends on wavelength, size of antenna aperture, and orbit altitude [10].

$$r \sim \left(\frac{\lambda}{D}\right)H$$

where r is spatial resolution, λ is wave length, D is antenna aperture size, and H is orbit altitude.

The formula shows that in order to gain high spatial resolution, it is necessary to increase the size of antenna aperture and reduce wavelength. Wavelength is determined by operating frequency range of RS electromagnetic spectrum, and it is longer as compared with wavelength range of optical systems. Thus in order to improve spatial resolution of SLR, the size of the antenna aperture must be increased. However it is hardly achievable since the overall dimensions of real aperture of noncoherent RS must be immense.

In this case synthetic aperture comes in handy since a large aperture is created when a relatively small antenna is moving over a targeted region. Such (coherent) RS utilize differences in time delay of the reflected signals and changes in Doppler frequencies over time to resolve targets. It should be noted that these differences do not depend on the distance to the target. That is why resolution of SLR with SAR doesn't considerably depend on ERS SC's orbit altitude.

Today there are different types of ERS space systems with SAR: *ERS*, *Radarsat*, *IGS-3R*, *Yaogan*, *COSMO SkyMed*, *TerraSAR-X*, *SAR-Lupe*, *TECSAR*, *ENVISAT*, *ALOS*, *SIR-X SAR*, and others (see Table 1.24).

The main technical characteristics of some of these systems are presented next [10].

European Earth Remote Sensing Space System ERS (European Remote Sensing Satellite)

ERS mission payload includes microwave-sensing equipment active microwave instrument (AMI) that is installed on both spacecraft (*ERS-1* and *ERS-2*) and ensures different operating modes. The mode for imaging underlying surface with the use of synthetic antenna aperture (AMI SAR image mode) has the following characteristics:

- Radiated power: 1270W;
- Radiation frequency: 5.3 GHz;
- Emission bandwidth: 15.5 ± 0.06 MHz;
- Polarization of transmitted and received waves: linear vertical;
- Pulse length: 37.1 μs;
- Spatial resolution: 30m;
- Height measurement accuracy of interferometry: 10m;
- Radiometric resolution: 2.5 dB (30m spatial resolution) and 1 dB (100m spatial resolution);
- Scanning swath width: 100 km at 23° incidence angle of EM waves at the center of swath.

Earth Remote Sensing Satellite ENVISAT-1

Remote sensing equipment of EAS *ENVISAT-1* includes the advanced synthetic aperture radar (ASAR), which is an improved RS with SAR that was used the *ERS* satellite series. The ASAR operates in C-band frequencies (3.9–6.2 GHz or 7.69–4.84 cm) and observes up to seven optional swaths along EAS flight path (of 100 km total width) under any weather conditions providing 30m resolution or 100m resolution on 400 km swath on single swath. Global monitoring mode provides 1 km resolution on a 400 km swath. Radiometric accuracy is 0.65dB and radiometric resolution is 1.5–3.5 dB.

Earth Resources Survey Spacecraft JERS-1 (Japan Earth Resources Satellite, Japanese Classification: Fuyo-1)

Radar station with synthetic aperture radar (real size of antenna array is 11.9×2.5m) is designed for all-weather imaging of Earth surface and coastal zones providing high resolution radar images. It has the following characteristics:

- Operating frequency: 1275 MHz;
- Spatial resolution: 18m;
- Swath width: 75 km, swath edge is 326 km to the right from the EAS track;
- Pulse power: 1.3kW;
- Pulse length: 35 μs.

Japanese Earth Resources Survey EAS Advanced Land Observation Satellite (ALOS)

The EAS is equipped with a high resolution radar station *VSAR* with the following characteristics:

- Operating frequency: 1.275 GHz;
- Spectrum range of emitted signals: 15 MHz;
- Spatial resolution limit: 10m (slant-range direction), 5m (along EAS track);
- Accuracy of RCS measurement: ±1 dB;
- Swath width: 70 km (high resolution mode), 250 km (low resolution mode).

Canadian Earth resources Survey Spacecraft Radar Satellite (Radarsat)

The size of the radar antenna array is 15×1.5m. EAS of the *Radarsat* series is equipped with a multifunctional synthetic aperture radar designed for all-weather Earth imaging, tracking of marine vessels and ice cover movement, as well as Earth terrain mapping and other purposes.

SAR of *Radarsat* spacecraft has the following main characteristics:

- Operating frequency: 5.263 GHz;
- Spatial resolution 9–100m (depending on operating mode);
- Polarization of transmitted and received signals: linear horizontal;
- Average radiation power: 300W;
- Pulse power: 5 kW;
- RF bandwidth: 11.6, 17.3, and 30 MHz;
- Sampling rate: 12.9, 18.5, and 32.3 MHz;
- Pulse length: 42 μs;
- Pulse repetition frequency: 1270–1390 Hz.

Table 1.24 Main Orbital Parameters and Technical Characteristics of Earth Remote Sensing Satellites with SAR

Spacecraft	SC Orbit Altitude, km	SC Orbit Inclination	Operating Wavelength, cm (Frequency GHz)	SAR Operating Mode	Spatial Resolution (Azimuth Direction), m	Swath Width (Frame Size), km
ERS-1	782 × 785	98.5°	(5.3)	—	30	100
ERS-2	782 × 798	98.54°				
ENVISAT-1	820	98.55°	7.69–4.84 (3.9–6.2)	7 optional swaths	30	100
				ScanSar	100	400
				Global monitoring	1000	400
JERS-1	567 × 569	97.7°	(1.275)	—	18	75
ALOS	700	98.1°	(1.275)	High resolution	5–10	70
				Low resolution	—	250
Radarsat	743	98.6°	(5.263)	Standard	28 × 25	500
				Wide swath	28 × 35	300
				High resolution	9 × 10	200
				ScanSar	30 × 35	300
				ScanSar	50 × 50	300
				ScanSar	55 × 32	500
				ScanSar	100 × 100	500
				Experimental	28 × 30	300
				Experimental	28 × 40	170
Almaz-1A	280	72.7°	(3)	—	15	30
Almaz-1B	400	—	3.5	—	5–7	20–35
			9.6	Fine	5–7	30–55
				Intermediate	15	60–70
			70	ScanSar	15–40	120–170
				—	20–40	120–170

COSMO SkyMed 1	614.4 × 633	97.86°	3.1 (9.6)	Depending on resolution required	less than 1 3–15 30 100	(10 × 10) 40 100 200
Lacrosse	676 × 696	68°	(9.5–10.5)	Fine ScanSar	less than 1 2–3 10–15	(2–4 × 2–4) (6–20 × 6–20) 100
SIR-C/X-SAR (Space Shuttle)	233 × 240	57°	(5.298) (9.6)	—	30 × 13–26	15–90 15–40
Osiris	600–800	90°	3	Telescopic —	8–10 3–5	— 30–50
TerraSAR-X	507.7 × 512.5	97.45°	(9.65)	Spotlight StripMap ScanSar Ultra-fine	1–2 3 16 0.5–1	(5–10 × 10) (30 × 50) (100 × 150) —
TanDEM-X	514	97.44	3.1 (9.6)	High Resolution SpotLight SpotLight StripMap ScanSAR	1 2 3 15	10 10 30 100

RS is capable of changing swath location relative to satellite track. Swath width ranges from 45 to 500 km according to required incidence angle and spatial resolution.

The Almaz Program (Russia)

Almaz is a program on the research of Earth resources with the help of spacecraft with SAR. Synthetic aperture radar *Almaz-1A* used two 1.5×15m waveguide antennas with two separate beams and had the following characteristics:

- Operating frequency: 3 GHz;
- Spatial resolution: 15m;
- Polarization of transmitted and received radiation: linear horizontal;
- Radiation power: 190W (pulse), 80W (average);
- Main pulse length: 0.07 and 0.1 μs;
- Pulse repetition frequency: 3 kHz;
- Beam width on the ground: 30 km;
- Swath width: 350 km;
- Radar imaging length along groundtrack: 20–240 km.

Another spacecraft of the *Almaz* program is *Almaz-1B*, whose onboard radar systemis designed for all-weather Earth surface observation. Vega Research and Production Enterprise designs the new-generation multifunctional radar system *Ekor-1B* for *Almaz-1B*. It consists of three subsystems that operate in various frequencies and modes:

- SAR-3: 3.5 cm wavelength, 5–7m ground resolution, 20–35 km swath width;
- SAR-10: 9.6 cm wavelength, 5–7m ground resolution, 30–35 km swath width (fine mode), 15m ground resolution, 60–70 km swath width (intermediate mode), 15–40m ground resolution, 120–170 km swath width (ScanSar mode);
- SAR-70: 70 cm wavelength, 20–40m ground resolution, 120–170 km swath width.

Earth Remote Sensing Module Priroda (Russia)

Module *Priroda* was a space platform equipped with different ERS equipment and was a part of the *Mir* Space Station since April 1996. The *Travers* radar station with synthetic aperture radar had the following main characteristics: operating wavelengths—9.3 and 23 cm, spatial resolution—50–150m, swath width—50 km.

Earth Remote Sensing Spacecraft COSMO (Italy, Spain, Greece)

ERS SC *COSMO* are equipped with SAR that operates at frequency of 9.65 GHz and provides 3 and 6–12m spatial resolution on swath of 40 and 100–120 km, respectively.

Dual-Purpose ERS SC COSMO-SkyMed (Italy)

The constellation of dual-purpose ERS SC *COSMO-SkyMed* consists of four satellites [18]. SC *COSMO-SkyMed* are equipped with synthetic aperture radar

SAR-2000 with 5.7×1.4m antenna. Radar operates in X-band wavelength (6.2–10.9 GHz or 4.84–2.75 cm) at the frequency of 9.6 GHz (3.1 cm wavelength). Different operating modes provide ground resolution of less than 1, 3–15, 30, and 100m on swaths of 10×10, 40, 100, and 200 km, respectively.

Radar Reconnaissance System Lacrosse

A typical example of military-oriented ERS SC is the *Lacrosse* system, constructed by Martin Marietta in the United States. Initial cost of the system design was $3 billion USD, with one spacecraft costing $0.5–1 billion USD. It was planned that the constellation would consist of six satellites. Frequency range of RS SC *Lacrosse* is likely to be within 9.5–10.5 GHz, ground resolution of radar image in fine mode is less than 1m, image size is $2–4 \times 2–4$ km; in ScanSar mode the resolution is 2–3m, image size is $6–20 \times 6–20$ km. In case swath width is 100 km, resolution will be 10–15m. *Lacrosse* conducts strategic reconnaissance and is designed for high resolution observation of small areas.

SIR-C/X-SAR

SIR-C/X-SAR was designed at the request of NASA and the space agencies of Germany and Italy. The equipment was designed by the Jet Propulsion Laboratory and Ball Communications (United States), Dornier (Germany), and Selena (Italy). The system consists of three RSs with synthetic aperture radars that operate in L- (1248 MHz), C- (5298 MHz), and X- (9600 MHz) bands. Pulse power is 4.3, 2.25, and 3.3 kW, respectively. Swath width is 15–90 km (SIR-C) and 15–40 km (X-SAR). Azimuth resolution is 30m, range resolution is 13 or 26m (both alternatives). Telescope mode provides range resolution of 8–10m.

French ERS SC Osiris on the Base of Radar 2000 Project

The customer is the Ministry of Defense of France. Contractors are Matra Marconi Space and Alcatel Espace. The SC has an S-band SAR that provides 3–5m resolution on a swath of 30–50km.

The German radar imaging satellite *TerraSAR-X* was placed in sun-synchronous orbit on June 15, 2007, and it presents an example of a modern commercial ERS SC designed for a wide range of applications, including military [12]. German Aerospace Center (DLR) covered the majority of expenses for the construction of *TerraSAR-X* and became a partner of Astrium GmbH, a German division of European concern EADS. Cost of the project was €130 million EUR. The share of DLR constituted €102 million EUR and the share of EADS was €28 million EUR. The mission payload of the satellite includes synthetic aperture radar *TSX-SAR*, a polarimetric multiband system with a mass of 394 kg and an active X-band phased-array antenna (PAA) (frequency 9.65 GHz) of high spatial and radiometric resolution comparable with that of aircraft SAR. The radar is developed based on technologies acquired by Europeans in the course of shuttle flights with the *SIR-C* and *SRTM* radars in 1994 and 2000. The size of onboard active PAA is $4.8 \times 0.8 \times 0.15$ mm and it consists of 384 transmit/receive modules.

The radar system carries out imaging in three main modes: Spotlight, Stripmap, and ScanSar. Spatial resolution (azimuth):

- 1–2m (Spotlight);
- 3m (Stripmap);
- 16m (ScanSAR).

Image size:

- (15–10) × 10 km (Spotlight);
- 30 × 50 km, acquisition length 1500 km (Stripmap);
- 100 × 150 km, acquisition length 1500 km (ScanSar).

Swath width: 463 (up to 622) km (Spotlight); 287 (up to 622) km (Stripmap), 287 (up to 577) km (ScanSAR).

TSX-SAR RS has several experimental modes:

1. Ultrafine mode which provides range resolution of less than 1m (up to 0.5m) with pulse bandwidth of 300 MHz.
2. Using a pair of PAA electronic modules, along track interferometry (ATI) mode allows receiving radio signals from two separate PAA subarrays, dual receive antenna (DRA) (each 2.4m long). After comparing signals from two subarrays, moving objects can be resolved. German Aerospace Center (DLR) has started designing two additional projects—*TanDEM-X* and *TerraSAR-X2*. The cost of the *TanDEM-X* project is €85 million EUR, and it's funded by a partnership program: €56 million EUR is funded by DLR, €26 million EUR is funded by Astrium EADS and €3 million EUR comes from other investors.

Considering the high cost of ERS SC with SAR, construction of light space-based high resolution SAR for small satellites can be viewed as a future field. Vega Research and Production Enterprise (Russia) designs light space-based SAR that will be launched by ballistic missiles removed from combat duty. Such ultra-high resolution SAR for small spacecraft will operate at 9.6 cm wavelength, cover the area of 500 km, and provide the resolution of:

- 2m in fine mode on swath of 10–20 km;
- 5–7m in lower resolution mode on swath of 50–100 km;
- 15m in low resolution mode on swath of 150–200 km.

The mass of radar equipment won't exceed 250 kg and power consumption will be no higher than 1300W.

The following main conclusions based on the facts stated above are given in [17]:

1. Requirements for SAR construction have a very specific and often contradictory nature since they depend on many factors. Radar ERS information for navigation purposes requires large area coverage and best possible resolution.

However, high resolution can be provided only on a small swath or image size. The ratio of swath width to resolution in azimuth direction is under 10,000 for most space-based SAR (i.e., if resolution is 5m, swath width can't be more than 50 km).

2. The review of ERS satellites with SAR shows that space powers and other countries have interest in such systems and they have been developing rapidly despite their extremely high cost. Further development directions of ERS satellites with SAR largely depend on data consumers but have theoretic and technical limits due to, for example, the diffraction limit of resolution, RS energy capacity, complexity of information processing algorithms and computer performance, transmission capacity, and other factors.

3. Considering the contradiction between acquiring high resolution and large swath width, the enhancement of SAR resolution can be seen as a justifiable development direction. It gives a wide range of opportunities. Moreover, this contradiction can be partially resolved by means of using wide-band signals instead of narrow-band ones with simultaneous usage of wide-angle apertures. Since theoretical limits haven't been reached yet, new methods and ways to improve resolution should be researched.

High-resolution SAR is especially important for acquiring radar imaging information for navigation and motion control purposes. It is obvious that this future field can be only developed with the use of recent advances in microelectronics by employing modern ECB.

This axiom is graphically demonstrated in Figure 1.19, which shows dynamics of financing of only civil space programs by six leading space powers from 1989 to 2007. The United States naturally has the leading position; the second position

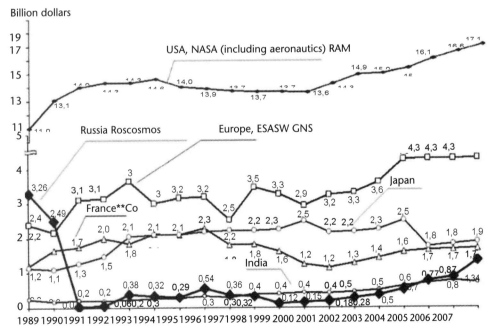

Figure 1.19 Financing of civil space programs by the leading world countries [19].

is occupied by the European Space Agency. Honorable third and fourth place are divided between France and Japan. Russia, which started rather actively in 1989, fought with India for less honorable second-to-last and last places in this space and financial race with varying success for the 16 years starting from 1991.

Thus, the amount of civil space programs financing of the United States, Russia, China, and India for the period from 2008 until 2011 is in Table 1.25.

1.7 The Effect of Space Radiation on SC

The main sources of radiation that affect SC body and equipment are the following [7]:

1. The so-called Earth's radiation belts that consist of two layers. The inner belt consists of high-energy protons and extends for at least 3.5 thousand km. High-energy electrons form the most dangerous part of the outer belt that extends for 15–25 thousand km.
2. Cosmic rays that are a flux of high-energy particles.
3. Solar ultraviolet and X-ray gamma radiation.
4. Solar particle radiation.

Solar activity has an 11-year cycle. During solar flares the level of solar activity increases 10–100 times as compared with solar quiet periods. Flares occur approximately every 10 days, however, astronauts can be warned against such hazards by observing the solar chromosphere and using modern forecasting methods.

Electromagnetic radiation is classified by wavelength into gamma radiation, X-ray, ultraviolet, light, infrared radiation, and radio wave radiation.

The ultraviolet region (1000–4000 Å) of solar spectrum accounts for 10% of the total energy and has a crucial impact. The direct effect of radiation is to change material's optical properties. The indirect effect is failure of certain SC chips and devices.

Table 1.26 shows average radiation doses of different sources of radiation.

The SC body considerably reduces the effect of electron radiation but protects far less from proton radiation; unfortunately, the SC body hardly reduces cosmic rays since they are high-mass and high-energy particles.

Apart from flares of high-level radiation during increased solar activity there's also variable radiation that is hard to predict, and constant radiation (constant

Table 1.25 Space Programs Financing

| Country | Amount of Financing, Billions USD | | | |
	2008	2009	2010	2011
United States	17,402	17,782	18,696	19,265
Russia	2,001	2,914	3,336	3,960
China	1,300	1,369	1,430	1,570
India	0.920	0.932	1,287	1,440

Table 1.26   Annual Space Radiation Doses

Source of Radiation	Radiation Dose, MeV/g	
	On the Surface	1 g/cm² Shielding
Inner radiation belt of Earth (protons)	10^{10}–10^{12}	10^7–10^9
Outer radiation belt of Earth (electrons)	10^{13}–10^{15}	10^6–10^8
Solar radiation	10^7–10^9	10^4–10^6
Cosmic rays	10^2–10^3	10^2–10^{3*}

* High-energy particles (heavy nuclei) are hardly absorbed by shield

background—protons, electrons, helium nuclei) that is significantly shielded by the main SC body.

Permissible radiation doses for various materials are given in Table 1.27. Comparison of Tables 1.26 and 1.27 shows that the radiation belts are an especially severe environment for SC operation. On the basis of Table 1.27, a number of general conclusions and recommendations can be made.

When permissible radiation dose is exceeded, plastics lose their mechanical and electric properties (transparent plastics lose transparency), ceramic materials and glass lose their mechanical, electric, and optical properties whereas rubber loses elasticity. As far as semiconductor devices are concerned, transistor amplification factor is reduced to such extent that amplifying and switching properties of transistors can be lost.

Therefore, plastics aren't used for coating of SC, especially those that pass through radiation belts.

Windows made of transparent plastic over optical equipment are to be shuttered. The shutters are to be opened only for active operation.

SC shielding materials should be chosen carefully since the level of radiation protection is directly dependent on the mass of shielding material.

In increased radiation environments the efficiency of photovoltaic cells (PV cells) is considerably reduced. They often degrade due to solar flares and lose efficiency when they operate in radiation belts for a long time.

Ultraviolet radiation (UVR) constitutes constant background of solar radiation. Though metals aren't affected by UVR, it damages optical equipment, therefore SC windows have quartz glass and protective shutters on the outside. The effect of

Table 1.27   Radiation Doses That Change Properties of Materials

Materials and Damaged Properties	Radiation Dose, MeV/g
Plastics (electric and mechanical properties)	10^7–10^9
Transparent plastic (transparency)	10^6–10^{10}
Elastomer (elasticity)	10^8–10^{11}
Glass (transparency)	10^5–10^{10}
Glass and ceramics (mechanical properties)	10^{11}
Quartz (transparency)	10^7–10^{11}
Semiconductors (electric properties)	10^8–10^{10}

UVR on plastic materials is analogous to that of ionizing radiation, hence screen vacuum thermal insulation (SVTI) methods are to be applied, but in this case it is necessary to prevent damage of plastics due to change of thermal regime of equipment operation.

Color pigments of enamel paint lose their properties due to UVR: TiO_2-based white paint turns yellow and optical properties are damaged. Even slight changes in optical properties of SC surfaces may change the thermal regime of onboard equipment operation. Therefore, only ceramic paint is used nowadays.

1.8   Micrometeoroid Effect on SC

Active effects of micrometeoroids on SC (holes) haven't yet been detected, probably due to relatively short operation periods of spacecraft to date. The influence of cosmic dust will be examined in detail in Chapter 1, Volume 2.

Micrometeoroids mainly cause erosion of optics. In low Earth orbit micrometeoroids cause optics erosion at an average rate of 200 Å/year. Away from Earth the effect is much smaller. Beyond the sphere of Earth's influence ($r \geq 1$ million km) the erosion rate is 1 Å/year. The research conducted on EASs has shown that near Earth there is a lot of micrometeoroid dust (dust particles with mass $m = 10^{-12}$ g) that causes optics erosion.

In order to prevent erosion, optics are placed perpendicular to the flight track and is shuttered (especially optical sensors for SC attitude control system). If SC flies away from the Earth, it is advisable to orient optical equipment towards Earth. There's a lot of data on micrometeoroid impact estimation, which is reflected in the graph on the number of meteoroids in Earth's vicinity depending on their mass (Figure 1.20).

The graph shows that there is a bigger number of meteoroids in low orbits of EAS due to a small dust cloud near Earth.

In order to calculate radius (in mm) of an impact crater, Bjork's formula can be used:

$$r = K(mV)^{\frac{1}{3}},$$

where m is meteoroid mass, g; V is meteoroid velocity, m/s, K is coefficient that is dependent on the material ($K = 1.09$ for aluminum pin striking aluminum wall, $K = 0.606$—iron striking iron, $K = 1.3$ lead striking lead, $K = 0.9$ aluminum striking iron).

Coefficient $K = 1$ can be taken for rough calculations.

The factor of wall penetration is the choice of wall thickness that exceeds crater radius 1.5–2 times (i.e., $\delta_{wall} = 1.5$–2).

The formula based on the hydrodynamic impact theory can be used to calculate the thickness of meteoroid shielding:

$$\frac{\delta}{d} = \frac{4}{3}\frac{1}{\sigma}\lg\frac{V_{int}}{V_f}$$

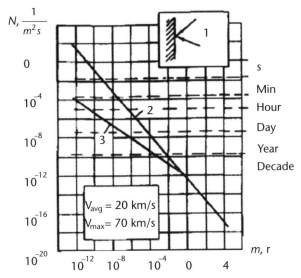

Figure 1.20 Relationship between impact frequency and meteoroid mass: 1—impact in one direction of meteoroid bodies with the mass > m; 2—for satellites in low orbits; 3—away from Earth.

where δ is wall thickness, d is diameter of incident particle; $\sigma = \rho_{wall}/\rho_m$ is the relationship between density of wall material and density of meteoroid material; $V_f > 1.5$ km/s is meteoroid finite velocity; V_{int} is meteoroid initial velocity.

As far as impact velocity of a meteoroid on a SC is concerned, average velocity $V_{avg} = 20$ km/s and maximum velocity $V_{max} = 70$ km/s are taken [7].

In space there is a constant background of meteoroid particles and flux of particles that occur in meteoroid streams. Meteoroid showers last for several days only; there are about 20 showers and they all are known.

Increasing thickness of the SC body isn't efficient for protection, since it makes SC heavier. Therefore, the protection should be developed in two directions:

1. Sectionalization of SC together with redundancy of equipment and support systems so that one section can operate in case the other breaks down (the major measure of protection).
2. Development of meteoroid shields. When penetrating a shield, a meteoroid breaks into a large number of small particles that do not pose danger to the SC hull, which allows considerable reducing thickness of the main wall.

The distance at which a shield should be placed to ensure best protection of SC skin can be calculated using the formula in Figure 1.21.

$$s = 2\delta_{skin}\left(\frac{\delta_{skin}}{\delta_{shield} - \rho_{shield}}\right)^{\frac{1}{2}}$$

where δ_{skin} is thickness of skin, mm; δ_{shield} is thickness of shielding, mm; ρ_{shield} is density of shielding material, g/cm3.

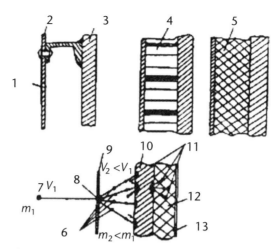

Figure 1.21 Meteoroid protection and schematic illustration of meteoroid disintegration: 1—venting; 2—shield; 3—skin; 4—honeycomb; 5—foam plastic; 6—meteoroid fragments; 7—meteoroid; 8—holes in shield; 9—shield; 10—skin; 11—fragments of skin; 12—sound and thermal insulation; 13—interior skin.

Example. For an aluminum shielding with $\delta_{skin}/\delta_{shield} = 10$ the distance is s (mm) $= 4\rho_{skin}$

The literature recommends that shield is made of beryllium that has high impact resistance. In case the same relationship is taken $\delta_{skin}/\delta_{shield} = 10$, the distance is $s \approx (5 - 6)\rho_{skin}$.

In order to protect essential parts of EAS with skin thickness $\rho_{skin} = 2$ mm, a shield $\delta_{shield} = 0.2 - 0.3$ mm is placed at a distance $s = 5\delta_{skin} = 10$ mm and the gap is filled with honeycomb core.

An inner shield is needed apart from outer shield since an impact can eject fragments from the skin's interior (Figure 1.21) that can damage elements of onboard systems. Inner shield that includes a thin metal layer can also serve as sound and thermal insulation.

1.9   The Problem of Space Debris in Earth Orbit

It was in the 1980s that the scientists first spoke of mass space pollution when the concentration of space debris in Earth orbit became so dense that ballisticians had to work hard in order to safely place this or that satellite in orbit. In recent decades the situation grew only worse. The number of debris pieces in near-Earth space is so big that it poses real threats to space stations and spacecraft that operate there. Earth's orbits are littered with abandoned satellites, used rocket stages, as well as garbage thrown into space and other debris.

In September 2012 the United States Space Surveillance Network estimated that about 23,000 objects larger than 5–10 cm are in orbit. According to this data it was calculated that in Earth orbit there is a total of about 750,000 objects bigger than 1 cm.

A considerable fraction of the cataloged objects emerged as a result of a well-known antisatellite missile test conducted by China in 2007 and also as a result of space accident in 2009 when two satellites accidentally collided.

The collision of objects from different orbits occurs at the speed of thousands of kilometers per hour.

What can it result in?

A collision at an orbital speed even with a tiny fragment can have devastating consequences. A bullet with a 1.2 cm diameter and 6.8 km/s speed penetrates almost half of an 18 cm thick aluminum block. Skin of a spaceship and rocket body is much thinner than such a block. A particle of 10 cm diameter can destroy a spacecraft.

Since each collision generates much more than two pieces, larger numbers of debris appear in orbit, which may result in disabling spacecraft, malfunction of various space systems such as communication, navigation systems, and other. Though the probability of debris destroying an orbital satellite is still small, such incidents have already occurred both on manned spaceships and orbital stations.

In 1983 the crew of the famous *Challenger* shuttle found a small pit on the spaceship windshield that was left after a collision with a foreign object. Though the pit was only 2.5 mm deep and wide, it made NASA engineers worry. After the spaceship landed, specialists carefully examined the pit and concluded that it was left by a paint particle detached from some other spacecraft.

The Soviet space station *Salyut-7* was also damaged by space debris: its surface was spotted with microscopic craters left after collisions with debris particles. *ISS* also had to maneuver and alter the course a number of times to avoid collisions with space debris. According to a documented case, on command from Earth astronauts once had to seek shelter in the emergency escape module of the *Soyuz* spacecraft until the danger passed.

Now there's a disastrous amount of debris in space. Considering that about fifteen hundred satellites will be launched in the next 10–15 years, the probability of collisions of separate fragments and spacecraft will dramatically increase. As a result orbital space can be densely occupied. It is natural to assume that a collision of a spacecraft with debris will create a larger number of debris. Some of new debris will collide with another spacecraft and the story will repeat itself. Thus, the numbers of debris in orbit will grow in geometrical progression. There may occur a real threat of losing satellites that provide people with means for space communication, TV broadcasting, navigation, meteorology, and research of Earth resources. Low Earth orbit will become too dangerous to place satellites and especially manned spaceships. Future space research with spacecraft in Earth orbit might even be precluded.

The most representative examples of collisions in space are as follows:

- December 1991: *Cosmos-1934* collides with a fragment of a disintegrated spacecraft *Cosmos-926*;
- July 1996: the French spacecraft *Cerise* collides with a stage of the French launch vehicle *Ariane*;
- January 2005: a stage of the American launch vehicle Thor collides with a stage fragment of the Chinese launch vehicle *CZ-4*;
- March 2006: spacecraft *Express-AM-11* collides with a piece of debris;
- February 2009: spacecraft *Iridium-25* collides with the satellite *Cosmos-2251*.

But that's in space.

Falling debris may cause consequences on Earth as well. It is known that objects with the size of more than 1 meter deorbit about once a week. Most of them burn up in atmosphere and the rest break into pieces and fall to Earth in the form of a metal shower. In the future mass meteor showers of artificial origin may have a nature of disaster. Since some spacecraft have nuclear power systems, their disintegration may cause radioactive contamination of the territories. As rare as they are, such cases still occur.

For example, at the end of October 1997 the satellite *Cosmos-954* suddenly depressurized and onboard systems broke down. Uncontrolled descent under the influence of the upper atmosphere led to deorbiting of the spacecraft and its falling in North-West of Canada on January 24, 1977. The major part of the spacecraft burnt up in the atmosphere, however it later turned out that fragments of radioactive space debris had scattered over the territory of several thousand square kilometers.

The cases of heavy spacecraft falling back to Earth:

- Uncontrolled deorbit of the space station *Skylab* (77 tons);
- Uncontrolled deorbit and falling on the territory of South America of the space station *Salyut-7* (~ 40 tons);
- Falling of the interplanetary space station *Mars-96* (~ 5 tons) due to unsuccessful launch;
- Controlled deorbit of orbital *Mir* Space Station (~130 tons);
- Falling of the spacecraft *Fobos-Grunt* (~ 9 tons) due to unsuccessful launch.

Space debris has already been under surveillance for a long period of time: the threat it poses is too real to ignore. Researchers are able to track particles of 10 cm size and bigger. However, smaller particles also pose a serious danger to spacecraft.

In order to prevent the worst scenario, groups of specialists from the leading space powers actively work on relevant technical solutions—designs of technical means for removal of space debris from Earth orbits are worked out, various alternatives for capturing and pulling of dead satellites, as well as for changing of their orbits and other measures are considered.

Among proposed manned missions there stand out such projects as *Phoenix* of the United States military agency DARPA and *CleanSpace One* of Swiss Ecole Polytechnique Federale de Lausanne (EPFL). These spacecraft are especially designed for capturing and destroying of space debris or adapting it to other space projects. There are projects for ground-based lasers capable of slowing down space debris, which results in their burning up in the atmosphere.

However, there aren't yet any economically acceptable methods of mass removal of space debris from space. Therefore in the near future most attention will be paid to control measures that involve prevention of creating new debris and explosions in orbit, placing spacecraft in graveyard orbit at the end of their operational life, atmospheric braking, and the like. As far as large objects such as *ISS* are concerned, it is expected that at the end of their usable life, they will return to Earth part by part. Satellites should end up not as debris in space, but come back to Earth.

1.10   The Use of Microelectronic Technologies for the Development of Space Microrocket Engines

Microelectromechanical system technology (MEMS technology) develops in the field of devices that convert chemical energy that releases in combustion into mechanical energy. One of the perspective application fields of such microsystems is in the development of engines of various designs for small aircraft and spacecraft (mass less than 500 kg) [20]. Small spacecraft that are utilized as Earth artificial satellites are divided into several classes, as shown in Table 1.28.

The technology of creating small space vehicles has rapidly developed in the past two decades. Figure 1.22 shows statistics on the number of Earth artificial satellites with the mass of less than 100 and 10 kg that were launched in the past two decades. It is notable that the number of small spacecraft grew exponentially in 1990–2013. The number of launched nanosatellites with the mass < 10 kg also grew as compared with the number of launched microsatellites with the mass < 100 kg. It is expected that such tendency will remain in the next decades.

Table 1.28   Classification of Small Spacecraft

	Mass, kg	Volume, L	Power, W
Small satellites	500–100	< 100	14–160
Microsatellites	100–10	< 15	6–14
Nanosatellites	10–1	< 0.5	2.4
Picosatellites	0.1–1		
Femtosatellites	0.01–0.1		

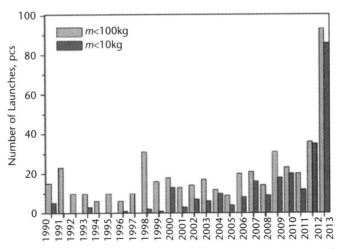

Figure 1.22   The number of micro (*m* < 100 kg) and nanosatellites (*m* < 10 kg) placed in orbit in 1990–2013 [21], according to data of SpaceWorks Enterprises, Inc.

The key advantage of small spacecraft is their low net cost. The reduction of weight-size parameters of spacecraft allows using less powerful vehicles for their launch into orbit or launching a large number of satellites at once. Reduction of satellite size leads to the reduction of their manufacturing cost and provides an opportunity for serial production which in turn decreases net cost. Simultaneous launch of a large number of uniform satellites allows creating large arrays and groups of devices that can cover a larger area of Earth surface together (e.g., for communications and research purposes. In recent years, there has been a growing use of microsatellites for observation and remote sensing purposes whereas the use of microsatellites for developing technologies has declined, according to SpaceWorks Enterprises, Inc. According to projections, commercial microsatellites will contribute the major part of satellites of such class, so that their number will significantly exceed the number of larger civil and military satellites.

1.11   Military and Special-Purpose Spacecraft

1.11.1   Missile Early-Warning System

A missile early-warning system (MEWS) is a special complex system designed for alerting a country's leadership about the use of a missile weapon by an adversary and identifying sudden attacks.

It is designed to detect missile attacks before missiles reach their targets. It consists of two layers: ground-based RS and a constellation of early-warning satellites.

MEWS is designed for automatic high accuracy detection of missile attack, alerting and providing command posts (CP) of the armed forces with information on launch of ballistic missiles (BM), missile attack, aggressor state, attacked regions, time to missile impact, scale of missile attack, the level of countermeasures, and current combat capability.

The system also serves as an essential source of data on the space situation for space surveillance system and space defense (SD) and the main source of commands for antiballistic missile (ABM) systems and complexes.

The previously mentioned fact plays an important role in the development of MEWS. The development of space and missile defense (SMD) systems and complexes—MEWS, space surveillance system (SSS), ABM, and SD that operate by the same algorithm, solve tasks interdependently, and complement one another—is the least expensive way to address the problems of missile warning, space surveillance, ballistic missile, and space defense.

In terms of combat capability MEWS serves as a source of information on test and combat training launches of national and foreign BM as well as information on military conflicts with the use of BM occurring in any part of the world.

MEWS includes:

- Spaceborne systems for detection of BM;
- Groups of above-the-horizon radar systems;
- Data transmission and operative command communications system;
- Command posts;

• Complexes for communication and display of warning information at informed CPs.

The analysis shows that meeting the requirements for MEWS in present conditions is a complex technical challenge.

It particularly concerns the requirement for high accuracy detection of missile attacks that includes detecting any types of missile attacks from any direction which means that MEWS is to provide assured detection of small-size warheads and precise prediction of impact location.

As for today neither Russia nor the United States has found an efficient spaceborne solution to this problem that meets the requirements for global, all-season, 24-hour and all-weather operation. One of the assured solutions to this problem is reinforcement of the ground echelon, which includes development of a group of powerful ground-based radar systems that operate in VHF and in the upper UHF band (*Daryal*-type) with the warhead vehicle range of more than 6,000 km.

The primary mission of an early-warning system is to detect a missile attack before the missiles reach their targets. A timely detection of an incoming strike would make it possible to determine the scale of attack and its origin, estimate potential damage, and choose an appropriate response. An early-warning system is absolutely necessary for implementation of a launch-on-warning posture, which assumes that a retaliatory strike would be launched before attacking missiles reach their targets.

Creation of an early-warning system that would be capable of fulfilling its mission is a very challenging task. To accomplish its mission, an early-warning system must be able to detect a missile as early as possible and provide reliable information about the scale of the attack. Since the system can issue a warning only when an attack is already underway, the time that is available for detection, assessment of the information, and generating an alert is extremely limited.

The extremely short-term period of time available for detecting an attack and making a decision about an appropriate response calls for very tight integration between MEWS and command and control systems. This requires an unprecedented degree of reliability of the early-warning system.

The important role that early warning systems play in nuclear command and control procedures and the unacceptably high cost of a potential error are the primary reasons why MEWS have been receiving so much attention in the recent years. The status of the Russian early warning system is the cause of the most serious concerns, since there are many visible signs of deterioration of that system.

The Soviet Union began the development of systems that would provide early detection of a ballistic missile attack in the early 1960s. The first two generations of early-warning radar stations (RS) deployed in the late 1960s to early 1970s were modifications of RS that were developed for space-surveillance and antisatellite systems. The primary mission of the early warning RS was to support missile defenses, which were still under development, rather than provide a warning required to achieve a launch-on-warning capability.

The concept of an integrated early-warning system that would include ground radars as well as satellites and that would be capable of providing the command and control of the strategic forces with the capability necessary for implementing a launch-on-warning option, did not appear until about 1972.This concept was a result of the effort that the Soviet Union undertook in the early 1970s in an attempt to streamline all its programs in the areas of missile defense, antisatellite warfare, space-surveillance, and early-warning.

The draft project, prepared in 1972, called for development of an integrated early-warning system that would include above-the-horizon and over-the-horizon RS, as well as space means. The early-warning satellites and over-the-horizon radars were supposed to detect launches of ballistic missiles during boost phase of their flight, thus ensuring maximum warning time. The satellites rely on infrared sensors that can directly detect radiation emitted by the missile plume. An over-the-horizon radar detects reflections of electromagnetic signal that it sends in the direction of a starting rocket.

Over-the-horizon RSs deployed on the Soviet territory were able to detect missile launches on the territory of the United States by using reflections of electromagnetic impulses from the Earth's ionosphere.

The draft project also called for deployment of a network of early-warning above-the-horizon RSs that were supposed to detect incoming missiles and warheads as they approach their targets. The RSs were intended to provide the important second echelon of early warning sensors, which were based on physical principles different from those deployed on satellites, thus providing more security to the entire system.

Another important role that was assigned to the early-warning radars was space surveillance. The project called for close integration of all existing and future radar facilities of MEWS and SSS that would provide the capability to track space objects.

The Soviet Union deployed all components of the system—above-the-horizon and over-the-horizon RSs, and early-warning satellites. The over-the-horizon RSs failed to live up to their expectations and did not play any significant role in early-warning system operations. The space echelon and above-the-horizon RSs deployment programs were more successful. However, both programs experienced considerable delays during their implementation and suffered serious setbacks at the time of the breakup of the Soviet Union. As a result, the current system is very far from the comprehensive, integrated, multilayered early-warning system envisaged by the original Soviet plans.

1.11.2 MEWS Ground-Based Echelon

Long-range early-warning radar stations (RS LR) and ground-based infrastructure (command posts, control centers, and the like) comprise MEWS ground echelon.

The construction of long-range early-warning RS began in 1954 when the government of the USSR adopted a decision to create a ballistic missile defense (BMD) system of Moscow. Its essential part was RS LR designed for detection and high precision locating of an adversary's missiles and warheads at a distance of several thousand kilometers.

The first early-warning RSs were constructed in 1963–1969. These were two RS of *Dnestr-M* type located in Olenegorsk (Kola Peninsula) and Skrunda (Latvia). The system was introduced into service in August 1970. It was designed for detection of ballistic missiles launched from the territory of the United States or areas of the Norwegian and North Sea. The main task of the system at that stage was to provide a ballistic missile defense system deployed around Moscow with information on missile attacks.

In 1967–1968, in parallel with construction of RSs in Olenegorsk and Skrunda, the Soviet Union began the construction of four RSs of *Dnepr* type (modification of RS *Dnestr-M*). Balkhash-9 (Kazakhstan), Mishelevka (near Irkutsk) and Sevastopol were chosen as construction sites. Another system was built at the Skrunda site in addition to the operating RS *Dnestr-M*. The stations were intended to increase the coverage provided by early-warning system to include the North Atlantic and areas in the Pacific and Indian oceans.

At the beginning of 1971 a command post of missile early-warning system was created on the basis of a command post of early detection.

Table 1.29 shows a list and main technical characteristics of MEWS RSs constructed in the USSR. The following families of RSs that have similar architecture and operation principle can be singled out: *Dnestr*, *Dnepr*, *Daugava*, *Daryal*, *Dunay*, *Volga*, and *Don*. Other designations such as *Hen House*, *Dog House*, *Cat House*, *Pechora*, and *Pill Box* are used abroad instead of national river classification.

The next important stage of the development of national and foreign MEWS RSs was transition to RS with phased arrays (PA), which was a breakthrough technological solution and which ensured the advances in microelectronic technology. *Daryal*, *Dunay*, *Volga*, and *Don* already had PA.

Figures 1.23, 1.24, and 1.25 show several elements of MEWS ground echelon. Figure 1.23 shows a photo of ground-based MEWS PA in Alaska, and Figures 1.24 and 1.25 show photos of national VHF band RSs *Voronezh-M* and *Pechora* RS with PA.

1.11.3 Phased Arrays

Phased arrays are the key engineering element of MEWS ground echelon and other ground- and space-based systems. A phased array (PA) is a group of antenna radiators in which the relative phases of the signals are set in such a way that the effective radiation pattern of the array is reinforced in a desired direction and suppressed in undesired directions (Figure 1.26).

Phase adjustment (phasing) allows a PA radar:

- To form (given various arrangement of antennas) the necessary antenna radiation pattern (ARP) (e.g., a directional or beam ARP);
- To steer a given antenna beam thus performing rapid (and often inertia-less) scanning—beam swinging;
- Shaping ARP to certain extent (i.e., changing beam width, controlling sidelobe ratio (sidelobe level), and the like). Wave amplitude of single antennas is also sometimes controlled.

These and some other features of PA as well as possibility of using modern automation and computational electronics equipment to control PA have caused their wide application in radio communication, radiolocating, radio navigation, and radio astronomy. PAs consisting of a large number of controlled elements form part of various ground-based (stationary and mobile), maritime, airborne, and spaceborne radio devices.

Application of such phased arrays has several advantages.

Table 1.29 List and Main Technical Characteristics of MEWS RSs Constructed in the USSR

Radar Station	Western Designation	Antenna Type / Wavelength Range	Number of Antenna Faces, Their Size and Azimuthal Sector Covered by One Face	Main Features
Dnestr	Hen House	Phased array, azimuth scanning by frequency modulation, no elevation scanning / 1.5–2m	2 faces / 200 × 20m / 30 degrees	Space-surveillance radar station
Dnestr-M	Hen House	Phased array, azimuth scanning by frequency modulation, no elevation scanning / 1.5–2m	2 faces / 200 × 20m / 30 degrees	Early-warning radar station. Dnestr-M is a modification of the space-surveillance Dnestr radar station
Dnepr	Hen House	Phased array, azimuth scanning by frequency modulation, no elevation scanning / 1.5–2m	2 faces / 200 × 20m / 60 degrees	Early-warning radar station. Modification of Dnestr/Dnestr-M design
Daugava	Pechora	Phased array / 1.5–2m	Transmitter 30 × 40m / ~60 degrees	A prototype transmitter station for Daryal radar stations
Daryal, Daryal-U, Daryal-UM	Pechora	Phased array / 1.5–2m	Transmitter 30 × 40m and receiver 80 × 80m separated by 0.5–1.5 km / ~110 degrees	Early-warning radar station
Dunay-3	Dog House	Continuous-wave phased array, azimuth scanning by frequency modulation / ~0.1m	Transmitter and receiver separated by 2.4 km / ~45 degrees	The radar station was built as part of A-35 Moscow ABM system; two similar radar stations are deployed back-to-back at one site
Dunay-3U	Cat House	Continuous-wave phased array, azimuth scanning by frequency modulation / ~0.1 m	Transmitter and receiver separated by 2.8 km / 51 degrees	The radar station was built as part of A-35 Moscow ABM system; two similar radar stations are deployed back-to-back at one site
Volga		Continuous-wave phased array, azimuth scanning by frequency modulation, / ~0.1 m	Transmitter and receiver separated by 3 km, ~50 degrees	Early-warning radar station
Don-2N	Pill Box	Phased array / ~0.01	4 faces / receiver antenna 6m in diameter and receiver antenna 10 × 10m / 90 degrees	Radar station of A-135 Moscow ABM system

Figure 1.23   Ground-based PA of missile early-warning system in Alaska.

Figure 1.24   VHF band MEWS RS *Voronezh-M* in Lekhtusi near Saint Petersburg.

Figure 1.25   Pechora radar station.

- An array consisting of *n* elements provides an *n* times increase in directivity (and, consequently, in gain) as compared with a single antenna and narrows the beam to determine antenna angular position more accurately for navigation and radiolocation purposes.
- A phased array manages antenna dielectric field and the level of radiated (received) power by placing independent amplifiers in parallel.
- An important advantage of phased array is rapid (inertia-less) observation by means of electrical beam scanning.
- Phased arrays have a number of constructive and technological advantages over other classes of antennas. Weight-size parameters of onboard equipment may be improved due to printed antenna arrays. The use of reflector antenna arrays reduces the cost of large radio telescopes.

Until the end of the 1980s the construction of such a system required a large number of devices; therefore, phased arrays controlled by electronics were mainly used for large stationary radars, for example, for massive *BMEWS* (ballistic missile warning radar) or the American Navy's smaller air-defense radar *SCANFAR* (modification of *AN/SPG-59*) installed on the American nuclear-powered guided missile cruiser *Long Beach* and nuclear-powered aircraft carrier *Enterprise*. Its further modification *SPY-1 Aegis* was installed on cruisers of *Ticonderoga* class and later on *Arleigh Burke*-class destroyers. A large radar *Zaslon* installed on the Soviet *MiG-31* interceptor and attack radar installed on *B-1B Lancer* are the only known cases of applying phased arrays on aircraft. Today it is installed on *Su-35* and *F-15*.

Such radars weren't installed on aircraft mainly due to their large weight since the first generation phased arrays utilized common radar architecture. While the antenna was modified, the rest remained unchanged, though computers were added to control phase shifters. This increased the weight of the antenna by the number of computer modules and created a greater load on electric power supply system.

Figure 1.26   Active phased array (APA).

Common classification of antenna arrays includes: (a) linear; (b) arc; (c) circular; (d) planar; (e) cylindrical; (f) conical (g) spherical; and (h) space-tapered arrays. According to excitation, antenna arrays are divided into systems with:

- Sequential feed;
- Parallel feed;
- Combined (sequential and parallel) feed;
- Space (optical or ether) excitation method.

Active phased array (APA) (Figure 1.26) is a type of phased array (PA) in which all or part of the elements are equipped with their own miniature microwave transmitters, dispensing with the single large transmitter tube of passive phased array radars. Each element of an active phased array is comprised of a module with a slot, phase shifter, transmitter, and, often, a receiver.

In case of a passive array, one transmitter with the power of several kilowatts feeds several hundred elements, each handling power of tens of watts. A modern microwave transistor amplifier also produces tens of watts. A radar with an active electronically scanned array has several hundred modules, each having the power of tens of watts, which together create a powerful main beam with a power of several kilowatts.

Though the final effect is identical, active arrays are more reliable, since failure of one transmit/receive element may distort array pattern which somewhat degrades performance of locator, but most of the antenna still remains operational. The catastrophic transmitter tube failures, which plague conventional radars, simply do not occur. A side benefit is the weight saving incurred by dispensing with the bulky high power tube, its associated cooling system, and its large high voltage power supply.

Another feature, which may be exploited only in active arrays, is the ability to control the gain of the individual transmit/receive modules. If this can be done, the range of angles through which the beam can be swept is increased substantially, and thus many of the array geometry constraints that plague the conventional phased array may be circumvented. Such arrays are termed *supergain arrays*. From published literature, it is unclear whether any existing or development arrays use this technique.

Alas, power dissipation and cost reduction are the key challenges that define the future development of APA.

Power dissipation is the first challenge. Due to shortcomings of modern transistor amplifiers—microwave range (MMIC)—the efficiency of the transmitter module is typically less than 45%. As a result, an APA emits a lot of heat that must be dissipated so as to prevent transmitter chips from overheating since GaAs MMIC chips are more reliable at low operating temperature. Traditional air cooling used in conventional computers and avionics is ill-suited for high density packaging of APA elements; therefore, modern APA have liquid cooling systems (American projects usually use polyalphaolefin (PAO), a coolant that is similar to synthetic hydraulic fluid). A typical liquid cooling system has pumps for circulating coolant through antenna channels to a heat exchanger that can be either an air cooler (radiator) or a heat exchanger in propellant tank with another fluid that passes through a coolant loop and carries away high temperature from the coolant tank.

Thus, the APA is more reliable than the conventional radar of a fighter aircraft with an air cooling system, although it has larger energy consumption and requires more intensive cooling. APA provides a much larger transmit power that is required for long-range target detection.

The other challenge is the cost of mass production of modules. A radar of a fighter aircraft typically requires from 1,000 to 1,800 modules; therefore, the cost of APA will be unacceptably high if one module costs more than one hundred dollars. The cost of the first modules was about $2 thousand USD, which excluded mass application of APA. However, the cost of such modules and MMIC chips is constantly decreasing due to the reduction of net cost of development and production, as will be examined in one of the following chapters.

Despite its drawbacks, an active phased array outperforms conventional radar antennas in almost all respects, providing greater tracking capacity and reliability although slightly increasing complexity and, probably, cost.

Figure 1.27 shows a standard structure of transmit/receive module, the main element of an APA.

The transmit/receive module is an APA's basic spatial element or signal processing.

The module includes an amplifier, and a circulator that has electrodynamic nonreciprocal characteristics. In order to ensure both receiving and transmission the device has to have both receive and transmit channels. Either a switch or a circulator isolates receiver from transmitter.

The receive channel includes the following devices:

- Limiter, which is either a surge arrestor or any other threshold device that provides overload protection;
- Low-noise amplifier that has two or more stages of signal amplification;
- Phase shifter, which is a device that delays signal in the channel for phase distribution across the aperture;
- Attenuator, which is a device that reduces signal amplitude for amplitude distribution across the aperture.

The composition of a transmit channel is similar to that of receive channel. Unlike the receive channel, the transmit channel doesn't have a limiter and its amplifier has lower noise requirements. However, transmission amplifier needs to generate higher power.

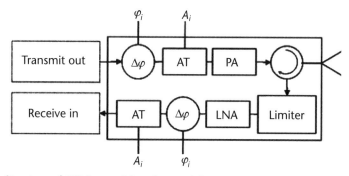

Figure 1.27   Structure of APA transmit/receive module.

1.11.4 MEWS Space Echelon

According to the project of the early-warning system, which was drafted in the USSR in early 1970s, the system was to include a space echelon in addition to the network of above-the-horizon and over-the-horizon RSs. Satellites were necessary to extend the capabilities of the early-warning system, for they were capable of detecting ballistic missiles almost immediately after launch.

The work on the space echelon of the early-warning system was assigned to the TsNIIKometa and the development of spacecraft was assigned to S. A. Lavochkin Design Bureau.

According to the design developed at TsNIIKometa, the space-based early-warning system, known as *Oko* or *US-KS*, included a constellation of satellites deployed on highly elliptical orbits and a command and control center near Moscow. The satellites were equipped with infrared and visible-spectrum sensors capable of detecting a burning missile motor against a background of space (but not against a background of Earth surface). The system began limited operations in 1978 and was placed on combat duty in 1982.

As the work on the *US-KS* system progressed, a set of requirements for a new space-based system, designated *US-KMO* (in the West this system is usually referred to, somewhat incorrectly, as *Prognoz*) was produced. This system was to provide coverage of possible submarine-launched ballistic missile launch areas from oceans, as well as from United States and Chinese territories. In order to do so, the satellites had to develop an ability to detect missile launches against the background of the Earth's surface. Certain problems that made the satellites suffer from explosive disintegration continued until 1984.

Missile launch detection system *Oko* designed for detection of launches from the US mainland included satellites *US-KS* in highly elliptical orbits with a control and data reception station and a launch site. The development of missile launch detection system was assigned to Chelomey Design Bureau in early 1960s. Preliminary design was prepared in 1962 and it included 20 satellites with the mass of 1400 kg placed in one polar orbit at the altitude of 3600 km.

LV *UR-200* was used to place the satellites in orbit. The satellites detected thermal radiation of first stage engine exhaust plumes.

On September 19, 1972, LV *Molnya* launched the first SC *Oko* (*Cosmos-520*) from the Plesetsk Cosmodrome. Another four launches were conducted in the following three years. The first two SC were equipped with infrared and television detectors. The third SC *Cosmos-665* had only television equipment. On December 24, 1972, it detected for the first time the launch of a *Minuteman* ballistic missile at night time. In 1976 *Cosmos-862* was placed into orbit and it had the USSR first onboard computer based on integrated microchips. A series of launches was performed since 1977 and as a result in 1978 a constellation of five satellites was formed that was the prototype of missile early-warning system.

It should be noted that at the beginning of program there were serious problems with reliability of the satellites. Only 7 out of 13 satellites launched in 1972–1979 operated more than 100 days. Later SC operation life was increased up to 3 years (American spacecraft *IMEWS-2* operated for 5–7 years). The deployment of *Oko* system started with the launch of four spacecraft in 1979 before the final outfitting of the ground segment station for control, data reception, and processing was finished. On April 5, 1979, the system was introduced into service, and in July of 1979 it detected start of launch vehicle from Kwajalein Atoll.

In 1980 six satellites were placed in elliptical orbit and the system was integrated into the missile early-warning system (MEWS). On December 30, 1982, the space-based system with six satellites was put on combat duty.

The satellites were equipped with an electronic self-destruct package that was activated if the satellite lost communication with ground control. Eleven out of 31 SC were lost due to this reason before 1983 (see Table 1.30). After 1983 it was decided not to activate self-destruct package in such situations, but it didn't considerably improve the failure statistics.

In 1984 SC *US-KS* of *Oko-S* system started operating in geostationary orbit. The satellite was placed at a longitude of 240° west and it surveyed the central part of the United States. The satellite observed rocket launches from the United States at the same angle as a satellite in highly elliptical orbit (HEO) while retaining its position relevant to Earth.

Introduction of geostationary satellites improved the reliability of the system. A satellite in GEO is capable of detecting launches even in case highly elliptical satellites aren't deployed. Though the quality of coverage and reliability of detection may suffer, the system doesn't go completely blind. LV *Proton-K* is used to place *US-KS* satellites in GEO and LV *Molnya-M* is used to place them in HEO. SC equipment detects infrared radiation of launched rockets only at the edge of Earth's visible disc (against atmosphere background) and transmits infrared image to Earth in real time. Eighty-six *US-KS* satellites were placed in orbit (four in GEO) in 1972–2002. The biggest number of launches was performed in 1984–1985 (eight each).

The launch of the second SC *US-KMO* (in the Russian language, *KMO* stands for continents, seas, and oceans) in February 1991 marked the beginning of deployment of the *Oko-1* system that is capable of detecting ICBM launches not only from the United States but also from other parts of the world. A feature of the system is vertical observation of missile launches against the background of the Earth's surface, which allows identifying flight azimuth of ICBMs. *Oko-1 (Prognoz)* with a satellite in GEO was introduced into service in 1996. The complete system is to include up to seven satellites in GEO and about four satellites in HEO. Eleven SC were launched from 1988 to 2004. SC operational life varies from 2 (SC *No. 8*) to 77 (*US-KMO No. 5*) months.

Flight tracks of the *Oko* system are shifted significantly to the West, which allows observing the territory of the United States at apogee while remaining in the Russian radio visibility zone. Television-vidicon cameras are used for observation in near IR and UV ranges.

Since one of the aims of this book is an attempt to understand the connection between reliability of SC and the utilized ECB, we shall analyze the following case.

As of January 2002, 86 first generation satellites were launched, 79 of them were placed in highly elliptical orbits (HEO). The remaining seven satellites were placed in geosynchronous orbits (GEO).

Three out of 79 launches in highly elliptical orbit failed. The remaining 76 satellites can be divided into three groups according to their operational life. The first group includes the satellites that operated for less than a year. Since such

Table 1.30 Russian Satellites of MEWS Space Echelon in 1972–1992

Satellite	NORAD Number	International Designation	Type	Launch Date (DD.MM.YY)	Launch Time (UTC)	Orbital Plane or GEO Station	Estimated End of Life DD.MM.YY	Comment
Cosmos-520	6192	1972-072A	HEO	19.09.72	19:19:03	4	Unknown	
Cosmos-606	6916	1973-084A	HEO	02.11.73	13:01:56	4	30.04.74	
Cosmos-665	7352	1974-050A	HEO	29.06.74	15:59:58	2	07.09.75	
Cosmos-706	7625	1975-007A	HEO	30.01.75	15:02:00	7	20.11.75	
Cosmos-775	8357	1975-097A	GEO	08.10.75	00:30:00	na	Unknown	Orbit was not stabilized
Cosmos-862	9495	1976-105A	HEO	22.10.76	09:12:00	5	15.03.77	Self-destructed
Cosmos-903	9911	1977-027A	HEO	11.04.77	01:38:00	7	08.06.78	Self-destructed
Cosmos-917	10059	1977-047A	HEO	16.06.77	04:58:00	9	30.03.79	Self-destructed
Cosmos-931	10150	1977-068A	HEO	20.07.77	04:44:00	2	24.10.77	Failed to reach the working orbit. Self-destructed
Cosmos-1024	10970	1978-066A	HEO	28.06.78	02:58:00	2	24.05.80	Moved off station in October 1979
Cosmos-1030	11015	1978-083A	HEO	06.09.78	03:04:00	4	10.10.78	Self-destructed. Orbit was not stabilized
Cosmos-1109	11417	1979-058A	HEO	27.06.79	18:11:00	9	15.02.80	Self-destructed. Orbit was not stabilized
Cosmos-1124	11509	1979-077A	HEO	28.08.79	00:17:00	4	09.09.79	Self-destructed. Orbit was not stabilized
Cosmos-1164	11700	1980-01 3A	HEO	12.02.80	00:53:00	9		Launch failure
Cosmos-1172	11758	1980-028A	HEO	12.04.80	20:18:00	9	09.04.82	
Cosmos-1188	11844	1980-050A	HEO	14.06.80	20:52:00	2	28.10.80	
Cosmos-1191	11871	1980-057A	HEO	02.07.80	00:54:00	4	16.05.81	
Cosmos-1217	12032	1980-085A	HEO	24.10.80	10:53:00	2	20.03.83	
Cosmos-1223	12078	1980-095A	HEO	27.11.80	21:37:00	7	11.08.82	
Cosmos-1247	12303	1981-016A	HEO	19.02.81	10:00:00	5	20.10.81	Self-destructed
Cosmos-1261	12376	1981-031A	HEO	31.03.81	09:40:00	6	01.05.81	Self-destructed
Cosmos-1278	12547	1981-058A	HEO	19.06.81	19:37:04	4	05.07.84	Self-destructed in December 1986

Table 1.30 Russian Satellites of MEWS Space Echelon in 1972–1992 (Cont.)

Satellite	NORAD Number	International Designation	Type	Launch Date (DD.MM.YY)	Launch Time (UTC)	Orbital Plane or GEO Station	Estimated End of Life DD.MM.YY	Comment
Cosmos-1285	12627	1981-071A	HEO	04.08.81	00:13:00	6	21.11.81	Failed to reach the working orbit. Self-destructed
Cosmos-1317	12933	1981-108A	HEO	31.10.81	22:54:00	9	26.01.84	Self-destructed
Cosmos-1341	13080	1982-016A	HEO	03.03.82	05:44:38	5	01.02.84	
Cosmos-1348	13124	1982-029A	HEO	07.04.82	13:42:00	9	22.07.84	
Cosmos-1367	13205	1982-045A	HEO	20.05.82	13:09:00	1	30.09.84	
Cosmos-1382	13295	1982-064A	HEO	25.06.82	02:28:00	7	29.09.84	
Cosmos-1409	13585	1982-095A	HEO	22.09.82	06:23:00	2	05.01.87	
Cosmos-1456	14034	1983-038A	HEO	25.04.83	19:34:00	4	13.08.83	Self-destructed
Cosmos-1481	14182	1983-070A	HEO	08.07.83	19:21:00	6	09.07.83	Failed to reach the working orbit. Self-destructed
Cosmos-1518	14587	1983-126A	HEO	28.12.83	03:48:00	5	01.06.84	
Cosmos-1541	14790	1984-024A	HEO	06.03.84	17:10:00	3	31.10.85	
Cosmos-1546	14867	1984-031A	GEO	29.03.84	05:53:00	1, 4	16.11.86	
Cosmos-1547	14884	1984-033A	HEO	04.04.84	01:40:04	7	23.08.85	
Cosmos-1569	15027	1984-055A	HEO	06.06.84	15:34:00	5	26.01.86	
Cosmos-1581	15095	1984-071A	HEO	03.07.84	21:31:00	8	19.08.85	
Cosmos-1586	15147	1984-079A	HEO	02.08.84	08:38:00	4	01.04.85	
Cosmos-1596	15267	1984-096A	HEO	07.09.84	19:13:00	9	26.11.86	
Cosmos-1604	15350	1984-107A	HEO	04.10.84	19:49:13	1	27.09.85	
Cosmos-1629	15574	1985-016A	GEO	21.02.85	07:57:00	4, 3, 1	16.01.87	
Cosmos-1658	15808	1985-045A	HEO	11.06.85	14:27:00	6	03.09.87	
Cosmos-1661	15827	1985-049A	HEO	18.06.85	00:40:26	na	21.10.89	Moved off station from the beginning of operation
Cosmos-1675	15952	1985-071A	HEO	12.08.85	15:09:00	8	18.01.86	

Cosmos-1684	16064	1985-084A	HEO	24.09.85	01:18:10	4	09.03.89	
Cosmos-1687	16103	1985-088A	HEO	30.09.85	19:23:00	2	30.09.85	Orbit was not stabilized
Cosmos-1698	16183	1985-098A	HEO	22.10.85	20:24:00	3	24.08.86	
Cosmos-1701	16235	1985-105A	HEO	09.11.85	08:25:00	8	23.11.87	Moved off station in December 1986
Cosmos-1729	16527	1986-011A	HEO	01.02.86	18:11:56	5	14.05.88	
Cosmos-1761	16849	1986-050A	HEO	05.07.86	01:16:47	3	23.10.88	
Cosmos-1774	16922	1986-065A	HEO	28.08.86	08:02:43	7	17.07.88	
Cosmos-1783	16993	1986-075A	HEO	03.10.86	13:05:40	1	03.10.86	Failed to reach the working orbit
Cosmos-1785	17031	1986-078A	HEO	15.10.86	09:29:18	9	16.01.91	Moved off station in December 1989
Cosmos-1793	17134	1986-091A	HEO	20.11.86	12:09:20	2	13.08.91	Moved off station in June 1990
Cosmos-1806	17213	1986-098A	HEO	12.12.86	18:35:36	5	20.11.88	
Cosmos-1849	18083	1987-048A	HEO	04.06.87	18:50:23	1	20.05.90	
Cosmos-1851	18103	1987-050A	HEO	12.06.87	07:40:28	6	23.11.89	
Cosmos-1894	18443	1987-091A	GEO	28.10.87	15:15:00	1	22.12.91	
Cosmos-1903	18701	1987-105A	HEO	21.12.87	22:35:42	8	11.11.92	
Cosmos-1922	18881	1988-013A	HEO	26.02.88	09:31:12	5	30.07.90	
Cosmos-1966	19445	1988-076A	HEO	30.08.88	14:14:54	3	14.12.90	
Cosmos-1974	19554	1988-092A	HEO	03.10.88	22:23:39	7	20.05.93	
Cosmos-1977	19608	1988-096A	HEO	25.10.88	18:02:31	6	12.07.90	
Cosmos-2001	19796	1989-011A	HEO	14.02.89	04:21:11	4	15.03.93	
Cosmos-2050	20330	1989-091A	HEO	23.11.89	20:35:44	9	08.10.93	
Cosmos-2063	20536	1990-026A	HEO	27.03.90	16:40:08	2	21.06.95	
Cosmos-2076	20596	1990-040A	HEO	28.04.90	11:37:02	1	30.10.92	
Cosmos-2084	20663	1990-055A	HEO	21.06.90	20:45:52	6	21.06.90	Failed to reach the working orbit
Cosmos-2087	20707	1990-064A	HEO	25.07.90	18:13:56	6	21.01.92	
Cosmos-2097	20767	1990-076A	HEO	28.08.90	07:49:13	3	30.04.95	
Cosmos-2105	20941	1990-099A	HEO	20.11.90	02:33:14	3	04.04.93	
Cosmos-2133	21111	1991-010A	GEO	14.02.91	08:31:56	4, 3, 2, 1, 4	09.11.95	Moved off station in February 1992

Table 1.30 Russian Satellites of MEWS Space Echelon in 1972–1992 (Cont.)

Satellite	NORAD Number	International Designation	Type	Launch Date (DD.MM.YY)	Launch Time (UTC)	Orbital Plane or GEO Station	Estimated End of Life DD.MM.YY	Comment
Cosmos-2155	21702	1991-064A	GEO	13.09.91	17:51:02	1	16.06.92	
Cosmos-2176	21847	1992-003A	HEO	24.01.92	01:18:01	6	13.04.96	
Cosmos-2196	22017	1992-040A	HEO	08.07.92	09:53:14	5	23.06.94	
Cosmos-2209	22112	1992-059A	GEO	10.09.92	18:01:18	1	16.11.96	
Cosmos-2217	22189	1992-069A	HEO	21.10.92	10:21:22	8	07.11.96	
Cosmos-2222	22238	1992-081A	HEO	25.11.92	12:18:54	1	03.12.96	
Cosmos-2224	22269	1992-088A	GEO	17.12.92	12:45:00	2, 1, 2	17.06.99	
Cosmos-2232	22321	1993-006A	HEO	26.01.93	15:55:26	4	04.06.98	

operational life is considerably shorter than satellite average operational life, it can be assumed that the reason for this is system failure. This group also includes satellites, the orbits of which weren't stabilized: *Cosmos-931*, *Cosmos-1030*, *Cosmos-1109*, *Cosmos-1124*, *Cosmos-1261*, *Cosmos-1285*, *Cosmos-1481*, and *Cosmos-1687*. The group consists of 21 satellites launched both at the beginning of the program and recently.

The second group is comprised of satellites that were launched before 1985. These satellites operated for 20 months on average. The third group of our classification consists of satellites that were launched after 1985 and the operational life of which was twice as long as that of the second group and constituted about 40 months.

Analysis of satellite operational life proves the assumption that in the middle of 1980s spacecraft were significantly upgraded, which resulted in increased operational life. Operational life was increased within the framework of the program that foresaw placing satellites in geosynchronous orbits (the first satellite was launched in 1984).

Cosmos-2232 had the longest operational life of 64 months among first generation satellites.

The number of second generation satellites is too small to make conclusions on their operational life. *Cosmos-2224* had the longest operational life of 77 months (which is longer than that of any other Eastern Bloc early-warning satellite). *Cosmos-2133* and *Cosmos-2209* operated for 56 and 50 months, respectively. Such operational life suggests that their operations were successful. *Cosmos-2282* ceased operations after 17 months, most likely because of a malfunction. *Cosmos-2350* ceased all maneuvers only after two months, which also indicates a failure.

The choice of observation geometry and, consequently, the highly elliptical orbits can be attributed to *the lack of proper infrared semiconductor sensors and microelectronic data processing capabilities* in the Soviet Union that are required for obtaining a capability of detecting rocket launches against Earth background. In absence of suitable IR-sensors, the Soviet Union had to create a system relying on a grazing-angle observation geometry, which allowed detecting launches against space background. Such conditions allowed the use of less sophisticated sensors but demanded more financial support from the government since constellations in highly elliptical orbits are to include *a larger number of satellites than constellations in geostationary orbits*.

Since the main subject of this book is space microelectronics issues, the previously mentioned examples from our recent past proves the importance of ECB in space projects.

1.11.5 Military Reconnaissance Satellites

A military reconnaissance (reconnaissance) satellite is an Earth artificial satellite (EAS) designed for Earth observation (television and photo survey) to ensure intelligence activities or a communications satellite deployed for reconnaissance applications. It is often referred to as spy satellite by journalists [22].

Functions of reconnaissance EAS:

- High resolution photography (imagery intelligence);
- Communications eavesdropping and detection of radio facilities (signals intelligence);
- Monitoring of nuclear test ban compliance;
- Detection of missile launches (missile early-warning system).

Available information on United States programs is mainly information on programs that existed up to 1972 and only small amount of leaked information on later programs is available. A few up-to-date reconnaissance satellite images have been declassified on occasion, or leaked, as in the case of *KH-11* photographs which were sent to *Jane's Defence Weekly* in 1985.

About 2,000 spent reconnaissance EASs produced in the USSR only are in high orbit, according to the Russian documentary *Secret Space*.

All reconnaissance EASs can be divided into three groups: imaging photographic, electro-optical and radar reconnaissance satellites.

Reconnaissance satellites were an important part of space programs of the USSR and the United States. After the first satellite was launched, the attention of S. P. Korolev was focused on lunar program whereas the United States aimed efforts at the development of the military reconnaissance program *Discoverer*.

The satellite launch plan developed in 1956 foresaw performing reconnaissance functions (observation of potential adversary's facilities from space) and detection of ballistic missile launches. During the Cold War the United States military space program was aimed at gathering intelligence information on the Soviet Union.

In 1954 the United States developed *Advanced Reconnaissance Systems* program, within the framework of which two projects of reconnaissance Earth artificial satellites (EAS) were implemented: *SAMOS* operated by the USF and *CORONA* designed for the purposes of CIA.

Discoverer satellites were designed for exercising methods of military space photographic reconnaissance (spy satellites). They were also used for researching the possibility of animal and human spaceflights. The launch of EAS *Discoverer-1* on

Figure 1.28   American radar reconnaissance satellite *Lacrosse* under construction.

February 28, 1959, marked the beginning of a series of launches (38 satellites) that were performed in a relatively short three-year period. *Discoverer-38* was launched the last on February 27, 1961. These EASs were equipped with special attitude control system as well as reentry equipment and were placed in polar orbits. Separation of reentry capsule and retrorocket firing started on command from the observing station on the Hawaiian Islands. The Americans weren't able to recover capsules for a long time. Reentry capsules were captured and searched for by the United States Air Force and Navy. Recovery of exposed film from satellites to Earth was performed under strict secrecy. Exposed film was for the first time successfully retrieved from *Discoverer-14* that was launched on August 18, 1960. After the reentry capsule separated during the satellite's seventeenth revolution, it was captured at the third attempt by transport aircraft *C-130* with the help of a special trailing device. Not the satellite itself but a small capsule (about 50 kg) was to be recovered; therefore, it was captured by an aircraft during parachute descent.

Thirteen out of 38 launches (1959–1961) were unsuccessful. Some of the capsules weren't captured by helicopters. All satellites placed in orbit by USF after

Figure 1.29   A model of German reconnaissance satellite *SAR-Lupe* on top of a *Cosmos-3M* rocket.

Discoverer-38 became classified. The information was declassified only in 1990s. The EAS was designated *CORONA*.

SAMOS and *MiDAS* were military programs of the United States. Within the *MiDAS* program (first launch on May 24, 1960) the use of early-warning satellites for detection of intercontinental missile launches was exercised. The system proved to be efficient in October of 1961 when it detected the launch of *Titan* ballistic missile from Cape Canaveral. Since the system was adjusted, the information was obtained only 90 seconds after launch. Reconnaissance satellite *Mercury* was successfully launched as a part of the *MiDAS* program. The satellite weighed less than 1100 kg and was designed for photographing the Earth's surface at the altitude of 160–200 km and research of human body functioning during spaceflight.

The National Reconnaissance Office, the National Security Agency, and the United States Naval Research Laboratory have recently declassified information on launches of satellites of *POPPY* series that were designed for radar observation of the Soviet Navy ships in 1962–1971. *POPPY* satellites were successors of *GRAB* satellites (first launched on June 22, 1960) that were launched in 1960–1962.

In October 2002 the United States declassified documents on reconnaissance satellites *KH-7* and *KH-9* (*CORONA*) [23]. The program *KH* (*Key Hole*) included a range of modifications: KH-7, -8, -9, -12, and other F1-30 (Figure 1.28). They operated for the purposes of CIA until the middle 1990s. The EAS *KH-11A* is credited with the capability of resolving objects with lateral size of less than 10 cm.

The operation of the first generation of satellites with equipment for close-up imaging started in June of 1963. *KH-7* satellites produced images of 0.46m resolution. In 1967 they were substituted with EAS *KH-8* (0.3m resolution) that operated until 1984. Satellite *KH-9* that provided wide coverage and the resolution of 0.6m was launched in 1971. It was the size of a railroad car and weighed

Figure 1.30   Three-ton spy satellite *KH-8*. The image was declassified in September 2011.

more than 9 tons. It had the equipment developed for the *Manned Orbiting Laboratory (MOL)*.

A considerable disadvantage of these space systems was connected with data transmission to Earth. First, there was a big time gap between imaging and delivering of photo data to Earth. Secondly, expensive equipment onboard the EAS became useless after separation of capsule with film. These problems were partially solved by equipping the satellites (starting from *KH-4B*) with several capsules with film.

The development of a system for electronic data transmission in real time was a comprehensive solution to the first problem. From 1976 until the end of program in 1990s the United States launched eight satellites of *KH-11* series that had electronic data transmission systems.

On February 11, 1965, the United States launched satellite *LES-1* of military communications satellite series that were designed for estimation of measures to reduce vulnerability of military satellites to the means of military space defense (at the time the USSR tested antisatellite systems). Satellite protection measures included substitution of solar cells with radioisotope power system, the use of attitude control system based on one-axis gyroscope, and the use of satellite-satellite communications link that allows dispensing with intermediate ground relay station during long-distance communication.

Satellites *LES* weighed about 450 kg and were about 3m long. They had a three-axis attitude control system that ensured pointing of some of the onboard antennas toward Earth and others toward the second EAS for the experiments with satellite-satellite communication. EAS's power supply was provided by two radio-isotope power systems that used plutonium-238 and had the initial power of 150W and the power of 130W after five years of operation. An antenna with parabolic reflector was used for communication with the other satellite. The satellite was equipped with devices for interference protection and stable radio communication with mobile small objects including aircraft.

Due to active operation of reconnaissance satellites, there emerged the second EAS generation of *CORONA* program [22]: *Ferret*, *Jumpseat*, SDS relay EASs, and *Spook Bird (CANYON)*.

CANYON satellites were launched from 1968 and placed in orbits close to geostationary orbit. They were aimed at the Soviet communications eavesdropping. At the end of 1970s they were replaced by EAS *Chalet* and *Vortex*.

Rhyolite and *Aquacade* satellites (geostationary orbit, 1970s) were used to track telemetry data of the Soviet ballistic missiles. In the 1980s they were replaced with EASs *Magnum* and *Orion* that were launched by a space shuttle.

A total of 145 launches of satellites with photographic reconnaissance equipment were conducted within the CORONA program, and 102 of them were successful.

It should be understood that this book can't present a complete picture of events since information on reconnaissance space equipment of the United States is classified as top secret and information in open access has a speculative and advertising (and sometimes misleading) character.

The United States was interested in the number of ballistic missiles in the USSR, location of cosmodromes in the North and in Kazakhstan, location of nuclear facilities, intercontinental ballistic missile submarines and their bases, as well as in many other objects of strategic importance.

Almost all the objects sent to space had dual purpose: applied scientific research and military. The examples are the United States *DMS* satellites and Soviet *Cosmos* EASs that were launched both as satellites and orbital stations.

Satellites of *DMS* series were primarily designed for the needs of military agencies, providing special strategic programs and command and control systems in various parts of the world with information. They produced high resolution images (in visible and infrared range) in real time being the only source of such data for coastal weather stations and weather ships of the United States Navy. Meteorological equipment delivered data on temperature, humidity, and atmosphere density vertical profile from the area covered by satellite. Both real-time and recorded meteorological data could be obtained from satellites. The satellites of the series were launched since early 1970s. On February 2, 1988, an advanced EAS *DMS-5D-2* was placed in orbit.

Lacrosse satellite that uses synthetic aperture radar came into operation at the end of 1980s [31] (Figure 1.28). This EAS provided resolution of 0.9m and could see through the clouds.

The development of satellite reconnaissance system (Figure 1.30) in the USSR started later than in the United States. On May 22, 1959, a decision was made to develop an orbital spaceship designed for reconnaissance and human spaceflight (the Resolution of CPSU CC and the Council of Ministers of the USSR No. 569-264cc). A manned spaceship (SS) *Vostok* and a photoreconnaissance SC *Zenit-2* were constructed. On April 26, 1962, the first television survey of Earth cloud cover was made by *Cosmos-4* satellite. It was a breakthrough in weather forecasting.

Spacecraft *Zenit-2* was the first Soviet reconnaissance satellite. The Armed Forces of the USSR adopted *Zenit-2* on March 10, 1964. The satellites of *Vostok-D* series had a bigger reentry capsule used to recover both film and camera unlike American satellites, the reentry capsule of which contained only film. From 1962 to 1968 the satellites of *Zenit-2,-4* series were used for photographic reconnaissance.

First generation satellites were put into the same orbits by the same launch vehicles as manned SSs *Vostok*. Duration of flights was eight days as a rule and the number of launches increased up to nine by 1964. The first program incident occurred at the end of 1964, when *Cosmos-50* exploded in orbit after eight days of flight. On January 19, 1968, the reconnaissance satellite *Cosmos-200* (*Tselina-O* type) was launched from the Plesetsk Cosmodrome. The satellites of *Zenit* series were equipped with an equipment complex that included a *SA-20* camera with focal length of 1m, a *SA-10* camera with focal length of 0.2m as well as *Baikal* photo-television equipment and *Kust-12M* reconnaissance equipment to transmit data via radio channel to visible ground receive stations. After the test flights (SC *Cosmos-4, -7, -9, -15*) two *SA-20* cameras were added to allow swath width of 180 km at the altitude of 200 km.

Thirteen launches of SC *Zenit-2* were performed during flight-development tests, and three of them ended up in launch vehicle failure (LV on the basis of ICBM *R-7*). A total of 69 launches of *Zenit-2M* were successfully performed from 1968 to 1979, and there was only one LV failure. Some 8–11 launches were conducted each year. SC *Zenit* had the following modifications (second generation): *Zenit-4* (1964–1970); SC *Zenit-4M* (*Rotor*) designed for advanced close look photoreconnaissance (1968–1973); *Zenit-4MK* (1969–1978); optical reconnaissance satellite *Zenit-4MKM* (*Gerakl*) (1977–1980).

SC *Zenit-6* was the following modification (1976–1980).

On July 12, 1963, the United States launched a new optical reconnaissance spacecraft with improved characteristics, *KH-7 Gambit*. The USSR developed a spacecraft of a new *Yantar* series, which was designated *Feniks* (developed by the Central Assembly and Design Engineering Bureau in Samara) when introduced into service. It was a prototype of the series of optical photoreconnaissance satellites: 11F622 *Yantar-1* for area survey and 11F623 *Yantar-2* for close look reconnaissance. At the time manned SC *Soyuz-R* for comprehensive reconnaissance was under development. It was replaced with transport spaceship (TKS) 11F727K-TK that supplied *Almaz* space station. Military and research spaceship 11F73 *Zvezda* was actively developed. However, none of these projects were taken to the stage of flight-development tests. Under the project Yantar, SC 11F624 *Yantar-2K* was developed for detailed photo reconnaissance. Based on this model, three new optical reconnaissance complexes were subsequently developed, which consisted of ultralong monitoring 11F650 *Yantar* 6K, operational detailed monitoring 11F661, and survey photographic monitoring 11F630.

The *Yantar-2K* complex (*Feniks*) was introduced into service in May of 1978. In terms of technical characteristics, it matched the American satellite *Big Bird* that had several reentry capsules. There were 30 launches of SC *Yantar-2K* on LV 11A511U *Soyuz-U* from 1974 to 1983. LV failed two times. The satellites exploded in orbit twice due to serious technical malfunction.

On the basis of *Yantar* there was created an electro-optical reconnaissance satellite *Neman* capable of converting photo images into digital signal and transmitting it to ground stations via radio channel.

In 1980 Production Association Arsenal started serial production of spacecraft of *Kobalt* type (modification of SC *Yantar-2K*) designed for observation and close look imaging of Earth surface (developed by TsSKB Progress, Samara). It was replaced with *Kobalt-M* spacecraft that had reentry capsule with film. Active operational life of such spacecraft constitutes about 60–120 days. On April 16, 2010, spacecraft *Cosmos-2462*, an optical reconnaissance satellite of *Kobalt-M* type, was successfully launched on the launch vehicle *Soyuz-U*.

In 1979–1984 there were 16 launches of SC *Yantar-4K1* (*Oktan*) and *Orlets* designed for high resolution photographic observation.

A new electro-optical reconnaissance satellite *Yenisey* with an operational lifetime of about a year was launched from the Baikonur Cosmodrome in August of 1994. It is a digital photographic reconnaissance satellite of fifth generation that provides close to real-time data transmission and presents a longer operational life modification of SC *Don* with 22 reentry capsules.

In June 1997 a photographic reconnaissance satellite of eighth generation 11F664 (*Yantar* series) was launched on LV *Proton-K* from the Baikonur Cosmodrome. In April of 2009 the ninth SC of the *Yantar* family (project 11F695M) was placed in orbit.

Cosmos-1426 launched on December 28, 1982, and marked the beginning of the fifth generation of the Soviet optical reconnaissance satellites with in real-time electronic data transmission. Unlike the fourth generation, the satellites of fifth generation have a narrow range of altitude and their orbits remain almost circular.

Flight endurance of these satellites is six to eight months. Normal operation of the photographic reconnaissance system of fifth generation foresees simultaneous operation of two satellites placed in orbits that have the distance of 910 between them. Introduction of long-lived satellites of fifth generation allowed reduction and in 1990 complete elimination of survey flights performed by the satellites of third generation. The last innovation of the Soviet optical reconnaissance program was SC *Cosmos-2031* launched in July of 1989.

Another field of space reconnaissance is signals intelligence. The development of signals intelligence systems began in August of 1960 when a task was set to develop in the interest of the Ministry of Defense of the USSR an experimental spacecraft *DS-K8* for exercising methods and means of parameter identification of radar signals produced by defense RSs. The first stage foresaw the development of unified SC *DS-U* and launches of two experimental spacecraft *DS-K40* that were performed in 1965–1966 but were unsuccessful due to launch vehicle failures. The second stage included the development of electronic intelligence SC of *Tselina* system that had microelements based equipment (Yuzhnoye Design Bureau, 1964).

SC *Tselina-O*, a nonoriented EAS with solar panels designed for survey electronic intelligence (*Cosmos-189*) was launched from the Plesetsk Cosmodrome in 1967. The system covered one and the same Earth area several times a day at different time. It had a three-month active operational life. Forty spacecraft of such modification were launched from 1968 to 1982.

SC *Tselina-D* was designed for detailed electronic measurements that included receiving, analyzing, and high precision georeferencing of electronic signal. This satellite belongs to the class of SC oriented in orbital coordinate system and features a more sophisticated complex of special and support equipment. Already during testing SC *Tselina-D* proved that it was able not only to detect radio objects and determine their location but also to identify their purpose, characteristics, and operational modes. A total of 71 SC were placed in orbit in 1970–1994, and two launches ended up in failure.

Tselina-R modification with equipment for observation of radio sources was developed in 1980s, and it ensured full-scale electronic reconnaissance.

The radar reconnaissance satellite *Almaz-T* (developed by NPO PM) with the resolution of 10–15m was launched in the USSR in 1981, and it can be considered a match of the United States *Lacrosse*.

Naval reconnaissance and targeting system (MKRTs) *Legenda* deserves special attention. The development of the world's first space system consisting of different types of reconnaissance SC for observation of the world's oceans started in early 1960s, and it was designed for the purposes of application of antiship strike weapons by ships and submarines of the Soviet Navy. MKRTs system utilized two types of spacecraft: radar reconnaissance satellite *US-A* (stands for *controlled active satellite*) and electronic reconnaissance satellite *US-P* (stands for *controlled passive satellite*). The head developer of the system was Design Bureau-1 (TsNIIKometa, Moscow). SC *US-A* and *US-P* were developed by Experimental Design Office No. 52 (NPO Mashinostroyeniya, the town of Reutov). At the same time the system for signal environment control *Tselina* that detected radiation in a wide range of frequencies was under development. However it wasn't until 1960 that the project was implemented due to the absence of decision at the Ministry of Defense on a single ordering party of space systems.

The nuclear power plant (NPP) *Buk* with electric power of 3 kW was used to supply the complex. Chemical sources were used to power several SC during the adjustment of onboard special complex. The first launch was performed in 1965. The active radar naval reconnaissance and targeting system with SC *US-A* was introduced into service in 1975 after successful flight tests. Electronic intelligence SC *US-P* was introduced into service in 1978.

The production of *US-A* ceased due to ban on SC powered by NPP in low orbits (orbit altitude 250–290 km). There was a total of 37 launches of SC *US*, two of which ended up in LV failure. Three to four launches were performed in 1975–1976 and 1981–1982. The first spacecraft were launched on LV of *Soyuz* series and on LV *Tsyklon-2* when introduced into service. On April 25, 1973, the satellite *US-A* fell on the territory of Canada due to failure of LV *Tsyklon-2*. The nuclear reactor pressure vessel stood the impact and nuclear contamination didn't occur; still, the Soviet Union had to pay $3 million (USD) to Canada.

SC *US-P* searched and identified surface targets not by means of radar illumination but by detecting their electronic emissions that are indicative of each type of ships. SC *US-P* were equipped with solar power plants and buffer storage batteries. A total of 37 SC of this type were launched in 1974–1991. The launches of SC *US-P* (and *US-A*) were performed only at the Baikonur Cosmodrome. The operating constellation of *US-P* SC was to include three spacecraft. Their orbits were phased in such a way that all the satellites flew along the same track with one day shift between one another.

At the end of their operational life SC left the working orbit. Spacecraft of 1975–1987 required small acceleration burn for deorbiting. The satellites remained in orbit for several years and were destroyed due to explosion of remaining propellant in propulsion system or of pressurized containers with buffer storage batteries (Figure 1.31).

At the end of this section we would like to mention *the role of military space reconnaissance of the United States in modern local conflicts* as an example of application of military reconnaissance satellites. The significance of this subject can be explained by the increasingly larger role of space reconnaissance in planning and

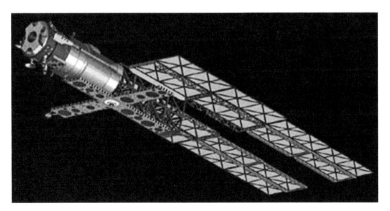

Figure 1.31 *SC US-PM.*

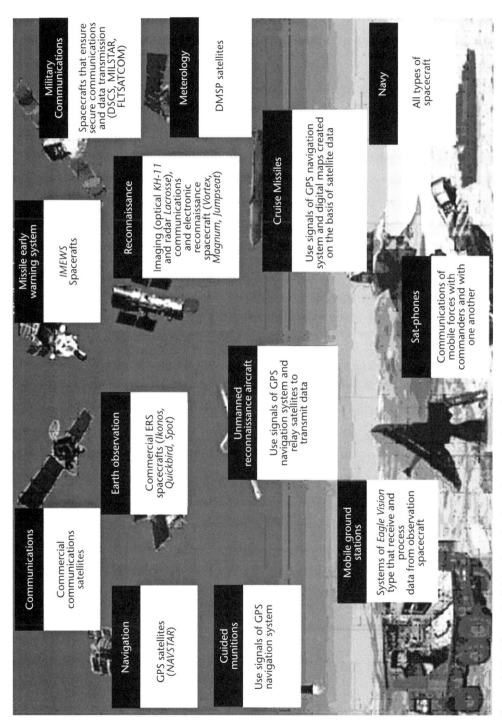

Figure 1.32   The main fields of application of space systems data.

conduct of combat operations after the Pentagon adopted the concept of network-centric warfare. Modern reconnaissance spacecraft are capable of detecting an adversary's activity at the stage of combat preparation. Powerful systems for data processing and transmission allow detecting, identifying, and destroying the target within the shortest time possible.

The Iraq War of 2003 is the most impressive example in terms of the use of data obtained from spaceborne systems. Figure 1.32 shows the main fields of application of spaceborne systems during the Iraq War.

To a certain extent this war became a proving ground that the United States Armed Forces used for testing new armament, including spaceborne systems. Different types of military and commercial observation satellites, navigation, and weather spacecraft as well as missile early-warning satellites were utilized. The constellation consisted of 50–59 military spacecraft of various purposes, 28 GPS spacecraft, and a large number of commercial SC for communications and Earth remote sensing, according to open sources.

During combat preparation the United States constellation wasn't enlarged. Military operations were supported by already operating spacecraft, which suggests that the United States have reached such position in space that a constellation deployed in advance and functioning at peacetime is capable of supporting military operations of such scale at any time and in any part of the world.

In the twenty-first century, data support to armed forces from space will remain one of key tasks of military space systems: providing low command level and potentially a soldier with space data. The consequences of information warfare can be compared with that of nuclear weapon development in the middle of the twentieth century.

Hands-on experience of armed forces living and operating in various strategic situations shows that today, operation of armed forces without space systems is problematic in peacetime and hardly possible during combat operations. Another possible military application of data from space reconnaissance systems could be the development of space support groups.

The Russian Army has a successful practice of application of space support groups at operational-tactical and tactical levels. The main task of these groups is evaluation of SC status and efficiency, formulation of proposals on the use of spacecraft for obtaining operational data, as well as providing different command levels with received data (reconnaissance, meteorological, navigation, and communication). Space support groups are one of the most perspective ways to eliminate the gap between potential capacity of spacecraft and their application for military purposes.

However, future development of such fields should be based on the recent advances of microelectronics, microsystems engineering, and power semiconductor electronics.

References

[1]	rossbel.ru/d/818179/d/belta_070514.doc.

[2]	www.cosmoworld.ru/spaceencyclopedia/news_arch/1039.doc.

[3] Bezborodov, V. G., S.V. Pushkarsky, M. A. Lukjaschenko, Organization and Application of Space Activities Results in the Russian Federation/FSBI, "Aviamettelekom of Roshydromet: New Technologies and Developments," *Meteospektr*, No. 1, 2014.

[4] www.realeconomy.ru/files/Federal%20space%20program.doc.

[5] www.tsenki.com/projects/federal-program.

[6] innovation.gov.ru/sites/defoult/files/documents/2014.

[7] Gushchin, V. N., B. M. Pankratov, A. D. Rodionov, *Basic Structure and Development of Spacecraft*, M: Mashinostroyeniye, 1992.

[8] Mikrin Y. A., N. A.Sukhanov, V. N. Platonov, et al., Design Concepts of Onboard Control Complexes for Automated Spacecraft, *Control Sciences*. No. 3, 2004, pp. 62–66.

[9] Kungurtsev, V. V., "Technology of Building Spacecraft Onboard Information Control Complexes on the Basis of Programmable Logic Integrated Circuits," *Information and Control Systems*, No. 5, 2006.

[10] Garbuk, S. V., V. Y., Gershenzon, *Earth Remote Sensing Space Systems*, M: IzdatelstvoA i B, 1997.

[11] Kashkin, V. B., A. I. Sukhinin, *Remote Sensing of the Earth from Space: Digital Image Processing*, M, Logos, 2001.

[12] Kupejko, A.V. "Germany Storms the Market of Geoinformatics," *NovostiKosmonavtiki*, No. 8 (295), 2007.

[13] Verba, V. S., et al. *Space-Based Earth Observation Radar Stations*, M.: Radiotechnika, 2010.

[14] dic.academic.ru/dic.nsf/enc-coliet/6661.

[15] geomatica.ru/pdf/2013-03/2013-02-16-36.pdf.

[16] Dvorkin, V.A., S. A. Dudkin, "Up-to-Date and Advanced Remote Earth Sensing Satellites," *Geomatics*, No. 12, 2013, pp. 16–36.

[17] Shpenst, V., Space-Based Radar Stations for Earth Remote Sensing, *Komponenty i Tekhnologii (Components & Technologies)*, No. 3, 2013, pp. 154–158.

[18] Kopik, A., "Italian Radar Reconnaissance Satellite Is in Space," *NovostiKosmonavtiki*, No. 8, 2007 p. 295.

[19] Suleymenov, Y. Z., Y. G. Kuljevskaya, G. G. Ulezko, E. A. Galantz, "State of Research in Kazakhstan According to Priorities of Science and Technology Development," *Space Research*, Kazakhstan National Center of Scientific and Technical Information, Almaty, 2008.

[20] Bayt, R. L., A. A. Ayon, K. S. Breuer, "Micropropulsion for the Aerospace Industry," *Sensors: Technology and Design*, February 2002.

[21] Cheah, K. H., J. K. Chin, "Performance Improvement on MEMS Micropropulsion System Through a Novel Two-Depth Micronozzle Design," *ActaAstronautica*, Vol. 69, 2011, pp. 59–70.

[22] http://proatom.ru/modules.php?name=News&file=print&sid=3299.

[23] Harendt, C., H. G. Graf, B. Hofflinger, E. Penteker, "Silicon Fusion Bonding and Its Characterization," *J. Micromech. Microeng.*, Vol. 2., 1992, pp. 113–116.

Selected Bibliography

Brovkin, A.G., B. G. Budygov, S. V. Gordeenko, et al., *Spacecraft On-Board Control*, edited by A. S. Syrova, M: MAI-Print, 2010.

Yetter, R. A., V. Yang, M. H. Wu, Y. Wang, D. Milius, et al., "Combustion Issues and Approaches for Chemical Microthrusters," *International Journal of Energetic Materials and Chemical Propulsion*, Vol. 6, 2007, pp. 393–424.

Kohler, J., J. Bejhed, K. F. Bruhn, U. Lindberg, K. Hjort et al., "A Hybrid Cold Gas Microthruster System for Spacecraft," *Sens. Actuators A.*, Vol. 97–98, 2002, pp. 587–598.

Kang, T.G., S. W. Kim, Y. H. Cho, "High Impulse, Low Power, Digital Microthrusters Using Low Boiling Temperature Liquid Propellant with High Viscosity Fluid Plug," *Sens. Actuators A.*, Vol. 97–98, 2002, pp. 659–664.

Lewis Jr., D. H, S. W. Janson, R. B. Cohen, E. K. Antonsson, "Digital Micropropulsion," *Sens. Actuators Λ.*, Vol. 80, 2000, pp. 143–154.

Zhang, K. L., S. K. Chou, S. Simon, "Development of a Solid Propellant Microthruster with Chamber and Nozzle Etched on a Wafer Surface," *J. Micromech. Microeng.*, Vol. 14, 2004, pp. 785–792.

Rossi, C., D. Briand, M. Dumonteuil, T. Camps, Phuong Quyen Pham, et al., "Matrix of 10 × 10 Addressed Solid Propellant Microthrusters: Review of the Technologies," *Sensors and Actuators A.*, Vol. 126, 2006, pp. 241–252.

Podvig, P. L., *History and the Current Status of Russian Early Warning*, Center for Arms Control, Energy, and Environmental Studies, Moscow Institute of Physics and Technology, 2002.

CHAPTER 2

Launch Vehicle and Spacecraft Failures and Accidents

2.1   Rocket and Space Technology Safety Issues

Today the rocket and space industry has become one of the most important and rapidly developing fields in the world economy, the value of which has long surpassed billions primarily due to commercialization of space projects (telecommunication services, multipurpose Earth surface monitoring, etc.). In the modern world, the prestige of a country is largely determined by its aerospace industry. It is obvious that advances in this field can be made only if high reliability and safety of military and civilian rocket and space technology is ensured. Considering that the commercial sphere is prevailing in the field, this factor has a growing economic importance since failures, accidents, and any contingency situations lead not only to little or no profit but to direct loss on the order of dozens and hundreds of millions of dollars. Space projects are generally considered to be the so-called mission critical applications (the failures of which result in both financial and human losses, collapse of long-term scientific, defense, and other significant state and commercial projects and programs).

Since currently there is no clear and standard terminology for defining such terms as *failure*, *fault*, *accident*, and *disaster*, henceforth we will refer to a *SC failure* as a nonfulfillment of designed mission functions by an element (block, system) at any stage of project implementation (launch, orbital injection, deorbiting, operation in open space) that hasn't resulted in breakdown (nonfulfillment) of project mission.

Fault means self-avoiding fault or single fault that may be avoided by operator's intervention or short-run disturbance of operability of onboard computers and then functioning will be restored without repair works.

SC accident shall be understood as a nonfulfillment of designed mission functions by an element (block, system) at any stage of project implementation (launch, orbital injection, deorbiting, operation in open space) that has resulted in nonfulfillment of project's mission.

SC disaster is generally considered to be an accident that has caused human or large-scale financial losses.

The history of failures and accidents goes hand in hand with the history of development of national cosmonautics. We will mention only the most prominent incidents next.

In 1957, testing began of the first Soviet intercontinental ballistic missile (ICBM) *R-7* designed by S. Korolev. The first prototypes didn't meet requirements in many areas and only two out of thirteen prototypes hit their targets. On April 27, 1958, an accident occurred to the launch vehicle (LV) *Sputnik* (*R-7* modification) with the third EAS. In 1959, accident occurred to the LV *Vostok*. In 1960, six accidents occurred to the LV *Vostok* and its further modification *Molnya*. In 1961, there were three accidents with the LV *Vostok* and *Kosmos-1*. In 1963, there were eight accidents with the LVs *Kosmos-1*, *Vostok*, and *Molnya*. On April 24, 1967, pilot-cosmonaut V.M. Komarov was killed at landing during testing of the spaceship (SS) *Soyuz* (*7K-0K* type serial No. 4 *Soyuz-1*). On June 30, 1971, the crew (G. Dobrovolsky, V. Volkov, V. Patsayev) of SS *Soyuz-11* (*7K-T* type) was killed during deorbiting. On December 19, 1978, the last stage *DM* of LV *Proton* operated in off-nominal mode. The biggest disaster occurred in March of 1980 when an explosion and fire during launch preparation of LV *Soyuz* with reconnaissance satellite resulted in death of 48 people. During the next 40 years, there were a number of other incidents (failures) that didn't have such tragic consequences.

The results of 2011, an anniversary year for Russian cosmonautics, were hardly satisfying since there were four launch vehicle accidents, and on top of that *Fobos-Grunt* got stuck in near-Earth space. Undoubtedly, the worst year for Russia was 1996, when 4 out of 24 LV launches ended up in a failure (it was also the year an ambitious automatic interplanetary station *Mars-96* equipped with cutting-edge mission payload would not fly to Mars). In comparison, the major space rivals had four unsuccessful launches each as a result of LV failure in 1995 and 1999 (total number of launches wass 30 and 31, respectively).

We will enumerate only the most prominent space failures of the anniversary year of 2011. On February 1, 2011, the spacecraft *GEO-IK-2* was placed in the wrong orbit and couldn't serve its purpose due to an accident of the *Briz-KM* upper stage (SW fault) during the launch of the *Rokot* launch vehicle. On March 4, 2011, SC *Glory* fell into the ocean due to failure of PLF jettison during the launch of the *Taurus XL* launch vehicle. Main payload SC *Glory2* and associated SC *KySat-1*, *Hermes*, and *Explorer-1* were lost. On August 17, 2011, the spacecraft *Ekspress-AM-4* wasn't placed in mission orbit and couldn't serve its purpose due to an accident with the *Briz-M* upper stage (SW fault) during the launch of the *Progress-M* launch vehicle. On August 18, 2011, SC *Shi Jian 11-04* fell due to an accident of the LV second stage during the launch of the *Long March 2C* launch vehicle. On August 24, 2011, transport cargo vehicle *Progress M-12M* fell due to an accident of the LV third stage during the launch of the *Soyuz-U* launch vehicle. On August 23, 2011, SC *Meridian-5* fell due to an accident of the LV third stage during the launch of *Soyuz-2-1B* launch vehicle. Moreover, on November 8, 2011, launch vehicle *Zenit-2SB* placed *AIS Fobos-Grunt* in support orbit. However, *Fobos-Grunt* failed to enter the departure trajectory to Mars and remained in support orbit due to a fault of the spacecraft propulsion system.

At the same time the number of launches of Russian SC has been growing in recent years—new launch vehicles and their stages as well as new SC designs, engines, and onboard systems are developed.

We shall introduce the results of statistical analysis in connection with the anniversary year of 2011 for the previous five years and compare achievements of the Russian Federation and other space countries that can be considered both its partners and its rivals.

Data on SC failures and accidents during the first 40 years of rocket and space technology (RST) development (1960s–1990s) was collected for the first fundamental overview [1, 2]. References [3, 4] are also worth mentioning since attempts were made to analyze causes of RST failures in terms of separate systems, their hardware, and software. To provide the material the authors of this book present the results of the original solid paper prepared by scientists and specialists of the KhAI National Aerospace University named after N. E. Zhukovsky and the State Space Agency of Ukraine as close as possible to the original text.

Figure 2.1 shows the total number of launches of launch vehicles that were manufactured in the following countries during this period: the United States, Russia, China, the EU, Japan, Ukraine, India, Israel, Iran, and Korea.

The absolute world leader in terms of the number and dynamics of launches during this period is Russia: while the number of launches in the United States first decreased (2008), then increased (2009), decreased again (2010), and then couldn't reach the level of 2007 even after an increase in 2011, Russia increased the number of launches each year despite the world economic crisis of 2008. China showed the same trend for growth (though lagging behind Russia two times in terms of number of launches), in 2010 and 2011 China caught up with and outperformed the United States.

Figure 2.2 shows dynamics of the number of LV launches over a longer period (1992–2011) in the five leading space countries (Russia, the United States, the EU, China, and Ukraine).

Here both the total number of launches and the number of unsuccessful launches are provided with a five-year breakdown.

It is rather complicated to analyze the data visually. Thus, the ratio of launch accidents to successful LV launches in China was the following: 4 of 13 in 1992–1996, 0 of 22 in 1997–2001, 2 of 29 in 2002–2006, and 2 of 59 in 2007–2011.

It is obvious that the proportion of launch accidents in China tended to decline after the first extremely unsuccessful period.

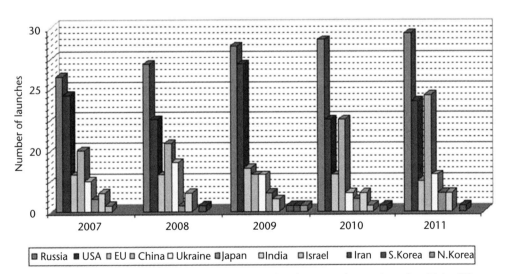

Figure 2.1   Orbital launches in 2007–2011 in countries that manufacture launch vehicles [2].

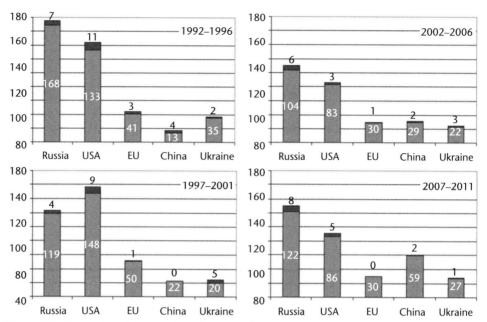

Figure 2.2 Dynamics of the number of successful and unsuccessful orbital launches in major countries that manufacture launch vehicles in 1992–2011 [2].

At the time Russia showed the following trend: 7 of 168; 4 of 119; 9 of 104, and 8 of 123. As compared with China, arithmetical relation of successful launches to unsuccessful ones in 2–4 periods of this space competition doesn't speak in favor of Russia.

Therefore, one of the aims of this book set by the authors is to thoroughly analyze available information and prove that one of the main causes of such failures and other unpleasant incidents of national space instrument making is a difference in approach to methods of designing, choice, and application of the microelectronic element base in SC used by Chinese and Russian SC developers. This chapter provides more detailed analysis results of SC failures and accidents statistics but for better understanding of the overall picture, it is useful to examine another two diagrams.

Figure 2.3 shows the dynamics of the total number of LV orbital launches for the period of 20 years prior to the anniversary year of 2011 in the same countries, and one obvious conclusion can be drawn—that China actively engaged in space race for world leadership: in 2010 China shared second and third place with another superpower, the United States, and now stands a good chance of permanently occupying the honorable second place (both in terms of total number of launches and the number of successful launches).

Figure 2.4 shows statistical data from Figure 2.1 in a way more convenient for analysis.

These examples allow us to draw a conclusion that the current reliability and safety of both world and Russian rocket and space technology leave much to be desired and require continuous analysis and decision making aimed at reducing risks of failures and accidents.

References [2, 4] analyze specific causes of accidents caused by failures of various devices, blocks, and systems of rocket and space technology (RST) with the

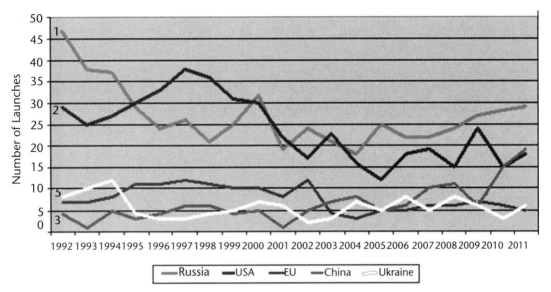

Figure 2.3   Dynamics of orbital launches in countries that manufacture launch vehicles [2].

main attention paid not to the impact of electronic component base (ECB) reliability, but mainly to reliability of computer onboard devices and their hardware and software. The references state the growing dependence of rocket and space complexes on properties of onboard computer control systems as well as its growing influence on safety. According to the analysis results of accident risk for the period of 40 years from 1960 to 2000 [2], every hundredth launch ended up in an accident due to software faults (flaws) and six out of seven failures of RST computer systems were caused by such faults.

It should be noted that the share of software-programmable and supported functions in aircraft and rocket and space complexes is constantly growing. The share of software-programmable functions of combat aircraft grew from 8% (*F-4*)

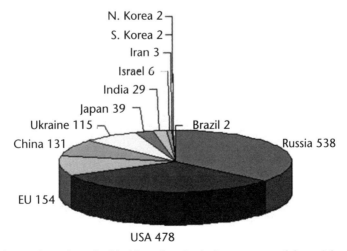

Figure 2.4   The total number of orbital launches (including unsuccessful ones) for the past 20 years (1992–2011) in countries that manufacture launch vehicles.

in the 1960s to 80% (*F-22*) in 2000, according to data published by the United States Department of Defense ten years ago [3]. LV and SC systems have similar trends. It is crucial to take this fact into account since software commonly causes failures even when traditional methods of backup (redundancy, majorization, time redundancy, and so on) are used. Failures caused by software faults can be managed only by means of application of multiversion design technologies and introduction of operational correction during the flight.

The aim of this chapter is to summarize and provide brief analysis of RST accidents that occurred in the first decade of the twenty-first century due to failures of LV and SC equipment, their computer systems, hardware, and software. This chapter is also based on the results of research that has been published in [2–6] in order to acquire a broad picture of 50 years of space age.

2.2   Analysis of Launch Vehicle Failure Causes

Analysis of failure causes of systems and devices of rocket and space technology is important today due to an obvious fact that modern rocket and space systems pose a huge potential threat to nature and mankind. While nuclear power plants (NPP) are generally considered to be the most dangerous man-made objects that can jeopardize global safety, products of rocket and space technology, launch vehicles (LV) in the first place, should be placed immediately after them. This dangerous group should include not only combat LVs equipped with nuclear warheads, but also the so-called peaceful LVs that are used for commercial and scientific research purposes.

The problem has three aspects. First, the rocket and space industry nowadays is rightfully one of the most important fields of the global economy, the annual value of which is estimated at the minimum of billions of dollars and this is a growing trend.

Second, the level of national safety and defense capacity is directly dependent on the level of rocket and space industry development. Third, a developed space industry historically determines the prestige of a state and its space image on a global level; thus, any faults, accidents, and failures of equipment taint the image of a country.

To grasp the main trends and causes of failures and accidents in this field, it is necessary to refer to statistical data in the first place. Unfortunately, there's still no reliable data on all accidents that have occurred in the main space countries—the USSR and the United States—as well as in other countries that are members of the so-called "space club": China, Japan, France, the United Kingdom, and others. Data on disasters, accidents, and failures of ballistic launch vehicles, military cruise missiles, and military satellites generally can't be found in public sources.

The broadest picture of rocket and space technology products accidents is provided and analyzed in [5], the main results of which we shall examine next, though the provided information may differ from that of other sources, as there isn't reliable enough data on all the accidents that occurred in the USSR, the United States, France, and the UK during the whole history of rocket and space technology development. Reference [5] show that it is useful to consider statistics of failures and accidents separately for three groups of products designed for space applications: LV, SC, and separate blocks and devices of LV and SC.

Data on LV and SC launches served as a primary basis for statistical analysis [4–6].

In 1957–2000 there were more than 4,000 LV launches, including the USSR (2,634), the United States (1,220), France (10), Japan (53), China (65), and India (9). As we can see, the number of LV launches in the USSR during this period was more than twice as big as that in the United States; therefore, the number of equipment failures was more statistically significant. Here, 120 out of 2,634 LV launches ended up in accidents that were registered from 1957 to 2000 in the USSR, according to open-source statistics.

Figure 2.5 shows statistical distribution of LV accidents in the USSR and the United States. This figure allows concluding that the number of LV accidents both in the USSR and the United States was unfortunately rather high at the first stages of space development (1961–1971), but in 15–20 years the efforts of scientists and engineers allowed reducing this number by an order of magnitude, though still each accident results in considerable financial losses, and negative economic and political consequences for a country that manufactures LVs. As we can see, the biggest number of LV accidents, 120 accidents, occurred in the USSR, as compared with 100 accidents in the United States; however, the total number of launches should be taken into account and for that period it constitutes 2,034 launches for the USSR and only 1,220 for the United States.

The maximum number of accidents occurred during the first stage of space development in 1961–1978 and afterward there were only isolated accidents, which is true both for Russian and American launches.

It should be noted that in 1961–1971 the United States was actively working in this field and conducted 578 launches, while the USSR had 621 LV launches, and since the number of accidents constituted 61 (10.5%) and 82 (13.2%) in the United States and the USSR, respectively, these values can be seen as statistically equal and suggest that there are some common causes of accidents.

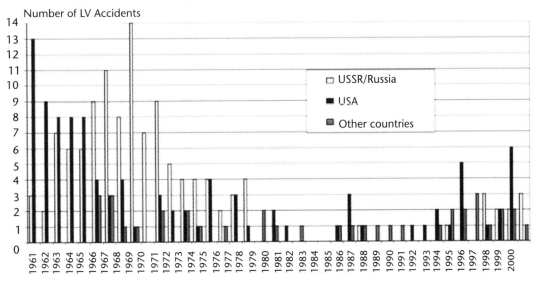

Figure 2.5   Histogram of launch vehicle accidents of the main manufacturing countries.

Reference [5] introduces the term *launch risk*, which estimates accident probability determined as the ratio of accident launches N_{ac} to the total number of successful N_{succ} and accident launches:

$$\text{Risk} = \frac{N_{ac}}{\left(N_{succ} + N_{ac}\right)} \qquad (2.1)$$

The average risk value was calculated by the authors [5] as a function of the total number of launches and accidents, rather than as arithmetic mean of three values for the countries. It should be noted that from the very beginning of space development there was a steady trend for the reduction of the number of LV accidents and the value of coefficient risk of accidents both for Russia and the United States, which was achieved largely by focused efforts of engineers and researchers on revealing conditions and causes of accidents that often result from failures of electronic systems' ECB.

Figure 2.6 shows that LV accident risk reduced from 30% at the beginning of the 1960s to zero in the middle of the 1980s. At the time, other countries fell behind the USSR and the United States in terms of RST development technology. The number of launches in other countries (excluding the USSR and the United States) did not exceed a dozen a year, but electronic control systems and LV designs underwent complex stages of running-in, and therefore risk values were rather high and fluctuating.

The second half of the 1980s saw an increase in both the number of LV accidents and associated risks. It can be partially explained by an increased number of commercial launches that emerged due to rapid development of telecommunication technologies. Though accident risk values are rather fluctuating, their maximum doesn't exceed the value of 0.167 (other countries in 1995, the United States in 1999). The average value of accident risk at the second half of the 1990s is fluctuating within the narrow range of 0.05–0.10. It seems that such trend of rocket and space

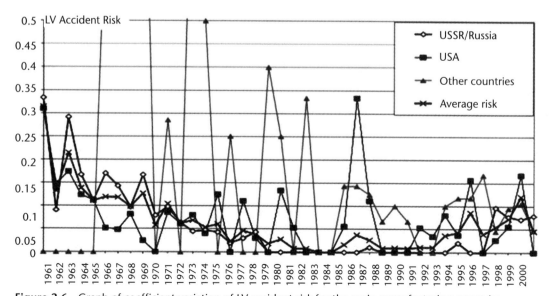

Figure 2.6 Graph of coefficient variation of LV accident risk for the main manufacturing countries.

industry is a steady one and will remain in the coming years. Some 5–10 accidents per 100 LV launches will continue to occur unless developers do not make active effort to detect and eliminate associated risk factors.

2.3   Analysis of Spacecraft Failures

Apart from launch vehicle accidents, the history of rocket and space engineering has seen a considerable number of disasters and accidents of space vehicles that were placed in planned orbits with the help of those launch vehicles. The history of such accidents is just as tragic: the consequences include human and multimillion financial losses, blows to the image of a space country, as well as the hazardous environmental implications of such disasters.

In order to grasp complexity of this problem, Table 2.1 [5] provides complete statistics of known data on SC failures that is grouped by countries manufacturing LVs that placed these SC in orbits (the USSR, the United States, and other countries).

Risk coefficient of SC accidents is calculated on the basis of one successful launch [5]:

$$\text{Risk}_{SC} = \frac{N_{SC\,ac}}{N_{succ\,l}} \tag{2.2}$$

A successful SC launch means not only that SC is placed in planned orbit, but also that it functions during the whole specified operation life.

Data from Table 2.1 is visually presented in Figures 2.7 and 2.8.

It should be noted that in the 1960s more SC accidents occurred in the Soviet Union (as compared with the United States), which was caused by difference in approach to organization of work of researchers, engineers, and designers in the United States and in the USSR. The difference between the USSR and the United States levelled off at the beginning of 1970s, and at the end of 1970s SC accidents essentially ceased. It can be explained by the fact that at the time technology for SC design, development, and operation approached certain level of perfection and most complex projects for interplanetary flights to the Moon, Mars, and Venus had already been completed. Meanwhile other countries operated fewer SC; therefore, absolute number of SC accidents as well as relative values of SC accident risks are in practice of zero value. Here it is worthwhile noting that at the moment of writing, the authors were uncertain in reliability of the information on a number of failures of space vehicles in the period from 1978 to 1985—*Ekran-02, Satcom-3, Insat 1A, SIRIO-2, GMS2, GOES3, Simphonics 1, 2*, and others—occurred while orbiting. Therefore, this time period is represented in these images as fault-free, despite the fact that the actual values of the incident risk coefficients were above zero during this period.

Unfortunately, the beginning of 1990s again saw an increase in SC accidents, which was directly connected with the development of telecommunication technology, sophistication of performed functions, as well as the use of new ECB. At the end of 1990s the risk values of SC accidents exceeded the values that were typical of the early space age and went beyond the critical value of 25%. Correspondingly, each fourth SC didn't make it to the end of the planned operation life during that

Table 2.1 Analysis of Spacecraft Failure Risks in 1961–2000

Year	USSR/Russia			U.S.			Other Countries			Total		
	Number of Successful Launches	Number of SC Accidents	Risk	Number of successful Launches	Number of SC Accidents	Risk	Number of Successful Launches	Number of SC Accidents	Risk	Number of successful launches	Number of SC accidents	Risk
1961	6	1	0.167	29	1	0.034	0	0	–	35	2	0.057
1962	20	2	0.1	52	0	0	0	0	–	72	2	0.028
1963	17	3	0.176	38	0	0	0	0	–	55	3	0.055
1964	30	2	0.067	57	1	0.018	0	0	–	87	3	0.034
1965	48	8	0.167	63	1	0.016	1	0	0	112	9	0.08
1966	44	0	0	73	2	0.027	1	0	0	118	2	0.017
1967	66	2	0.03	59	1	0.017	2	0	0	127	3	0.024
1968	74	4	0.054	45	0	0	0	0	–	119	4	0.034
1969	70	3	0.043	40	0	0	0	0	–	110	4	0.036
1970	81	0	0	29	1	0.034	4	1	0.25	114	2	0.018
1971	83	5	0.06	32	1	0.031	5	0	0	120	6	0.05
1972	74	0	0	31	0	0	1	0	0	106	0	00
1973	86	3	0.035	23	2	0.087	0	0	–	109	5	0.045
1974	81	2	0.025	24	1	0.042	1	0	0	106	3	0.028
1975	89	2	0.022	28	0	0	8	0	0	125	2	0.016
1976	99	4	0.04	26	0	0	3	0	0	128	4	0.031
1977	98	4	0.041	24	2	0.083	2	0	0	124	6	0.048
1978	88	1	0.011	32	0	0	4	0	0	124	1	0.008

Year												
1979	87	0	0	16	0	0	3	0	0	106	0	0
1980	89	0	0	13	0	0	3	0	0	105	0	0
1981	93	0	0	18	0	0	7	0	0	123	0	0
1982	101	0	0	18	0	0	2	0	0	121	0	0
1983	98	0	0	22	0	0	7	0	0	127	0	0
1984	97	0	0	22	0	0	10	0	0	129	0	0
1985	98	0	0	17	0	0	6	0	0	121	0	0
1986	91	0	0	6	0	0	6	0	0	103	0	0
1087	95	1	0.011	8	0	0	7	0	0	110	1	0.009
1988	90			12	0	0	14	0	0	116	0	0
1989	74			18	0	0	9	0	0	101	0	0
1990	75			27	1	0.037	14	1	0.071	116	2	0.017
1991	59			18	1	0,056	11	1	0.091	88	2	0.023
1992	54			28	0	0	13	0	0	95	0	0
1993	47			23	0	0	9	0	0	79	0	0
1994	48		0	26	1	0.038	15	1	0.067	89	2	0.022
1995	32		0	27	2	0.074	15	1	0.067	74	3	0.041
1996	25		0	33	1	0.03	15	0	0	73	1	0.014
1997	28	7	0.25	37	2	0.054	21	3	0.143	86	12	0.14
1998	24	5	0.208	34	9	0,265	19	2	0.105	77	16	0.208
1999	26	3	0.115	30	7	0.233	17	2	0,118	73	12	0.164
2000	35		0	28	1	0.036	20	2	0,1	83	3	0.036

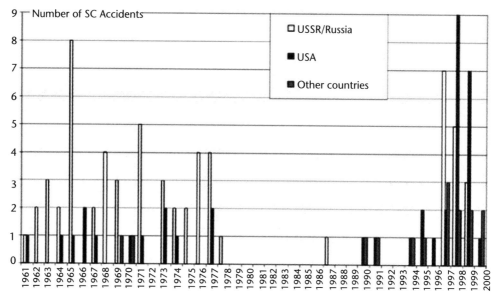

Figure 2.7 Histogram of spacecraft accidents of the main manufacturing countries (1961–2000).

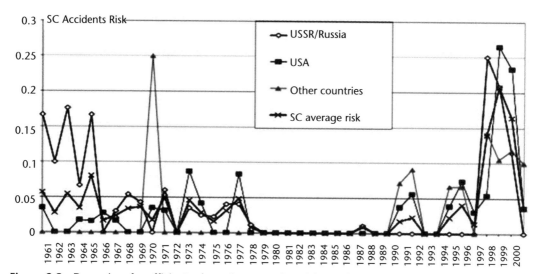

Figure 2.8 Dynamics of coefficient values of spacecraft accident risk (1961–2000).

period. This fact naturally cannot but disturb the developers and customers since such failures have a very high cost and result in unpleasant consequences both for the manufacturing country and the country (territory) that involuntarily became a hostage of the accident situation.

Figures 2.9 and 2.10 [2] show comparative analysis of the number of accidents and risk values of LV and SC. Apart from mean values of SC and LV accident risks, Figure 2.10 also shows value of total risk of SC mission implementation that is made up of SC accident risk and LV accident risk.

The formula from [2] is used to determine the risk value of SC mission implementation.

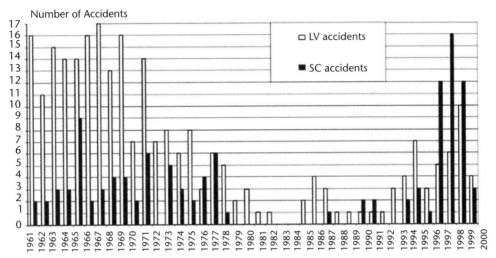

Figure 2.9 Histogram of the total number of launch vehicle and spacecraft accidents (1961–2000).

Probability of successful mission implementation is calculated with the equation:

$$R_{\text{mis impl}} = 1 - \text{Risk}_{\text{mis imp}} \tag{2.3}$$

While

$$R_{\text{mis impl}} = R_{\text{LV}} \cdot R_{\text{SC}} = \left(1 - \text{Risk}_{\text{LV}}\right) \cdot \left(1 - \text{Risk}_{\text{SC}}\right) \tag{2.4}$$

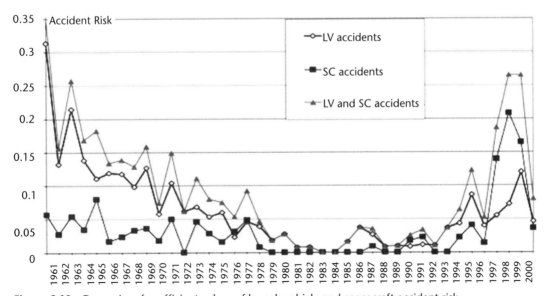

Figure 2.10 Dynamics of coefficient values of launch vehicle and spacecraft accident risk.

Therefore,

$$1 - \text{Risk}_{\text{mis impl}} = \left(1 - \text{Risk}_{\text{LV}}\right) \cdot \left(1 - \text{Risk}_{\text{SC}}\right) \qquad (2.5)$$

And finally

$$\text{Risk}_{\text{mis impl}} = \text{Risk}_{\text{LV}} + \text{Risk}_{\text{SC}} - \text{Risk}_{\text{LV}} \cdot \text{Risk}_{\text{SC}} \qquad (2.6)$$

The previously mentioned data [2, 6] allows concluding that coefficient value of LV accident risk was twice as big as that of SC accident risk until the middle of 1970s. Then values of risks leveled off and starting from the second half of the 1990s the risks of SC accidents exceeded those of LV accidents. Therefore, at the modern stage of RST development a major problem of SC mission is not placing SC in orbit, but ensuring flawless SC operation during the whole planned operation life.

2.4 Analysis of Failure Causes of Rocket and Space Technology Products

Considering specific character of this book's subject, it is necessary to provide a thorough analysis of trends of coefficient values of accident risks due to failures of various components of LV and SC.

Table 2.2 shows temporal dynamics of changing proportion of failures that resulted in LV and SC accidents over a period of 10 years. A range of common causes of LV and SC failures is provided separately. It should be noted that unfortunately

Table 2.2   Main Reliably Ascertained Causes of RST Failures

Accident Causes	1960s Quantity	%	1970s Quantity	%	1980s Quantity	%	1990s Quantity	%
Failures and explosions of launch vehicle (LV) stages	136	79	60	66	38	90	31	29
Spacecraft (SC) failures	9	5	9	10	0	0	0	0
Propulsion system (PS) failures	6	3	5	5	1	2.5	10	10
Radio equipment (RE) failures	2	1	2	2	1	2.5	7	7
Upper stage (US) failures	3	2	1	1	1	2.5	6	6
Power supply systems and cable network (PSS) failures	2	1	1	1	0	0	9	9
Control system (CS) failures	16	9	14	15	1	2.5	24	23
Onboard computer hardware (HW) failures	0	0	0	0	0	0	6	6
Onboard computer software (SW) failures	0	0	0	0	0	0	10	10
Total	*174*		*92*		*42*		*103*	

the authors [2] couldn't differentiate types of SC failures for 1960–1980s due to lack of information.

Results of analysis of data contained in Table 2.2 are shown in Figure 2.11–2.14, where failure causes correspond to Table 2.2.

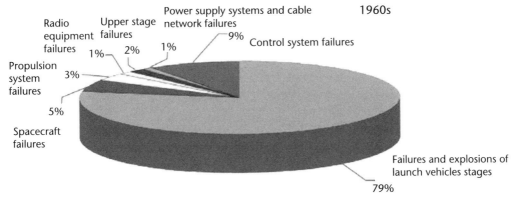

Figure 2.11   Distribution of causes of rocket and space technology failures in 1960s.

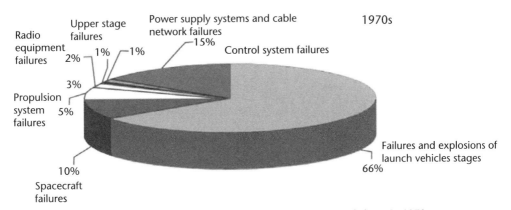

Figure 2.12   Distribution of causes of rocket and space technology failures in 1970s.

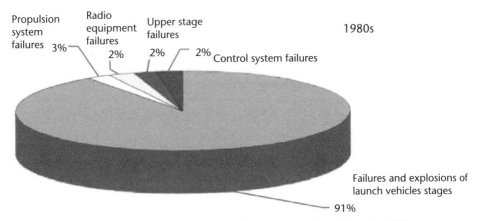

Figure 2.13   Distribution of causes of rocket and space technology failures in 1980s.

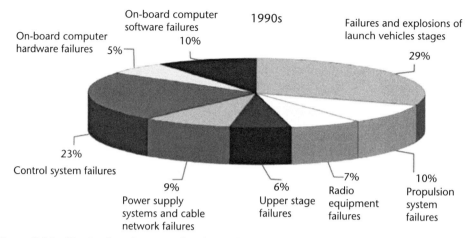

Figure 2.14 Distribution of causes of rocket and space technology failures in 1990s.

In order to determine value of accident risk coefficient due to failure of a certain component during flight the authors [2] propose using the following formula:

$$\text{Risk}_i = \frac{N_{\text{acc } i}}{\left(N_{\text{succ } 1} + N_{\text{acc LV}}\right)} \tag{2.7}$$

In order to determine accident risk due to failure of a certain component during space flight, the authors [2] recommend using a simple formula:

$$\text{Risk}_i = \frac{N_{\text{acc } i}}{\left(N_{\text{succ } 1} + N_{\text{acc LV}}\right)} \tag{2.8}$$

The major accident risks due to failures of various components of RST products that reflect current trends are shown in Table 2.3. It has been taken into account that in 1990s there were 816 successful launches and 44 LV accidents, according to official data.

Table 2.3 Accident Risks due to Failures of RST Components Based on Statistics for 1990s (In Terms of One LV Launch)

Accident Cause	Number of Failures	Accident Risk
Failures and explosions of launch vehicle (LV) stages	31	0.036
Propulsion system (PS) failures	10	0.012
Radio equipment (RE) failures	7	0.008
Upper stage (US) failures	6	0.007
Power supply system and cable network (PSS) failures	9	0.01
Control system (CS) failures	24	0.028
Onboard computer hardware (HW) failures	6	0.007
Onboard computer software (SW) failures	10	0.012

Naturally, values of risk coefficients provided in Table 2.3 could be made more specific if differentiation of accident risks of specific LV and SC components were taken into account, but in general given values provide a real picture.

The simple analysis given here shows that LV stages are especially susceptible to failures; the second hardly honorable place is occupied by electronic control systems. However, today new risks emerge that are connected with the use of computer equipment for onboard control systems. Failures of onboard computer software (SW) together with failures of propulsion systems share third and fourth places among all failure causes.

On average each hundredth launch (once a year on average) ends up in an accident due to SW failures. It should be noted that hardware (HW) failures are twice as rare as that of SW. Table 2.4 shows data on failures of onboard computer HW and SW from 1991 to 2000.

An obvious conclusion can be drawn from these facts is that in order to understand mechanism and causes of RST failures it is necessary to conduct research in the following fields:

1. Collection, processing, and analysis of statistical data on RST failures and accidents with the use of regression/correlation analysis of statistical data, which allows making prognosis for future trends of RST development;
2. Specification of mathematical models for risk estimation with regard to both LV and SC components and specific consequences of accidents;
3. Comparative analysis of accident causes, in particular during introduction of new information and other technologies together with other critical systems, such as nuclear power, transport, and so on [7, 8];
4. Development of new methods and means for improving reliability and safety of control system (CS) and onboard computers on the base of specialized ECB designed for space application [9].

Table 2.4   Launches of Launch Vehicles in 2000–2009

Year	Russia	USA	EU	China	Ukraine	Japan	India	Other Countries	Total
2000	32	28	12	5	7	1	0	0	85
2001	19	22	8	1	6	1	2	0	59
2002	25	17	12	5	1	3	1	1	65
2003	21	23	4	7	3	3	2	1	64
2004	18	16	3	8	7	0	1	1	54
2005	26	12	5	4	5	2	1	0	55
2006	23	18	5	6	7	6	1	0	66
2007	22	19	6	10	5	2	3	1	68
2008	24	16	6	11	8	1	3	0	69
2009	27	24	7	6	6	3	2	3	78
Total	237	195	68	63	55	22	16	7	663

2.5   Trend Analysis of Launch Vehicle Accident Risks in 2000–2009

Just like in previous sections, open sources including publications in periodicals, information agencies' news, relevant websites, and sections [9], in the first place, were used for the analysis.

The aim was to determine the cause of each incident: component, block, and type of failure.

Another aim was to analyze risk trends of accidents occurring due to failures of various LV and SC components, including hardware and software of onboard and ground computer systems.

Data on failures and accidents of military ballistic and cruise missiles were not examined.

The authors of [2], which is mainly referred to in this chapter, rightly point out that none of the available sources can provide systemized data selection on accident situations, and since there is still no complete and reliable data on all the disasters, additional measures were taken to ensure reliability of initial data. In particular, the principle of data cross-validation was implemented, which includes selective verification of data on incidents from different sources.

The major problem of verbal data vectoring (date of incident, its significance, LV/SC, cause of incident) was to determine cause of an accident or contingency situation. In this case the principle of multiple analysis was used which takes into account not only data from various sources but also expert opinion. It should be emphasized that statistics on LV launches and failures is rather well-ordered and there are several diverse credible sources for its verification whereas data on spacecraft failure are sketchy. Data on SC failures and accidents provided in articles [2, 4, 6] were obtained on the basis of analysis of different-type information sources.

Data on the number of LV launches in manufacturing countries for the analyzed period are given in Table 2.4. The table also reflects the launches registered by the UN Commission on Space Research (COSPAR) and North American Aerospace Defense Command (NORAD).

During this period Russia and the United States conducted the biggest number of LV launches. Ukraine is among top five after the European Union and China. However, the total number of LV launches during this decade reduced by 25% as compared with 1990–1999 (663 against 891 launches [3]). Production of LVs in Russia, Ukraine, and the United States reduced by 30% and by 22% in the European Union. At the same time number of LV launches trended upward in Japan (1.5 times more launches), China (1.7 times), and especially in India (2.6 times).

The "Other Countries" column in Table 2.4 reflects LV launches of such countries as Brazil, Iran, Israel, North Korea, and South Korea that were insignificant in number.

A feature of Ukrainian LV launches is that Ukraine doesn't have its own cosmodrome and has to conduct launches from cosmodromes in other countries.

Ukraine launches its *Zenit-3SL* launch vehicles as part of international project *Sea Launch* (32 launches in 2000–2009) as well as *Zenit-3SLB*, *Dnepr*, and *Tsyklon-3* from the cosmodromes of the Russian Federation within the projects *Land Launch* and *Kosmotras* (23 launches in 2000–2009).

In particular, a successful launch of LV *Tsyklon-3* was conducted from the Plesetsk Cosmodrome in 2009. This LV was operated since 1969. There were 122 launches altogether, 7 of which had a contingency situation. LV of the *Tsyklon* series is one of the most reliable ones (94.3% of successful launches), exceeding in this parameter such LVs as *Ariane*, *Delta*, and *Proton*, [8] notes.

The first launch of LV *Tsyklon-4* of the fourth generation from Brazilian cosmodrome within a joint project of Ukraine and Brazil *Alcantara Cyclone Space* was planned for 2012 but wasn't conducted due to political aspects.

In 2000s apart from *Tsyklon-3* such launch vehicles as *Tsyklon-2* and *Atlas II* (no accidents were registered during history of their operation) and *Ariane-4* were taken out of operation. New LVs such as *Soyuz-FG*, *Atlas V*, *Delta IV*, *Soyuz-2*, and *Proton-M* were put in operation.

Table 2.5 provides a known data on analysis results of LV failure risks in 2000–2009. It shows aggregated data on LV failures grouped by manufacturing

Table 2.5 LV Launches in 2000–2009 That Resulted in an Accident (ACC) or Satellite Delivery into a Wrong Orbit (WO)

Year	Type of LV Failure	Russia	USA	EU	China	Ukraine	Japan	India	Other Countries	Total
2000	ACC	1	0	0	0	2	1	0	0	4
	WO	0	1	0	0	0	0	0	0	1
2001	ACC	1	1	0	0	0	0	0	0	2
	WO	0	0	1*	0	0	0	1*	0	2
2002	ACC	1	0	1	1	0	1**	0	0	4
	WO	1*	0	0	0	0	0	0	0	1
2003	ACC	0	1	0	1	0	1	0	1	4
	WO	0	0	0	0	0	0	0	0	0
2004	ACC	0	0	0	0	0	0	0	1	1
	WO	0	1*	0	0	2	0	0	0	3
2005	ACC	4	0	0	0	0	0	0	1	5
	WO	0	0	0	0	0	0	0	0	0
2006	ACC	1	3	0	0	1	0	1	0	6
	WO	0	1*	0	0	0	0	0	0	1
2007	ACC	1	1	0	0	1	0	0	0	3
	WO	0	1	0	0	0	0	1	0	2
2008	ACC	0	1	0	0	0	0	0	1	2
	WO	1	0	0	0	0	0	0	0	1
2009	ACC	0	2	0	0	0	0	0	1	3
	WO	1	0	0	1	0	0	0	1***	3
TOTAL	ACC	9	9	1	2	4	3	1	5	34
	WO	3	4	1	1	2	0	2	1	14

Notes:

* Due to LV failure, the satellite was placed into a wrong orbit and wasn't able to be corrected. The satellite couldn't serve its purpose.

** During launch of the Japanese LV *H-11A 2024*, the satellite *DASH* didn't separate from the second stage whereas the satellite *MDS-1* was successfully delivered in orbit. The launch can be classified as partially successful.

*** Half of the payload wasn't placed into the desired orbit. The reported cause is failure during PLF jettison: one of fairing halves didn't separate from the stage, which resulted in considerable lack of speed. Some SC that are placed into wrong (lower) orbits are capable of orbit correction by means of their propulsion system and propellant reserve.

countries. LV failures led to accidents (LV explosion, fall, considerable trajectory errors, self-destruction) as well as to delivery of SC into wrong orbit, which is generally considered to be a partial failure.

Orbit correction by a satellite obviously leads to considerable reduction of its operational life. For example, on July 12, 2001, due to early engine cut-off of LV *Ariane-5* second stage, the satellites *Artemis* and *BSAT-2B* were placed into a much lower orbit than intended (geostationary orbit).

The navigation satellite *Artemis* was able to transfer to geostationary orbit whereas communication satellite *BSAT-2B* remained in medium Earth orbit and couldn't serve its purpose. Such LV launches shall be considered unsuccessful rather than partially successful. Furthermore, in 6 out of 14 registered cases when SC were placed into a wrong orbit (see Table 2.5), the satellites weren't able (didn't have the capacity) to make orbit correction. As a result such LV launches were regarded as unsuccessful.

Figure 2.15 shows variation of launch vehicle launches risks in 2000–2009 (the category "Other Countries" includes Japan and India as compared with Table 2.5). Risk value (probability of failure) was calculated as relation of unsuccessful LV launches $N_{LV fail}$ to the total number of LV launches $N_{LV launch}$ by the formula:

$$\text{Risk}_{LV} = \frac{N_{LV\,fail}}{N_{LV\,launch}} \tag{2.9}$$

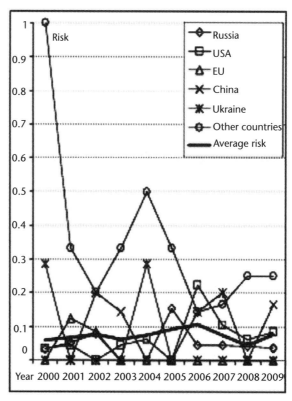

Figure 2.15   Variation of LV launches risks in 2000–2009.

Table 2.6 Comparison of Risks of Launch Vehicle Failures in 1990–1999 and 2000–2009

Year	Number of Launches	Number of Failures	Risk	Year	Number of Launches	Number of Failures	Risk
1990	117	1	0.009	2000	85	5	0.059
1991	89	1	0.011	2001	59	4	0.068
1992	96	1	0.010	2002	65	5	0.077
1993	82	3	0.037	2003	64	4	0.063
1994	93	4	0.043	2004	54	4	0.074
1995	81	7	0.086	2005	55	5	0.091
1996	76	3	0.039	2006	66	7	0.106
1997	91	5	0.055	2007	68	5	0.074
1998	83	6	0.072	2008	69	3	0.043
1999	83	10	0.120	2009	78	6	0.077
Total	891	41	0.046	Total	663	48	0.072

Over the past decade, the value of total risk of LV launches has fluctuated within the range of 0.04…0.10 and constituted 0.072 on average, including accident risk of 0.051 and risk of delivery in wrong orbit of 0.021. The highest value of risk (0.106) was registered in 2006, when 7 LV failures occurred at the total of 66 launches. This trend seems to be stable and will remain in years to come.

Consequently, companies working in the rocket and space industry will have to put up with losing of 4–10 rockets per 100 launches.

Table 2.6 provides the results of comparison of LV failure risks in the 1990s and 2000s [2]. As compared with 1990–1999, the risk of launch vehicle failure during the analyzed decade increased by more than 1.5 times (Table 2.6), whereas the total number of launches reduced approximately by 25%, as mentioned earlier.

The cause of such an obviously negative trend is transition from highly reliable but not efficient enough LVs *Tsyklon-2* and *Atlas II/III* to new improved LVs, the elements of which haven't yet been proven, according to the authors [2].

The analysis of LV reliability shall be examined independently. Thus, Table 2.7 shows operational characteristics of LVs that were used most frequently during this period. LVs *Tsyklon-2*, *Tsyklon-3*, *Dnepr*, *Rokot*, and *Start* were based on converted intercontinental ballistic missiles.

It follows from Table 2.7 that the following launch vehicles were absolutely reliable:

- LVs *Tsyklon-2* and *Atlas II/III*, which were taken out of operation;
- LV *Soyuz-FG*—modified LV *Soyuz-U* with new engines of first and second stages that feature special injectors for better mixing, which allows increasing load capacity for smooth delivering of three cosmonauts into orbit.

For any customer, a key parameter for reliability is the number of past successful launches. American LV *Delta II* and Russian LV *Proton-K* are leaders in these terms: the latest accident occurred in 1997 and 1999, respectively. The latest accidents of LVs *Proton-M* and *Zenit-3SL* occurred in 2007. Table 2.7 does not cover

Table 2.7 Operational Reliability of Launch Vehicles

No.	LV	First Launch	Last Launch	Latest Accident Launch	Number of Launches				Reliability, %
					Total	Past Successful	Accident	Partially Successful	
1	Tsyklon-2	August 6, 1969	June 25, 2006	—	106	106	0	0	100 (100)
2	Atlas II (III)	December 7, 1991	February 3, 2005	—	70	70	0	0	100 (100)
3	Soyuz-FG	January 20, 2001	October 7, 2010	—	32	32	0	0	100 (100)
4	Atlas V	August 21, 2002	September 21, 2010	—	23	23	0	1	100 (95.65)
5	Delta IV	March 11, 2003	May 28, 2010	December 21, 2004	13	9	0	1	100 (92.31)
6	Soyuz-2	November 8, 2004	October 16, 2010	—	8	8	0	1	100 (87.50)
7	Delta II	February 14, 1989	September 14, 2010	January 17, 1997	147	92	1	1	99.32 (98.64)
8	Proton-M	July 7, 2001	October 14, 2010	September 6, 2007	47	30	1	3	97.87 (91.49)
9	Soyuz-U	May 18, 1973	September 10, 2010	October 15, 2002	715	45	18	1	97.48 (97.34)
10	Ariane-4	January 22, 1990	February 15, 2003	December 1, 1994	116	73	3	0	97.41 (97.41)
11	Ariane-5	June 4, 1996	August 4, 2010	December 11, 2002	52	38	3	1	94.23 (92.31)
12	Tsyklon-3	June 24, 1977	January 30, 2009	December 27, 2000	122	4	5	2	95.90 (94.26)
13	Kosmos-3M	May 15, 1967	April 27, 2010	October 27, 2005	425	10	20	8	95.29 (93.41)
14	Zenit-3SL(B)	March 28, 1999	November 30, 2009	January 31, 2007	34	10	2	1	94.12 (91.18)
15	Dnepr	April 21, 1999	June 21, 2010	July 26, 2006	16	9	1	0	93.75 (93.75)
16	Proton-K	March 10, 1967	February 28, 2009	October 27, 1999	310	44	26	9	91.61 (88.71)
17	Rokot	November 20, 1990	September 8, 2010	October 8, 2005	17	6	2	0	88.24 (88.24)
18	Zenit-2M	March 13, 1985	June 22, 2009	September 9, 1998	38	7	5	0	86.84 (86.84)
19	Start	March 25, 1993	April 25, 2006	March 28, 1995	7	5	1	0	85.71 (85.71)

Notes: Data is valid as of the time of writing (October 2010). Parenthetical values of reliability are given without including partially successful launches (placing into wrong orbit, delivery of part of payload).

the launches of the multiple re-entry vehicles of the space shuttle series (Columbia, Atlantic, Discovery, Challenger, Endeavor, and Enterprise), as in this section under consideration were only single-use rocket launchers, although the subject of the shuttle reliability analysis is of a significant interest for researchers. The authors address this subject in the third book of this series, *Space Electronics*.

2.6   Analysis of SC Failure Trends in 2000–2009

Aggregated data on SC launches and failures are provided in Table 2.8. SC launch is considered to be successful provided that SC is placed into desired orbit and operates during the whole planned operation life.

The authors utilize the following classification of SC failures [2]:

- LV accidents (LV ACC), including failures during separation of SC from LV;
- SC placement in wrong orbit (LV WO);
- Failures at SC deployment in orbit (FDO) including nondeployment of solar panels, stabilization failures, and orientation failures;
- Operational failures of onboard equipment (FOE) and mechanical faults and damages (MFD).

If placed in wrong orbit, some SC are capable of correcting it. However, if guaranteed operation life of communication satellites constitutes 12–15 years, the use of resource and propellant of its propulsion systems reduces SC operation life by 2–4 times on average.

A total of 149 out of 1060 launched SC failed over the past 10 years. Some 59% of them failed due to LV accident and SC placement in wrong orbit, 12% failed due to SC failure immediately after placement in orbit, and 29% failed due to fatal failures of onboard equipment and mechanical damage during guaranteed operation life.

Over the analyzed period the biggest number of SC was lost in 2006, with an accident of LV *Dnepr* that occurred on July 26, 2006, accounting for 18 out of 23 lost SC.

It is notable that a quarter of registered failures of onboard equipment was managed by means of redundancy and maintenance procedures (e.g., software upgrade) via commands from flight control center.

Figure 2.16 shows graph of variation of SC various risk components in 2000–2009 with regard to data from Tables 2.5 and 2.6 as well as integrated risk (not including SC equipment failures) calculated by formula:

$$\text{Risk}_{\text{SC}} = \frac{N_{\text{LV ACC}} + N_{\text{LV WO(FF+PF)}}}{N_{\text{SC launch}}} + \frac{N_{\text{FDO}} + N_{\text{FOE(FF+PF)}} + N_{\text{MFD(FF+PF)}}}{N_{\text{SC launch}}} \qquad (2.10)$$

where N_{ACC} is number of SC lost as a result of LV accident; $N_{\text{LV WO (FF+PF)}}$ is number of SC placed in wrong orbit due to LV failure; N_{FDO} is number of SC failures

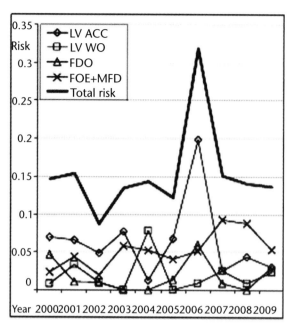

Figure 2.16   SC failure risks.

occurring in orbit immediately after launch; $N_{FOE\,(FF+PF)}$ is number of fatal and partial failures of SC onboard equipment; $N_{MFD(FF+PF)}$ is number of fatal and partial failures due to SC mechanical faults and damage; and $N_{SC\,launch}$ is total number of SC launched in 2000–2009.

The conducted analysis shows that values of SC launch risk are rather fluctuating as compared with LV risks examined earlier. Mean value of SC integrated risk is 0.15, which is twice as big as risk of LV failure. The highest risk of SC failure falls on 2006 due to the accident of LV *Dnepr* with 18 satellites onboard. Furthermore, in recent years there has been a growing trend of failures of SC onboard equipment. Statistics of SC failures intentionally doesn't include numerous failures and breakdowns at the International Space Station (ISS). Table 2.5 also doesn't include data on failures occurring in 2000–2009 on SC launched before 2000.

Comparative analysis of statistics of SC launches and failures (not including failures occurring due to LV failures) in 1990–1999 [2] and over the analyzed decade (Table 2.9) shows that the total number of SC launched in 2000–2009 increased by 1.25 times as compared to 1990–1999.

Considering that in 2000–2009 the number of LV launches decreased, it can be concluded that the number of multiple launches (one LV launches 1.6 SC on average) has increased, therefore increasing the cost of LV failure.

As far as risk of SC failure caused by onboard equipment is concerned (including unsuccessful deployment of SC in orbit after launch), it has increased from 0.059 to 0.073 as compared with 1990s.

Moreover, it should be noted that 7% of satellites are lost during or immediately after placement in orbit which accounts for 40% of the total number of SC failures.

Table 2.8 Statistics of Spacecraft Launches and Failures

Year	Number of SC Launches	Number of Failures by Types										Total Number of Launches
		LV ACC	LV WO			FOE			MFD			
			FF	PF	FDO	FF	PF	MF	FF	PF	MF	
2000	130	9	1	0	2	0	0	1	0	0	0	15
2001	91	6	2	1	2	5	2	3	0	0	0	23
2002	103	5	1	0	1	2	2	0	0	3	1	15
2003	104	8	0	0	0	0	3	5	0	0	0	16
2004	77	1	3	3	0	0	0	1	0	0	0	8
2005	74	5	0	0	1	2	1	1	0	0	0	11
2006	116	23	1	0	7	5	0	3	0	0	0	43
2007	119	3	0	3	1	4	3	2	2	3	0	21
2008	114	5	0	1	0	2	1	0	0	1	1	11
2009	132	4	1	2	4	2	0	1	0	0	0	16
TOTAL	1060	69	9	10	18	22	12	17	2	7	2	179

Notes: The following indications are used in the Table: LV ACC: SC lost as a result of LV accident; LV WO: SC is placed in wrong orbit due LV failure; FDO: failure at deployment in orbit and inability to establish contact with SC immediately after placement in orbit; FOE: onboard equipment failure; MFD: mechanical faults and damage of SC design elements; FF: fatal failure, PF: partial failure; and MF: managed failure.

Table 2.9 Comparison of Spacecraft Failure Risks in 1990–1999 and 2000–2009

Year	Number of Launches	Number of Failures	Risk	Year	Number of Launches	Number of Failures	Risk
1990	116	2	0.017	2000	130	9	0.069
1991	88	2	0.023	2001	91	5	0.055
1992	95	0	0	2002	103	3	0.029
1993	79	0	0	2003	104	6	0.058
1994	89	2	0.022	2004	77	4	0.052
1995	74	3	0.041	2005	74	4	0.054
1996	73	1	0.014	2006	116	13	0.112
1997	86	12	0.14	2007	119	12	0.101
1998	77	16	0.208	2008	114	10	0.088
1999	73	12	0.164	2009	132	11	0.083
TOTAL	850	50	0.059	TOTAL	1060	77	0.073

2.7   Analysis of LV and SC Accident Causes in 2000–2009

The authors of the fundamental work [2] propose to classify all failures leading to accidents and other contingency situations by special types separately for SC and LVs.

Thus, the following types of LV failures can be distinguished:

- First stage (FS);
- Second stage (SS);
- Third stage (TS);
- SC separation mechanisms (SC SM);
- Upper stage (US);
- Computer control system hardware (HW);
- Software (SW).

The following eight types are distinguished according to types of spacecraft systems, equipment, and components:

- Radio equipment (RE);
- Software (SW);
- Power supply system (PSS);
- Mechanical fault or SC design damage (MFD);
- Hardware (HW);
- Gyroscopic devices (GD);
- SC propulsion system (PS);
- Personnel error (PE).

Tables 2.10 and 2.11 systemize causes and consequences of LV and SC failures registered in 2000–2009. It should be noted that the analysis doesn't include the space shuttle *Columbia* disaster that occurred on February 1, 2003, and was caused by damage of thermal protection system.

Though this chapter provides more or less systemized analysis of RST accidents occurring only over the first decade of the twenty-first century, the beginning of the second decade shows that unfortunately there will be information for analysis in the future as well. Unfortunately, comparison of LV and SC accidents occurring in the last decade of the past century and the first decade of the present century shows that total risk of accidents and failures hasn't reduced, but instead increased unacceptably.

Table 2.10   Causes and Consequences of Launch Vehicle Failures

Failure Location	LV Accident	SC Placement in Wrong Orbit	Total
FS	9	2	11
SS	7	4	11
TS	5	4	9
SC SM	5	1	6
SW	5	1	6
US	1	2	3
HW	1	0	1
Total	33	14	47

Table 2.11   Causes and Consequences of Spacecraft Failures

Failure Location	Fatal Failure	Partial Failure	Managed Failure/ Fault	Total
RE	20	11	0	31
SW	3	2	16	31
PSS	15	3	2	20
MFD	5	8	2	15
HW	3	0	3	6
GD		2	3	5
PS	2	2	0	4
PE	1	0	0	1
TOTAL	49	28	26	103

It should be emphasized that risks of accidents caused by onboard computer systems failures have increased. Failures and faults of software and hardware (20% and 6%, respectively) occupy the second place (after radio equipment failures) among causes of SC failures. Furthermore, part of radio equipment failures can be connected with hardware failures, primarily with aspects of ECB choice and application. At the same time each seventh LV failure is caused by software defect whereas failure of onboard control system hardware isn't typical of LV. Three main groups of LV failure causes can be identified for the countries with well-developed rocket and space industry (primarily Russia, the United States, and China): early engine cut-off of different LV stages; problems with separation of stages, boosters, upper stages, fairings, and platforms placed in orbit; and penetration of foreign particles in critical elements of propulsion systems and electrical mechanisms.

In practice it is hard to determine precisely, but it is very likely that early cut-off of propulsion systems is less connected with PS abnormalities and more so with failures and faults of onboard control systems. Failures of the second group are likely to be connected with defects of pyrotechnic electrical devices of LV design as well as with absence of timely start commands from onboard control systems or less likely with defects of cable connections. The third group of failures suggests possible defects of manufacturing and assembling of LV elements and design imperfections. In any case understanding of underlying causes shall reduce LV accident risks.

2.8   Analysis of Computer System and Software Failures

As Figure 2.17 shows, 13% of LV failures over the analyzed period were caused by software of various control systems, whereas it accounts for 20% of SC failures, and 6% of SC failures are the result of hardware failures of onboard computers (Figure 2.18).

At the same time only 6% of SC fatal failures were caused by SW failures (same as by HW failures), whereas 15% of LV fatal failures were resulted by the same cause. Such discrepancy can be explained by the fact that SC onboard computers change to the so-called safe mode when SW defects emerge. This allows the control

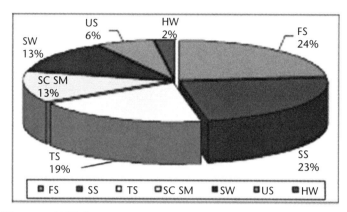

Figure 2.17   Diagram of LV failure causes.

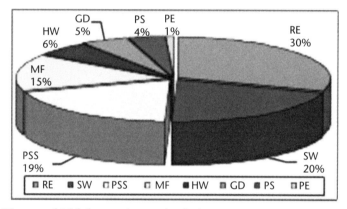

Figure 2.18   Diagram of SC failure and operation fault causes.

center to detect the defect and eliminate it by means of SW updating via a backup communication channel. Obviously, LVs don't provide such an opportunity.

Failures of control system software lead to accidents (or placement of SC in wrong orbit) every 110th LV launch on average, which is in compliance with calculations made earlier for the 1990–1999 period.

However, it should be noted that this value can be considerably higher in practice. The reason is that early engine cut-off is indicated as a cause of one-third of LV stages and upper stage failures. The cause is formulated in such a way when there are abnormalities in operation of propulsion system or faulty commands for early cut-off issued by onboard control system that is responsible for navigation, stabilization, propellant management, separation of LV stages, and payload.

This confirms the data on the registered facts of failures of the SC onboard computers for the period from 1991 to 2000 [2], represented in Table 2.11.

The accident of LV *Zenit-3SL* that occurred on March 12, 2000, is of particular interest to Russian specialists. The investigative committee that looked into the disaster cause found that the accident occurred due to a logical error in a program algorithm of the ground automated launch processing system. A command required for closing the valve of the pneumatic system of the second stage wasn't issued. Loss of gas pressure in the system that ensured operation of second stage vernier engine exceeded 60%, which led to off-design operation mode and engine cut-off at 461st second of the flight.

The automated launch processing system was developed by specialists of rocket and space complex Energia (Russia). It is noted that the defect was induced by software upgrade. Moreover, it was the second time that this defect showed in launch processing system, according to Yuzhnoye State Design Bureau. (On September 9, 1998, the first defect led to accident of LV *Zenit-2* that was part of *Globalstar* program.)

In other words, the LV *Zenit-3SL* accident could have been prevented if investigation had found the true cause of LV *Zenit-2* accident and it had been eliminated. Should that have been the case, the reliability of LV *Zenit-3SL* would constitute 97% exceeding that of LV *Ariane-5*. Direct consequences of a software defect include reduced credibility value of the LV *Zenit-2/3SL* and the reputation of the *Sea Launch* project as well as the loss of clients and profit.

However, software isn't a only source of failures, but a flexible mechanism for the management of design drawbacks and the failures of other SC systems. For example, the Japanese *Daichi* (*ALOS*) satellite for Earth remote sensing produced low quality images due to insufficient stabilization and a high level of noise, and this problem was solved by means of software programmable functions of correction and digital filtering on the satellite.

Another example is the case of SC *Hayabusa* (Japan), which was launched for researching the Itokawa asteroid. As propellant of maneuvering rocket engines was used up early (after two out of three gyroscopes of stabilization system broke, propellant was used excessively to keep the spacecraft flight direction), Japanese scientists were able to reprogram spacecraft control systems so that SC orientation and stabilization were controlled by exhaust thrust created by the use of xenon of ion main propulsion system. In summary several important conclusions and recommendations can be made.

On one hand, extended functions and possibilities of electronic onboard systems raise requirements for their reliability. On the other hand, they lead to an increased number of defects and complexity of their detection, especially in the context of operation system introduction and multitasking. An illustrative example is emergence, detection and correction of a defect of onboard control system software of the Mars rover *Mars Pathfinder*. The defect was connected with the problem of priority inversion occurring when access of the threads to the information bus was blocked in OS VxWorks. A detailed description of the case can be found in [10].

Analysis shows that this problem wasn't detected in due time since it was caused by a rare set of circumstances that hadn't been simulated during ground testing. Nevertheless, this defect showed on the second day after *Pathfinder* landed on Mars and occurred repeatedly. At the same time, the developers of the *Mars Pathfinder* software found and implemented mechanisms for recovery after failures and faults induced by various problems, most of which did not occur during the mission.

Therefore, development of RST control systems requires pretest analysis of significance of potential failures and affecting factors in order to provide efficient recovery mechanisms and ensure that test profile corresponds with actual operation conditions. It is useful to apply mathematically sound formal methods of design

Table 2.11 Onboard Computer Failures from 1991 to 2000

Failure Date	Launch Date	Type of Rocket Launcher	Type of Spacecraft	Place of Failure (Rocket Launcher/ Spacecraft)	Failure Reason (Hardware/ Software)	Country of Origin	Failure Cause
02.05.95	03.04.95	Pegasus	Orbcomm FM-2	Spacecraft	Hardware	U.S.	Disruptions in the onboard processor. Memory cleaned and functionality restored
08.09.97	20.02.86	Proton 8K82K	Orbital station Mir	Spacecraft	Hardware	USSR	Onboard computer failed. Complex orientation was disrupted; some onboard systems were deactivated.
02.01.98	23.09.97	Cosmos-3M	FAISAT-2V	Spacecraft	Hardware	U.S.	Onboard computer developed faults. Solar cell power capacity reduced, unstable operation of the onboard equipment in the Earth's shadow.
30.05.98	20.02.86	Proton 8K82K	Orbital station Mir	Spacecraft	Hardware	USSR	Central onboard computer failed. As a result, the orientation systems of the complex stopped. Due to power shortage, many onboard systems were shut down. Temperature in the living quarters of the complex dropped.
04.07.98	18.12.93	Ariane-4.44L	DBS-1	Spacecraft	Hardware	U.S.	Failure of the SCP control processor. Control was automatically switched over to the backup processor, and the pace vehicle proceeded operation with no consequences to the serviced clients.
20.12.99	18.12.99	Atlas-2AS	Terra	Spacecraft	Hardware	U.S.	Onboard computer failed. NASA specialist succeeded in restoration of the onboard computer operation on January 2, 2000.
19.07.91	17.07.91	Ariane-4.40	Orbcomm-X	Spacecraft	Software	U.S.	Software failure of the SC power control system 40 hours after deployment in orbit.
01.05.95	03.04.95	Pegasus	Orbcomm FM-1	Spacecraft	Software	U.S.	SV control system software disruptions. Software correction performed.
04.06.96	04.06.96	Arian 5	4 satellites of the type Cluster F	Rocket launcher	Software	EU	Operand overflow resulted in an erroneous command, which forced autodemolition of the rocket launcher.

17.08.97	20.02.86	*Proton 8K82K*	Orbital station Mir	Spacecraft	Software	USSR	Due to an error in the onboard computer program, repeated docking of the freight vehicle *Progress M-35* and the orbital station Mir failed.
20.07.98	18.10.89	*Space shuttle OV-104 Atlantis No. 5*	Space probe Galileo	Spacecraft	Software	U.S.	Anomalous behavior was detected in one of two subsystems responsible for reception of the commands from Earth. The spacecraft proceeded to error-protection mode. On July 23, 1998, the specialists of the Jet Propulsion Laboratory in Pasadena, CA, managed to rectify the faults by sending the debugging program to the probe, which replaced the faulty elements in the software of the onboard computer.
27.08.98	27.08.98	*Delta-3*	Galaxy-10	Rocket launcher	Software	U.S.	In the 70th second of the flight, a disruption occurred in the control system software and the rocket started to veer from the plotted course. When the deflection from the course exceeded the permitted limits, the rocket was destroyed by a command from Earth. The revised software will be installed on all subsequent copies of the rocket launcher *Delta-3*.
09.09.98	09.09.98	*Zenith-2*	Globalstar (12 pcs)	Rocket launcher	Software	Ukraine	Logic error in the algorithm of the ground automated system of the prelaunch preparation. Failure in issuing the command for closure of the pneumatic system valve of the second stage, which resulted in the failure of both the channel of the control system and in emergency termination of the flight.
27.11.98	18.10.89	*Space shuttle OV-104 Atlantis No 5*	Space probe Galileo	Spacecraft	Software	U.S.	Two disruptions occurred in the software within six hours of each other, resulting in the loss of some information that the scientists hoped to collect about Jupiter and the planet's satellites.
30.07.99	20.02.86	*Proton 8K82K*	Orbital station Mir	Spacecraft	Software	USSR	During a routine experiment, the computing machine CCM-1 failed due to an error in compiling the program. The complex's orientation was disrupted, but was restored within four days after the incident.
23.09.99	11.12.98	*Delta-2-7425*	Spacecraft Mars Klimate Orbiter	Spacecraft	Software	U.S.	Navigational error due to feet and inches not being converted to the metric system. The station went through Mars's atmosphere and burned up.
12.03.00	12.03.00	*Zenith-3SL*	ICO F1	Rocket launcher	Software	Ukraine	Logical error in the program algorithm of the ground automated system for prestart preparation. No command was issued for closure of the pneumatic system valve of the second stage, which resulted in engine shutdown.

and verification of hardware and software in general, such as event-B and model checking in accordance with ECSS standards.

The potential character of this field is proven by the fact that NASA has a special website dedicated to this topic and has conducted annual symposiums on application of formal methods since 2009 [11].

2.9   Analysis of Onboard System Failures at the International Space Station in 2000s

A stand-alone consideration is commanded by the theme of the space stations. A great many readers are aware that the currently operational international space station is of the twelfth series in this class of the space vehicles. They have their own classification: manned and unmanned ones, single and multimodule, general and special purpose.

The chronological succession of the single module space stations looks as follows:

1. A series of the orbit stations of the USSR:
    - *Salut-1* (manned, long time operation);
    - *Station—DOS-1*, 1971;
    - *DOS-2*, 1972—failed to be put into orbit;
    - *Salut-2*, 1973—a part of the project Almaz, decompression of the body;
    - *Kosmos 557 (DOS-3)*, 1973—control is lost after deployment into orbit;
    - *Salut-3 (OPS-2)*, 1974–1975;
    - *Salut-4 (DOS-4)*, 1974–1977;
    - *Salut-5 (OPS-3)*, 1976–1977.
2. Orbital station of the American *SkyLab*, 1973–1979.
3. Orbital station of the Chinese *Tiangun*, 2011.

The astronautics history so far knows only two manned multimodule stations—these are the Russian Mir (*Salut-8* or alias *DOS-6*) 1986–2001 and the international space station MKS (operational since 1998); 15 states contribute to its creation and operation. In order to analyze how onboard control systems affect reliability of rocket and space technology, it is worth looking at the biggest incidents that occurred at the International Space Station in 2000–2009.

On February 21, 2000, the main computer out of three onboard computers that performed command and control functions in the ISS United States segment stopped functioning for a short period of time. Contact with the station was restored only by switching to the backup onboard computer. The fault was caused by software defect, specialists assume.

On April 25, 2001, all three onboard computers of the ISS United States segment failed and contact with Mission Control Center in Houston broke. It turned out later that there was a failure of computer storage device and files on hard drive were damaged (today solid-state drives are used instead of hard drives to store data).

On February 4, 2002, ISS lost control for several hours due to failure of onboard electronic system that relays commands from Russian sensors that determine position of the station to U.S.-manufactured gyroscopes. As a result voice communication with the Earth and attitude control system at ISS failed.

Three months later (May 21, 2002) there was a three-hour shutdown of all life support systems and almost all of the scientific equipment. The fault was caused by a defect of one of the onboard computers. On June 12, 2007, failure of an onboard control system in the Russian segment caused improper operation of attitude control thrusters, equipment for oxygen generation and carbon dioxide removal, as well as other life support systems failed to turn on. Furthermore, automatic reset of the central computer generated false fire alarm. Investigation into fault causes showed that the underlying cause was condensation on electrical contacts that led to short circuit and a command for disconnection of power supply of the main and redundant computers. Therefore, even multiple redundancy of onboard computers cannot help to avoid a failure if there are defects of SW or ECB (responsible for hardware failures).

2.10   Methods of Ensuring Onboard Equipment Reliability of Spacecraft of Long-Life Operation

Useful life (UL) of SC that comprise modern space systems and complexes constitutes from 10 to 15 years, which allows reducing the number of SC launches for deployment and maintenance of various orbital groups.

One of the main indicators of SC reliability is its faultless operation. Reliability of any product is embedded at the stages of designing and manufacturing, and then it shows during testing and normal operation under practical operation conditions. Consequently, aspects of reliability shall be taken into consideration at all the stages of SC life cycle: during designing, development of engineering documentation, ground development of prototypes, flight tests, and normal operation [12, 13].

Stages of Design and Development

Design of space systems' equipment begins with preparation of a statement of work (SOW).

Probability of faultless operation (PFO) of onboard systems during operation life in accordance with their purpose shall comply with the so-called SC normative reliability budget. Table 2.12 shows typical values of reliability indicators of onboard systems' basic elements of communication, telecommunication, and navigation SC with useful life of 15 years [13].

It should be noted that such PFO values shall be ensured at the last year of useful life (fifteenth or twenty-fifth year).

Basic set of equipment for such SC shall be designed with the necessary level of backup that ensures required PFO; any single failure of an element or interconnect circuits, or any unauthorized command sequence shall not result in onboard electronics (OE) or SC failures.

Indicator of OE failure is nonfulfillment of at least one function specified in SOW when all backup sets, blocks, and circuits are used. The following requirements

Table 2.12 Probability of Faultless Operation of Onboard Systems' Basic Elements of Designed SC [13]

Elements of Onboard System	PFO*, min
Payload (relay)	0.91
Onboard control complex	0.958
Orientation and stabilization system	0.951
Power supply system	0.962
Correction system	0.930
Thermal control system	0.992
Solar panel mechanisms	0.9998
Antenna mechanisms	0.9999
Platform in general	0.8
SC in general	0.72

* As of the end of UL 15, 25 years.

are imposed on SC in terms of durability: SC UL shall constitute a minimum of 15 years, including:

- Testing and acceptance in orbit—0.25 year (2160 hours);
- Operation according to designated purpose (operational UL)—15 years (131,490 hours).

General requirements for OE durability (resource) are indicated in SOW and include specific particular requirements for OE resources both in stand-by and intermittent modes of operation with indication of specific operation resource (10% reserve for stand-by mode) and regard to preset minimum number of switching-on (for intermittent mode) during operation according to designated purpose during SC UL and testing at manufacturing facility, maintenance, and at technical complex.

OE shall comply with required technical and operational characteristics (preserve them) for 18 years starting from the moment of acceptance and ensure:

- Shelf life (3.5 years), including SC production cycle (1.5 years) and storage of accepted SC (2 years);
- Resource for all kinds of equipment testing within shelf life—4380 hours;
- UL as a part of SC according to designated purpose—15 years.

Required reliability parameters at minimum costs of equipment production ensure:

- Development and implementation of requirements for quality of electrical, electronic, and electromechanical (EEE) parts;
- Application of design choices that are perspective from the point of view of modern methodology of analysis and guarantee of OE reliability.

In order to ensure quality of applied EEE parts, they are selected for future use during testing at test facilities (TF) so as to reduce the number of failures caused by parts with hidden defects [14].

Functional analysis, analysis (calculation) of reliability, types of consequences and significance of failures, analysis of worst-case scenarios (as well as analysis of electric and thermal load on components, resource, and shelf life), and safety analysis are conducted to ensure reliability of OE [15].

A program for control of critical elements is developed based on research results. At design stages it is possible to embed, substantiate, and estimate future OE reliability; therefore, special attention is paid to verification models of required OE reliability at these stages.

OE of SCs of long-lived operation is regarded as complex systems that are characterized by structural redundancy. In such systems, binary variable can be used for OE status [16]:

$$S(T_{OE}) = \begin{cases} 1 & \text{in case equipment (system) remains operational during } T_{OE} \\ 0 & \text{otherwise} \end{cases}$$

(2.11)

where T_{OE} is operational life of onboard equipment.

The main indicator of OE reliability is faultless operation during specified UL. The analysis of specific OE probability of faultless operation includes examination of reliability structure diagram (RSD) formed out of a finite set of sequential and parallel RSDs of certain type. Thus, character of OE with parallel structures is defined by method and diagram of switchover.

In case of parallel reliability, function of onboard equipment PFO takes the following form:

$$P(T_{OE}; \Theta) = 1 - P\{\max_N (X^{(1)}, X^{(2)}, \dots X^{(N)}) < T_{OE}; \Theta\}$$

(2.12)

where N is the number of parallel elements in onboard equipment; $X^{(j)}$ is random value of operation life of j element (circuit) of onboard equipment ($j = 1, 2, \dots, N$); $\max_N (X^{(1)}, X^{(2)}, \dots X^{(N)})$ is function of reliability that corresponds with maximum N value; and Θ is parameter vector (set of parameters that determine time distribution of faultless operation).

In case stand-by redundancy is used, general function of PFO takes the following form (on condition of absolute reliability of switchers and failure indicators) [13]:

$$P(T_{OE}; \Theta) = P\left\{\sum_{j=1}^{N} X^{(j)} > T_{OE}; \Theta\right\}$$

(2.13)

Alternative connections of elements in OE RSD include different types of sequential and parallel connection of OE elements (circuits). Mathematical models of reliability calculation for OE and its electronic components (functional devices, blocks, circuits) that can be either in active or passive (stand-by) status are widely spread. Exponential failure law is applied.

In order to calculate reliability $P(t)$ of OE with such components, the following relations can be used [13]:

- Nonredundant block or sequential circuit of elements (location of single failure):

$$p(t) = e^{\lambda t} \tag{2.14}$$

- Block with m/n redundancy with blocks in parallel mode:

$$p(t) = \sum_{i=2}^{n-m} C_n^i \left(1 - e^{-\lambda t}\right)^i \left(e^{-\lambda t}\right)^{n-1} \tag{2.15}$$

- Block with m/n redundancy with blocks in stand-by mode:

$$p(t) = e^{-m\lambda t} \left[1 + \sum_{i=1}^{n-m} \frac{\left(1 - e^{-\lambda_{xp} t}\right)^{i-1}}{i!} \prod_{j=0}^{i-1} \left(j + m \frac{\lambda}{\lambda_{xp}} \right) \right] \tag{2.16}$$

where t is time of block operation; λ is failure rate in active mode; λ_{xp} is failure rate in standby mode ($x\lambda p =/ \lambda 10$); n is number of identical parallel blocks; and m is number of operating blocks that determine performance of circuit and those backed up by other $n - m$ blocks.

The use of highly reliable EEE parts allows applying reduction factor for failure rate reference values of EEE parts that form OE [16]. One of the approaches to quantitative estimation of reduction factor K_{TF} that defines value of operational failure rate λ_o more precisely is proposed in papers [17, 18].

Stages of Ground Experimental Development

Ground experimental development (GED) of OE and SC shall provide development and verification of requirements for technical characteristics and reliability specified in SOW under conditions similar to operation conditions. Methodology of development includes:

- More demanding modes of development and acceptance;
- Strict sequence of development (from simple to complex) of SC element levels equipment (units, assemblies, onboard equipment) → onboard system → SC (platform and payload module), as well as stage-by-stage development (component, system level, preliminary, exploratory, and flight tests);
- Gradation of development and test norms. Gradation of development norms means that the higher the development level is, the lower the qualification margin of destabilizing factors applied during development (temperature, number of thermal cycles, test duration, mechanical stress, and so on). Thus, maximum norms and development modes are applied to SC equipment.

The following types of tests are conducted at the level of OE component development in accordance with complex experimental development program:

- Laboratory development tests (LDT);
- Design related tests (DRT);
- Preliminary tests (PrT);
- Specific exposure tests (if necessary);
- Endurance (if necessary).

Requirements for Test Norms

Tested SC OE with UL of 10–15 years developed by NPO PM undergoes development and preliminary tests for the following external exposures:

- Mechanical (modes in accordance with SOW);
- Thermal (thermal vacuum), including cyclic tests with temperature range extended by ±10°C relative to operational range, at least 20 thermal cycles, 4 of which are thermal vacuum ones;
- Triple cold start impact (at minimum negative temperature with 10°C margin;
- Pressure decrease for OE with voltage 100V and more, cold start, and pressure decrease can be combined with thermal (thermal vacuum) tests.

OE flight units further undergo the following types of tests during manufacturing and acceptance [13]:

- Burn-in testing lasting at least 300 hours and including 10 thermal cycles with the thermal range extended by ±10°C;
- Mechanical stress (modes in accordance with technical specification);
- Thermal (thermal vacuum) exposure in thermal range extended by ±5°C;
- Pressure decrease for OE with voltage of more than 100V (at the stage of acceptance test of the first OE flight unit, further flight units are tested if necessary regarding follow-on development);
- Single cold start test (at minimum negative temperature with 5°C margin);
- Verification test for faultless OE operation in enabled state for the past 100 hours of operation.

In general, OE ground development is a process with feedback at each level that aims at ensuring efficiency of OE development during all tests and development activities [19]. Figure 2.19 shows structure of SC and its components development with regard to stage-by-stage testing and applied models [13]. Modern SCs with specified quality and reliability indicators are manufactured in shortest period of time determined by contract and at minimum costs; therefore, ground development requires constant optimization.

NPO PM implements perspective approaches to optimization of SC and OE development [20] and applies the following methods of optimization:

- Nomenclature reduction of produced equipment for development of OE and its elements by means of integrating different types of tests on a single piece of equipment;

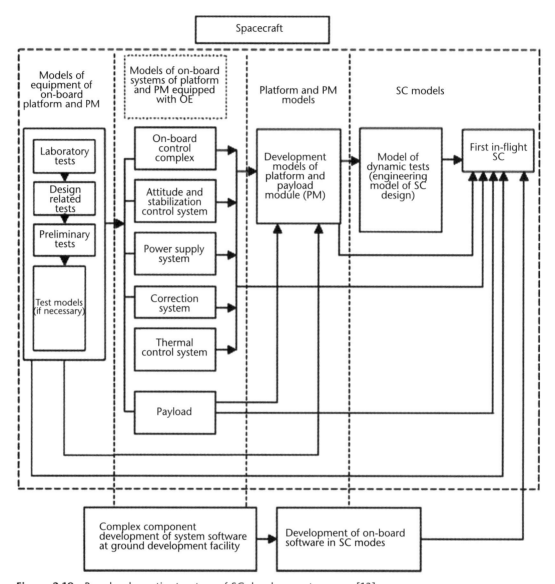

Figure 2.19 Rough schematic structure of SC development process [13].

- Reduction of work costs by means of substitution of physical OE development models with program models and the use of equipment and software designed within the framework of various projects (interproject unification);
- Reduction of development period and costs by means of decreasing stages of SC component testing (for example, the stage of laboratory testing, special and endurance tests if verified by calculations);
- Integration of some development tests with flight unit approval tests;
- Optimization of development and production stages (elimination of less informative tests on the basis of estimation of OE design features; integration of test types that provide similar results of effect on OE or that become more informative when integrated; elimination of redundant tests that identify the

same OE properties under different types of exposures; substitution of equipment testing in general with testing of its components, fragments, or models; reduction of test time (optimization of exposure time) including the use of rapid methods; equivalent replacement of one test type with another; replacement of test categories: qualification, periodic, and type tests with screening tests).

Justification criterion of any optimization is comparative assessment of initial and optimized parameters. Reference [21] proposes an approach to quantitative estimation of efficiency of SC ground development optimization in accordance with all the tests at all levels of development based on its comparison with reference or acceptable efficiency level.

OE reliability is determined during production testing and operation by detecting failures that belong to different categories. There are the following types of OE failures: design, production, and operation failures; EEE failures.

Parameters of OE faultless operation in terms of any type of failure can be estimated by means of a model that reflects structure of OE production testing as well as by the number of failures due to a specific cause (except for operational ones) during stages of production testing.

Production testing of OE and its components includes:

- Incoming control and additional testing of EEE batches at TF;
- First OE switching-on after manufacturing and adjustment;
- Burn-in testing;
- Predelivery tests;
- Acceptance tests;
- Endurance, type, periodic tests (if necessary);
- Testing in assembly with onboard systems;
- Testing in assembly with SC (undocked and docked).

In order to determine faultless operation in terms of EEE failures, tests additionally include results of incoming control and additional EEE tests at TF (screening, diagnostic nondestructive control, sampling physical destructive analysis).

General function of OE reliability according to results of ground development for random discrete values (i.e., the number of OE failures due to different causes at different stages of testing) is as follows:

$$p_{OE} = \prod_{i=1}^{r} F_i\left(K, M, N_0^*, n_{TF}^*, n_{i,j}, p_{i,j}, p_{TF}^*\right) \tag{2.17}$$

where F_i is function of faultless operation in terms of i-type failure; K is number of produced OE of certain type; M is number of types of OE production tests; N_0^* is initial amount of batches of certain EEE types that are received for incoming control at TF (EEE failures detection); n_{TF}^* is number of defect and potentially unreliable EEE parts of a certain batch that have been discarded at TF (EEE failures detection); p_{TF}^* is efficiency of tests of EEE batches at TF (EEE failures detection); n_{ij} is number of OE failures due to i-type cause at j-type test; p_{ij} is efficiency of j-type test for i-type failure; and r is number of failure causes (design, production).

Methods to determine OE total reliability and confidence intervals in case of independent tests and OE failures due to various causes are proposed in [22].

Stages of Flight Tests and Normal Operation

General verification process of OE and SC required characteristics includes flight tests and OE inspection during SC acceptance in orbit. Efficiency of flight tests shall be ensured so that compliance of OE with SOW requirements is verified and there are no design and production failures and faults connected with fundamental design errors and insufficient development of OE production technology and testing. The following aspects are taken into account as well:

- Results of OE operation in orbit (failures, operating hours, timeouts, values of key parameters);
- Calculated and experiment-calculated estimates of reliability of OE and its elements obtained at design stages and specified according to test and operation results;
- Amount of development tests (number of samples, operating hours, OE failures);
- Types and modes of OE and SC development tests;
- Results of normal and accelerated OE endurance tests;
- Results of diagnostic and predicting control of EEE parts used for SC OE.

In order to estimate and predict OE reliability parameters during flight tests and normal operation, a software and methodical complex has been developed that includes a system of analyzed and predictable parameters of OE and SC reliability and technical state, algorithms and methods of analysis and prediction of their performance during orbital flight with the use of parametric, nonparametric, and hybrid models for processing small amounts of various statistical data. The studied complex of methods of stage-by-stage OE reliability provision at different stages of life cycle has been implemented during production of SCs *SESAT*, *Ekspress-AM*, *GLONASS-K*, *Ekspress-1000* platform, which has proven that it is possible to manufacture spacecraft with a useful life of 10–15 years and reliability factors at the level of best domestic and foreign samples.

References

[1]   Zheleznyakov, A.B., *Rocket Downfall During Launching*, Saint Petersburg, Russia: Sistema, 2003.

[2]   Kharchenko, V. S., V. V. Sklyar, and O. M. Tarasyuk, "Rocket-Space Risk Analysis: Evolution of Sources and Tendencies," *Radio-Electronic and Computer Systems*, Vol. 3, No. 3, 2003, pp. 135–149.

[3]   Hansen, M., and R. F. Nesbit, "Report of the Defense Science Board Task Force on Defense Software," Defense Science Board, Washington, DC, 2000.

[4]   Tarasyuk, O. M, V. S. Kharchenko, and V. V. Sklyar, "Safety of Airspace Technique and Reliability of Computer Systems," *Aviation and Space Technique and Technology*, Vol. 9, No. 1, 2004, pp. 66–80.

[5] Gorbenko, A. V., S. A. Zasukha, V. I. Ruban, O. M. Tarasyuk, and V. S. Kharchenko, "Safety of Airspace Technique and Reliability of Computer Systems: 2000–2009," *Aviation and Space Technique and Technology*, Vol. 78, No. 1, 2011, pp. 9–20.

[6] Kharchenko, V. S., V. V. Sklyar, and O. M. Tarasyuk, "Rocket-Space Risk Analysis: Evolution of Sources and Tendencies," *Radio-Electronic and Computer Systems*, Kharkiv: National Aerospace University "KhAI," Ed. 3, 2003, pp. 135–149.

[7] Aisenberg, Y. E., and M. A. Yastrebenetsky, "Comparison of the Safety Principles for the Control Systems for Launchers and Nuclear Power Plants, *Space Science and Technology*, Vol. 8, No. 1, 2002, pp. 55–60.

[8] Kharchenko, V. S., M. A. Yastrebenetsky, and V. V. Sklyar, "New Information Technology and Safety Aspects of Information Control Systems of Nuclear Power Plant," *Nuclear and Radiation Safety*, No. 2, 2003, pp. 18–29.

[9] Labenskiy, V. B., "Applying of Correlation-Regression Analysis for Tasks Planning in Rocket-Space Industry," *Automation and Information Sciences*, No. 4, 2001, pp. 101–110.

[10] Jones, M. B., "What Happened on Mars?," Microsoft Corporation, http://research.microsoft.com/enus/um/people/mbj/mars_pathfinder/.

[11] NASA, "The First NASA Formal Method Symposium," http://ti.arc.nasa.gov/events/nfm09/

[12] Interfax, "Accident Caused by a Fault," http://www.interfax.ru/print.asp?sec=1446&id=169821.

[13] Patraev, V. Y., and Y. V. Maksimov, "Methods of Reliability Control of an On-board Equipment of Space Vehicles of Long-Lived Operatio," *Izvestiya vysshikh uchebnykh zavedeniy. Priborostroenie (Journal of Instrument Engineering)*, Vol. 51, No. 8, 2008, pp. 5–12.

[14] Urlichich, Y. M., and N. S. Danilin, *Quality Management of Space Radio-Electronic Equipment Under the Conditions of Open Economy*, M.: Maks Press, 2003.

[15] Instructional Guidelines 154-24-2001, *Conducting Analysis to Ensure Reliability Of Equipment, Systems and Spacecraft*, Zheleznogorsk: NPOPM, 2001.

[16] Patraev, V. Y., "Spacecraft Reliability Model, SAKS2004," Theses of International Research and Practice Conference, Krasnoyarsk: Siberian State Aerospace University, 2004.

[17] Fedosov, V. V., and V. Y. Patraev, "The Reliability Growth of the Spacecraft Electronics on Application of Radio Electronics, Subjected to Additional Screening Tests in Specialized Technical Testing Centers," *Aerospace Instrument Making*, No. 10, 2006, pp. 50–55.

[18] Fedosov, V. V., and V. Y. Patraev, "Estimation of Destructive Physical Analysis Towards Reliability Performance Microelectronics Products Installed on the Board of Spacecraft, *Aerospace Instrument Making*, No. 1, 2008, pp. 37–40.

[19] Quality Management System, *Stages of Ground Experiment Development of Products: Types of Screening and Control Tests, General Requirements*, Zheleznogorsk: NPO PM, 2007.

[20] Patraev, V. Y., and Y. V. Maksimov, "Optimization of Testing Spacecraft with Useful Life of 10–15 Years," *Dvojnyetekhnologii (Dual Technology)*, No. 3, 2004, pp. 66–80.

[21] Patraev, V. Y., and Y. V. Maksimov, "Assessment of Communications and Navigation Satellites Experimental Test Development Optimization Efficiency," *Materials of the Russian Research and Technology Conference Dedicated to the 40th Anniversary of Launch of the first SC Glonass*, Krasnoyarsk: SibSAU, 2007, pp. 35–40.

[22] Patraev, V. Y., and Y. V. Maksimov, "Estimation of Reliability of Onboard Equipment by Results of Additional Screening Tests of Completing Electro-Radio Elements and Production Tests of the Onboard Equipment," *Aerospace Instrument Making*, No. 8, 2006, pp. 46–49.

CHAPTER 3

Microwave Electronics for Space and Military Applications

3.1   Basics of Microwave Electronics

Before discussing the peculiarities of gallium arsenide and its application, we should remember the basics of microwave technology. As is well known [1–3], the range of ultrahigh frequencies is the frequency range of electromagnetic radiation (100–300 GHz) located in the spectrum between ultrahigh television frequencies and far infrared frequencies. This frequency range corresponds to a wavelength of 30 cm to 1 mm, so it is also called decimeter and centimeter waves. In English-speaking countries it is called the microwave range—meaning that the wavelength is very small compared to the wavelengths of conventional radio broadcasting, having an order of several hundred meters.

As according to the wavelength, the radiation of microwave range is intermediate between light radiation and conventional radio waves, and it has some properties of both light and radio waves. For example, it spreads the same as the light in a straight line and is covered by almost all solid objects.

Much like the light it is focused and is distributed in the form of the beam, and it is reflected. Many radar antennas and other microwave devices are like enlargements of optical elements such as mirrors and lenses.

At the same time, microwave radiation is similar to radio waves of broadcast band in the sense that it is generated by analogous methods. One may apply to microwave radiation the classical theory of radio waves, and it can be used as means of communication, being based on the same principles. But thanks to the higher frequencies it gives better opportunities of information transmission, which improves communication efficiency. For example, one microwave beam can simultaneously carry several hundred calls. The similarity of microwave radiation with light and the increased density of the carried information proved to be very useful for radar and other spheres of technology.

The use of microwave radiation in modern space and military equipment can be reduced to the following main directions.

Radar. Decimeter and centimeter range waves remained the subject of pure scientific curiosity before the Second World War, when there was an urgent need for new and effective electronic means of early detection. Only then the intensive studies of the microwave radar began, although its principal feature was demonstrated as early as 1923 in the Scientific Research Laboratory of the U.S. Navy.

The essence of radar is that the short, intense pulses of microwave radiation are emitted, and then some part of the radiation, which has returned from the desired remote object (a ship or an aircraft), is registered.

Communication. Radio waves of microwave range until recently were widely used in communications technology. In addition to various military radio systems [4, 5], there are numerous commercial lines of microwave communications worldwide. Since these radio waves do not follow the curvature of the earth surface, but extend in a straight line, these communication lines are usually composed of repeater stations that are installed on the tops of hills or radio towers with approximately 50 km intervals. Parabolic or horn antennas, mounted on towers, receive and transmit microwave signals further. At each station the signal is amplified by an electronic amplifier prior to retransmission. Since microwaves allow narrowly focused receiving and transmission, we don't need large electric power consumption for transmission.

Although the system of towers, antennas, receivers, and transmitters may seem expensive, in the long run it all pays off thanks to the big informational capacity of microwave communication channels. The cities of the United States more than 20 years ago were interconnected by a complex network of more than 4,000 microwave relay links forming a communication system that stretched from one ocean coast to the other. The channels of that network were capable to pass thousands of phone calls and numerous television programs simultaneously.

Communications satellites. The system of relay radio towers, which was required for the transmission of microwave radiation over long distances, was based, of course, only on the ground. Other methods of retransmitting were necessary for intercontinental communication. Here the connecting artificial Earth satellites (AES) came to help; while being deployed into a geostationary orbit, they can serve as the microwave communications relay stations [5–7].

An electronic device, called active-relay AES, receives, amplifies, and retransmits the microwave signals, transmitted by ground stations. The first experimental AES of this type (*Telstar, Relay*, and *Syncom*) have successfully carried out the retransmission of television broadcasts from one continent to another in the early 1960s. On the basis of this experience, commercial satellites of intercontinental and internal communication have been developed. Satellites of the last intercontinental series *Intelsat* were put in different points of the geostationary orbit in such a way that their coverage areas provide service for the subscribers worldwide. Each *Intelsat* satellite of the latest versions provides customers with thousands of high-quality communication channels for simultaneous transmission of telephone, television, fax signals, and digital data.

Scientific research. Microwave radiation has played an important role in studies of the electronic properties of solids. When such a body is in the magnetic field, the free electrons in it start to rotate around the magnetic lines of force in the plane perpendicular to the magnetic field direction. Frequency of rotation, called a cyclotron, is directly proportional to the magnetic field strength and inversely proportional to the effective mass of the electron. Effective mass determines the acceleration of the electron under the influence of any force in the solid state circuits. It differs from free electron mass, which determines the acceleration of the electron by action of any force in vacuum. The difference is due to the presence of the forces of attraction and repulsion, with which the atoms and other electrons influence the electron in a chip. If a solid body in a magnetic field receives microwave radiation, this radiation is strongly absorbed when its frequency is equal to the electron cyclotron frequency. This phenomenon is called cyclotron resonance; it allows measuring the effective mass of the electron. Such measurements have given a lot of valuable information on the electronic properties of semiconductors, metals, and metalloids.

Microwave radiation plays an important role in space rescarch. Astronomers have learned a lot about our own galaxy, exploring radiation with a wavelength of 21 cm emitted by hydrogen gas in interstellar space. Now we can measure the speed and determine the direction of motion of the galaxy arms, as well as the location and density of the areas of hydrogen gas in space, and more.

The key elements of all the microwave devices are sources of microwave radiation [7, 8].

Rapid progress in the field of microwave technology is largely associated with the invention of special electro vacuum devices—klystron and magnetron, which are capable of generating large amounts of microwave energy.

Generator on conventional vacuum triode, used at low frequencies, is very inefficient in the microwave range.

Two main disadvantages of the vacuum tube as a microwave generator are finite transit time of an electron and interelectrode capacity. The first relates to the fact that the electron takes some (albeit small) time to fly between the electrodes of the vacuum tube. During this time, the microwave field has time to change its direction into reverse, so that the electron is forced to turn back before reaching the other electrode. As a result, the electrons oscillate in the lamp without any use, without giving away their energy in the oscillating circuit of the external circuit.

Magnetron. The magnetron, which was invented in UK before World War II, doesn't have these drawbacks, because a completely different approach to the generation of microwave radiation was taken as the basis—the principle of cavity resonator. The same way an organ pipe of this size has its own acoustic resonance frequencies, the cavity resonator has its own electromagnetic resonances. The walls of the cavity act as an inductance, and the space between them acts as the capacity of some resonance circuit. Thus, cavity resonator is similar to the parallel low-frequency generator with separate capacitor and an inductor. The dimensions of the cavity resonator are chosen, of course, so that the desired ultra-high resonance frequency matches this combination of capacitance and inductance [2].

The magnetron commonly has several cavities arranged symmetrically around the cathode in the center. The device is put between the poles of a strong magnet. The electrons, which are emitted by the cathode by the action of the magnetic field, are forced to move along circular trajectories. Their speed is such that they cross at the periphery at the fixed time the open slots of the resonators, giving their kinetic energy and exciting the oscillations in the resonators. Then again, the electrons return to the cathode, and the process is repeated. With this device the time of flight and the interelectrode capacitance does not interfere with the process of generation of microwave energy.

The magnetrons can be made a larger size, and then they give powerful microwave pulses of energy. But the magnetron has its drawbacks. For example, the resonators for very high frequencies are so small that they are difficult to manufacture technologically, and the magnetron itself due to its small size may not be powerful enough. Besides that, the magnetron needs the heavy magnet, and the required mass of the magnet increases with the power of the device. It is therefore evident that powerful magnetrons are not suitable for spacecraft onboard installations.

Klystron. This electro-vacuum device, based on a slightly different principle, does not require an external magnetic field. The klystron electrons move in a straight line from the cathode to the reflecting plate and then back. At the same time they cross an open gap of cavity resonator in the shape of a donut. The control grid and the resonator grids group the electrons into separate clusters, so that the electrons cross the gap of the resonator only at certain moments of time. The gaps between the clusters are aligned with the resonance frequency of the resonator so that the kinetic energy of the electrons is transmitted to the resonator, whereby strong electromagnetic waves are set therein. This process can be compared with the rhythmic sway of initially fixed swing.

First klystrons were quite low powered devices, but later they broke all records of magnetrons as the high power microwave generators. Klystrons were created, delivering up to 10 million watts of power per pulse and up to 100 thousand watts in continuous mode. The klystrons system of research linear particle accelerator produces 50 million watts of microwave power in a pulse.

Klystrons can operate at very high frequencies; however, their output power is usually less than one watt. It is known from the literature that the options of klystron design are intended for high output power in the millimeter range [3, 8].

Klystrons can also serve as microwave signals amplifiers. To do this, the input signal should be supplied to the cavity resonator grid, and then the density of the electron clusters will vary in accordance with this signal.

Traveling wave tube (TWT) [1, 2, 6]. TWT is another electro-vacuum device for the generation and amplification of electromagnetic waves of microwave range. It is a thin evacuated tube inserted in a focusing magnetic coil. Inside the tube there is a slowing down coil-wire. The electron beam goes along the axis of the coil wire, and the amplified signal wave runs along the coil wire. Diameter, length, and pitch of the coil wire, and the velocity of the electrons, are selected in such a way that the electrons give some part of their kinetic energy to the traveling wave.

Radio waves travel with the speed of light, while the velocity of the electrons in the beam is much smaller. However, since the microwave signal is forced to follow the coil, the speed it moves along the axis of the tube is close to the electron beam velocity. Therefore, the traveling wave for a long time interacts with the electrons and is amplified by absorbing their energy.

If the lamp does not receive an external signal, the random electrical noise at a certain resonance frequency is amplified and TWT of traveling wave works as a microwave generator, but not as an amplifier. TWT output power is significantly less than of the klystrons and magnetrons at the same frequency. However, TWTs allow tuning over a wide frequency range and can be used as highly sensitive low-noise amplifiers. This combination of properties makes TWT a valuable device of microwave technology.

Flat vacuum triodes [3]. Although klystrons and magnetrons are more preferred as the microwave generator, technological and design improvements in some degree restored the important role of vacuum triodes, especially as amplifiers at frequencies up to 3 GHz.

The difficulties associated with the transit time are eliminated due to the very small distances between the electrodes. Undesirable interelectrode capacitances are minimized, since the electrodes are made net-shaped, and all external connections are made on large rings that are outside of the lamp. As is customary in the microwave technology, the cavity resonator is applied. Resonator tightly encloses the lamp, and ring connectors provide contact over the entire circumference of the resonator.

Gunn diode generator [1–3]. The first such semiconducting microwave generator was proposed in 1963 by J. Gunn, who worked for the IBM Thomas J. Watson Research Center. Currently, these devices provide power of only some milliwatts at the frequencies less than 24 billion hertz, but within these limits they have distinct advantages over low-powered klystrons.

As Gunn diode is a single crystal of gallium arsenide, it is more stable and durable, in principle, than the klystron, which should have a heated cathode to create a flow of electrons and requires high vacuum. In addition, Gunn diode operates at a relatively low supply voltage, whereas klystron needs bulky and expensive power supply sources with a voltage of 1,000 to 5,000 volts.

Conventional channels for transmitting microwave range waves are shaped as waveguides. A waveguide is a carefully treated metal tube of rectangular or circular cross-section, inside which the microwave signal goes. To put it simply, the waveguide directs the wave, causing it to be reflected from the walls. However, in fact the wave propagation along the waveguide is an extension of oscillation of the electric and magnetic fields of the wave, as well as in free space. Such waveguide propagation is possible only provided that its dimensions are in relation to a specific frequency of the transmitted signal. Therefore, the waveguide is accurately calculated, processed, and intended for only a narrow range of frequencies. It transmits other frequencies poorly or doesn't transmit them at all.

The higher the frequency of the wave, the smaller the size of its respective rectangular waveguide is; after all, these dimensions are so small that it greatly complicates manufacturing and reduces transmitting limiting power. Therefore, the development of circular waveguides has been initiated (circular cross section), which may be large enough even at high microwave frequencies. The use of a circular waveguide is constrained by some difficulties. For example, a waveguide should be straight; otherwise, its efficiency reduces. Rectangular waveguides are easily bent; they can obtain the desired curved shape, and that does not affect the propagation of the signal.

Solid components. Solid components, both semiconducting and ferrite, play an important role in microwave technology. Thus, germanium and silicon diodes are used for detecting, switching over, straightening, frequency conversion, and amplification of microwave signal [3, 5, 8].

The special diodes are also used for amplification—they are called varicaps (with a managed capacity)—in the scheme, called parametric amplifier. Widely spread amplifiers of such kind are used to amplify very small signals, as they almost do not make the intrinsic noise and distortions.

The ruby maser is also a solid microwave amplifier with a low noise level. This maser, whose action is based on quantum mechanical principles, amplifies microwave signal due to transitions between the levels of the internal energy of atoms in the ruby crystal. Ruby (or other suitable material of the maser) is immersed in liquid helium, so that the amplifier operates at extremely low temperatures (only a few degrees higher than the temperature of absolute zero). Therefore, the level of thermal noises in the scheme is very low, so that the maser is suitable for radio astronomy, radar, and other ultra-sensitive measurements, which need to detect and amplify very weak microwave signals.

Ferrite materials, such as magnesium oxide and iron-yttrium-iron garnet, are widely used for the manufacture of microwave switches, filters, and circulators. Ferrite devices are controlled by magnetic fields, and a weak magnetic field is enough for controlling the flow of high-power microwave signal. Ferrite switches have the advantage over mechanical—they have no moving parts to wear, and switching is carried out very quickly.

A typical ferrite device is the circulator. Acting like a roundabout, the circulator provides a signal following only on certain paths connecting the various components. Circulators and other ferrite switching devices are still used to connect several microwave components of the system to the same antenna.

The tunnel diode is also used in microwave technology. It is a semiconductor device operating at frequencies up to 10 GHz. It is used in generators, amplifiers, frequency converters, and switches. Its operating capacity is small, but this is the first semiconductor device capable of working effectively at such high frequencies.

3.2   Structure and Properties of Gallium Arsenide

In previous chapters, we have considered the problem of silicon technology and silicon microelectronic devices. The main drawback of silicon as the main raw material for the manufacturing of microelectronic products compared to other compound semiconductor materials is a relatively low mobility of charge carrier, which limits the ability to work at high frequencies. Gallium arsenide is a more complex material than silicon; it is two-component, and technologically it is more difficult to work with, but it has higher mobility of charge carriers in comparison with silicon. The carrier mobility is the parameter that largely determines the operating frequency of the integrated circuit. The limit of the operating frequency of standard silicon processors today lies in the range of about 10 GHz, and gallium arsenide can operate in the range of 100 GHz to 1 THz. One can form different heterostructures on gallium arsenide, which increases the charge carrier mobility. Therefore, silicon is the material for the manufacture of chips, designed to work at low frequencies and with digital devices, and gallium arsenide is mostly for work with analog signals at very high frequencies [9].

Today monolithic circuits based on gallium arsenide are needed in any radio-electronic devices with radio transceivers, including mobile phones and radar stations. Chips based on gallium arsenide and gallium nitride are widely used in the field of wireless connections.

Today, more than 90% of all semiconductor devices are manufactured on silicon worldwide. At the same time, more than 50% of scientific and technical publications in the field of semiconductors and semiconductor devices are dedicated to the study of $A^{III}B^V$ type compounds. In recent years compounds of this class have received wide use as the material for the manufacture of various semiconductor devices.

Active investigation of the properties of these compounds in recent years has led to the discovery of a number of new physical phenomena, creation of innovative electronic devices, and significant contribution to the development of solid state physics.

Thus, gallium arsenide (GaAs) is one of the main semiconductor materials belonging to the compounds of $A^{III}B^{V}$ class, which, thanks to the successful combination of physical properties, takes the second place (after silicon) in its significance in modern electronic technology.

Speaking of GaAs crystal structure, it should be noted that the majority of compounds of $A^{III}B^{V}$ type, including gallium arsenide, are crystallized with zinc-blend structure. The unit cell of this structure contains two atoms, A and B, and is repeated in space so that each component forms a face-centered cubic lattice. In the case of GaAs the structure can be represented as interpenetrating face-centered lattice of Ga and As atoms, which are shifted relative to each other by a quarter of the main diagonal—see Figure 3.1(a).

Three major crystal planes of classical lattice GaAs are shown in Figure 3.1(b). Each As atom on the surface {100} has two bonds with Ga atoms from the underlying layer. Two other links are free. The plane {110} contains the same number of atoms of Ga and As. Each atom has only one connection with the underlying layer. Atoms on the surface {111} have three bonds with Ga atoms of the lower layer; the fourth link remains free.

The distance between the nearest neighboring atoms is 0.244 nm and equals to the sum of the atomic radius As (0.118 nm) and Ga (0.126 nm). The lattice constant is 0.565 nm.

It should be emphasized that due to the partial heteropolarity, the bonds in GaAs are much stronger than homopolar bonds in Si, or Ge. This leads to low lattice vibration amplitude (and, consequently, to a larger mobility) higher melting point and a wider bandgap.

This difference in chemical bonds is distinctly shown during mechanical shearing of the crystals. In diamond, for example, the crystals cleave substantially along the {111} planes. As shown earlier, in GaAs {111} superimposed planes, which are formed by different atoms (the plane of the gallium atoms and the plane of the arsenic atoms), the electrostatic interaction between these planes makes the shearing

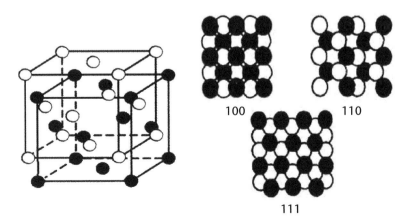

Figure 3.1   (a) Structure of GaAs, and (b) main crystalline lattice planes.

difficult. GaAs crystals readily cleave along the planes {110}, which contain the same number of atoms of gallium and arsenic.

3.3   Comparative Characteristics of GaAs and Si Properties

The number of physical properties of GaAs makes it one of the most interesting materials for semiconductor microelectronics technology. Some basic properties of this material are presented in Table 3.1 [9].

First, it should be noted that the main advantages of GaAs as the base material for semiconductor devices are defined by structural features of its energy bands. Figure 3.2 shows the classic band diagrams of GaAs and Si. The band diagram is constructed in a conventional manner. The x-axis shows the values of the wave vector for a number of its directions in the Brillouin zone; the vertical axis shows the value of the energy of electronic states.

Significant differences between GaAs and Si are in the nature of dependence of the conduction bands energy to the wave vector. Due to large bandgap the intrinsic concentration of electrons and holes in GaAs is less than in Si, so theoretically GaAs can have a very high value of resistivity. This allows us to use such material as a dielectric in ICs designed to work in centimeter and millimeter wavelength ranges, as well as for insulation structures in digital ICs.

Also, the larger bandgap allows the creation of devices capable of operating at higher temperatures than silicon ones.

As mentioned earlier, one of the most important properties of the material is high electron mobility (six times higher than silicon) in electric fields of low strength, which potentially creates microwave devices with improved performance. The small value of the minority carrier lifetime and greater (than that of silicon) bandgap make GaAs a very promising material for radiation-resistant devices and ICs.

To create the heterostructure of $A^{III}B^{V}$ materials on GaAs substrates, designed for manufacturing high-quality instruments, numerous researchers have developed

Table. 3.1   Basic Properties of GaAs

External view	Dark gray cubic crystals
Molecular mass	144.64 atomic unit
Lattice constant	0.56533 nm
Crystal structure	Zink-blende type
Melting point under n.c.	1513 K
Band gap under 300 K	1.424 electron volt
Electrons, effective mass	0.067m
Light holes, effective mass	0.082m
Heavy holes, effective mass	0.45m
Holes mobility under 300 K	400 cm²/ (V•s)
Electrons mobility under 300 K	8500 cm²/ (V•s)

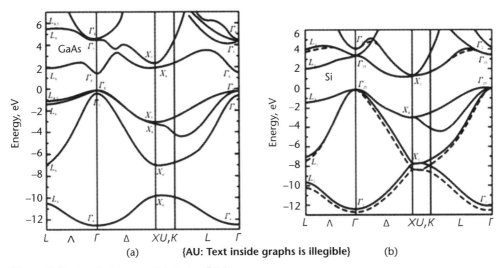

Figure 3.2   Band diagrams: (a) GaAs, (b) Si.

rather complicated methods of epitaxial films growing. At the same time the optical properties of such heterostructures open real prospects of creating elements of digital, microwave, and optical devices on a single GaAs crystal.

However, there are reasons that impede the practical implementation of the benefits that can be gained in the use of GaAs compound in microelectronics technology. The main drawback of gallium arsenide is that it is a two-component compound, and in connection with that it is necessary to lower the maximum temperatures during the production processes, thereby preventing dissociation of surface structures composition. Doping by diffusion process, which found wide application in the manufacture of silicon devices, appeared to be almost unacceptable in transition to GaAs. Gallium arsenide does not have natural, stable, easy-to-form oxide. The ability of silicon to form such oxide became an important factor in the creation of technology of production of the first silicon MOS transistors. GaAs surface is also more susceptible to the effects of various chemicals used in industrial processes that require in some cases the development of a fundamentally new approach to the practical implementation of these processes. Furthermore, GaAs is very brittle and prone to fracture material in the case of its procession.

3.4   Microelectronic Devices Based on GaAs

Various device structures on GaAs and A1GaAs that are used when creating analog and digital ICs have a variety of purposes. The most commonly used element in the development of both digital and analog ICs, is the basic element, which is a GaAs field effect transistor.

All devices on the basis of GaAs can be divided into several classes, the main of which are shown in Figure 3.3.

We'll consider these in more detail.

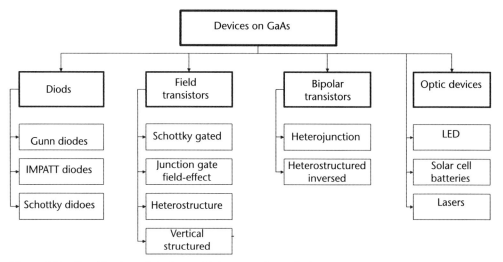

Figure 3.3   Classification of devices, which are made on GaAs.

3.4.1   Diodes Based on GaAs

GaAs-based diodes are made of three types. Among them, Gunn diodes and avalanche-transit diodes for use in the microwave band and millimeter band waves, as well as Schottky diodes, are widely used in space application electronics as varactors, mixers, and schemes to change the signal levels.

Gunn Diodes

The Gunn diode was invented by John Gunn back in 1963. This type of semiconductor diode is used for generating and converting waves in the microwave range. Unlike other types of diodes, the Gunn diode action principle is based not on the properties of p-n-junctions, but on its own bulk properties of the semiconductor.

Schematic representation of the construction of the device is shown in Figure 3.4(a). The diode includes an active region of L length, in which the concentration of the dopant mixture is equal to n, and two ohmic contacts. The dependence of the electron transfer rate to the intensity at which a negative differential conductivity in such structure appears is 3.2 kV/cm, the peak value of the electrons velocity in this case equals 22 • 10^7 cm/sec.

The physical mechanism of the device operation is the following. Under the condition $nL > 10^{12}$ cm^{-2} in the cathode area dipole domains of space charge are formed, which move to the anode and disappear there. The frequency of the diode generation is determined by domain transfer through the active region. Depending on the connection type, nL value, and bias voltage value, other modes of diode operation are also known. The most common use of Gunn diodes are high-frequency signal generators and amplifiers based on them.

Historically, Gunn diodes were the first GaAs-based diodes suitable for practical application in microwave devices. Nevertheless, they received only limited use in IC manufacturing. Prior to the development of GaAs FETs manufacturing technology,

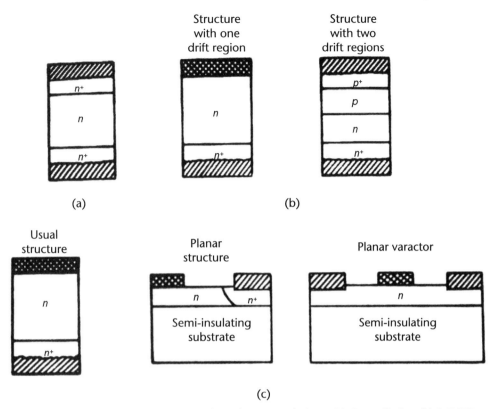

Figure 3.4   Sketches of diode structures based on GaAs designs: (a) Gunn diodes; (b) IMPATT diodes; (c) Schottky barrier diodes.

numerous groups of researchers took efforts aimed at using Gunn diodes as active elements of digital ICs.

Another possible area of} Gunn diodes application in integrated circuits is generation of signals in millimeter wavelength range.

IMPATT Diodes

In IMPATT diodes as well as in Gunn diodes the existence of the region of negative differential conductivity in the current-voltage characteristics of the device is used. However, the mechanism of occurrence of this region in IMPATT diodes differs from that of Gunn diodes. Usually IMPATT diodes consist of Schottky barrier or p-n-junction, shifted in the opposite direction before the avalanche, the drift region, and the ohmic contact. Figure 3.4(b) shows schematically the structure of IMPATT with one and two drift regions. In diodes with a drift region carriers occur near Schottky barrier and the electrons move toward the ohmic contact. In diodes with two drift regions, carrier generation occurs near the center of the structure, then the electrons move toward the ohmic contact, and the holes move toward p^+ contact. Negative resistance appears due to the existence of finite values of the time of formation of avalanches and the carriers drift. These times determine the current change delay through the structure with respect to the voltage in it and create conditions for the appearance of diode negative differential conductance in microwave and millimeter wavelength ranges.

The most promising area of IMPATT application in ICs is manufacturing millimeter wavelength signal generator for analog circuits. IMPATT diodes, in the same manner as Gunn diodes, are conveniently used to generate microwave signals, but their use as amplifiers is much more difficult because it is necessary to provide isolation of input and output circuits using circulators or hybrid branches. The inclusion of such elements in microwave IC is not appropriate. Therefore, IMPATT diodes are widely used as a relatively powerful signal generator.

Schottky Diodes on GaAs

Schottky diodes on GaAs are an important element of modern analog and digital ICs. It is largely determined by the relative ease of manufacture of contacts to gallium arsenide. Figure 3.4(c) shows three types of Schottky diode structures that may be implemented in IC manufacture.

The first type is a regular vertical structure, which is basically similar to that of IMPATT with a single drift region. This structure is most often used in the manufacture of discrete diodes, designed to work as varactors, varicaps, or mixers. It can be used in the manufacture of ICs on semi-insulating GaAs substrates, which leads, however, to a certain complication of the process cycle due to the need to provide access to the contact areas from the substrate side.

An example of the second type Schottky diode structure is planar structure, used in the production of diodes, designed to offset the level of signals and perform logic functions in digital ICs. The value of direct voltage drop of such diodes is determined by their area and the current value.

The third type of Schottky diode structure is used in the manufacture of tunable varactors for integrated FET-based generator circuits. The use of diodes of this type allows for greater capacitance variation limits than conventional varactors, and correspondingly allows increasing the frequency tuning range of the generators. Diodes capacitance variation limits expansion is explained by the fact that the change of reverse bias value affects a change in depletion region thickness and in the diode effective area.

3.4.2 Field Transistors

On the basis of GaAs FETs are also made. These include varieties of Schottky gate transistors and *p-n* junction transistors, FETs with heterojunction, permeable base, and vertical structure. Sketches of FETs structures are schematically shown in Figure 3.5.

Despite these structural differences, the operating principle of all these types of transistors is based on the main carrier's current value control in the device channel via the voltage applied to the gate. These FETs can be manufactured as devices operating either in enhancement mode or in depletion mode. Channel region in transistors operating in enhancement mode (normally closed transistors) at zero

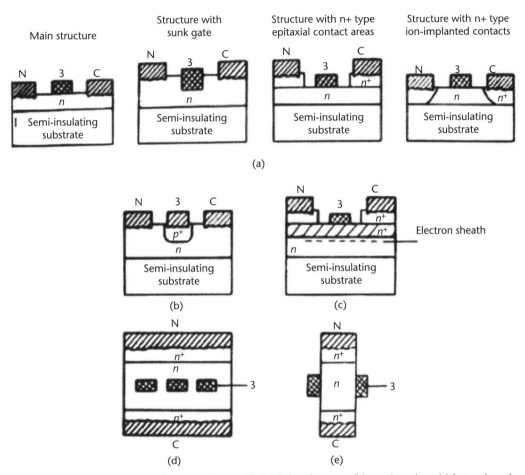

Figure 3.5 Vertical sections of FETs on GaAs with (a) Schottky gate, (b) p-n junction, (c) heterojunction, (d) permeable base, and (e) vertical structure.

potential on the gate is fully depleted, and to ensure the channel conductivity it is necessary to use a direct bias. In transistors operating in depletion mode (normally open transistors), the current flows in the channel even with zero bias on the gate, and it is required to apply a reverse bias to reach the cutoff mode and block the channel.

In digital ICs field effect transistors operating both in depletion and enrichment modes are commonly used, whereas in monolithic microwave ICs only devices of the first type are used. The transistors operating in enhancement mode have lower power consumption compared to depletion-type transistors, but still their frequency and noise properties are significantly worse.

Both in digital and analog microwave ICs, chip developers most commonly use field effect transistors with Schottky gate. FETs with *p-n* junction are typically applied in creating low power consumption devices, which at the same time have lower speed performance.

Let us consider in more detail these basic elements.

Field-Effect Transistors with Schottky Gate and p-n Junction

Typical structures of FET with Schottky gate and *p-n* junction with the planar arrangement of the source, gate and drain regions are fabricated on semi-insulating GaAs substrates are shown in Figure 3.5(a) and (b), respectively. The active region of these structures may be formed directly by ion implantation in a substrate, during epitaxial growth or ion implantation into the epitaxial layer. For several reasons operation of utilizing ion implantation directly into the substrate is becoming more and more common in the manufacture of digital ICs and microwave range FET-based ICs. This is due to good reproducibility of shop-proven process, the possibility of carrying out selective doping, maintaining high quality of the surface, and the possibility of organizing mass production of inexpensive products.

It follows then that the differences in FET structures are associated with the use of either conventional or recessed-gate, as well as with the methods of creating +/– source and drain contacts, which can be made by epitaxial methods or by making a source-drain implant (Figure 3.5(a)). Such structures with recessed-gate, with or without +/– source and drain contacts are now widely used in the manufacture of both high-power and low-noise FETs. The properties of recessed-gate areas and source and drain ohmic contacts determine to a large extent the frequency parameters and breakdown voltages of FETs. In the manufacture of digital ICs, the use of devices with planar geometry is preferable, as this minimizes the issues associated with coating of steps in the metal-insulator system when creating a multilevel metallization.

Currently, there has been completed a significant number of studies dealing with simulation of FET operation and electronic circuits on their base. The very first FET model using gradual-channel approximation has been offered by Shockley. Later, his students developed a much more sophisticated model.

Figure 3.6 shows the standard results computer simulation of Schottky field transistor operation conducted on the basis of method proposed by Fresnel. The figure shows the electron density distribution of electrostatic potential and electric current throughout the transistor structure. The solid lines in the figure show the electron density value related to the concentration of doping of the active layer. Dashed lines represent electrostatic potential distribution. Arrows indicate electric current distribution. We can clearly see the existence of the depletion region under the device gate. Increasing the density of equipotential lines in the area between the gate and the drain corresponds to the existence of high electric fields and confirms the experimentally observed fact that the device failure is associated with the electrical breakdown in this area.

Field Effect Transistors Based on Heterostructures

The heterojunction FET layout is similar to the conventional FET geometry, but the properties of materials used are quite different. Figure 3.5(c) shows that a heavily doped A1GaAs layer is in the heterojunction FET gate area, and the electron transfer occurs in the undoped GaAs area adjacent to this layer. Since in this case the intensity of electron scattering by impurities is significantly reduced, the designers can obtain very high electron mobility, in particular at a temperature of 77 K.

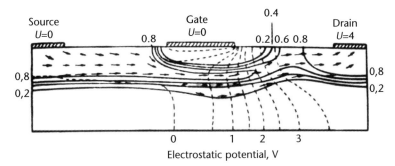

Figure 3.6   Results of computer simulation of Schottky field transistor operation.

Typically, the designers obtain the source structures for manufacture of hetero-junction FET designers using the molecular beam epitaxy. To facilitate the creation of ohmic contacts after growing of undoped GaAs and A1GaAs layers, a n^+ layer is grown during such IC manufacturing workflow. In this case the A1GaAs layer can be used as an etch-stopper while removing the n^+ layer between the drain and source areas. The presence of these two A1GaAs layers separated by a thin GaAs layer allows forming of both depletion-mode and enhancement-mode FETs on the same wafer.

At the initial stages of heterojunction FETs design, the researchers had already optimistically assumed that new digital ICs would be created on their basis. However, heterojunction FETs also proved to be promising for the microwave frequencies applications. Recent studies have shown that the heterojunction FETs are used both in digital and microwave range ICs.

Vertical Structure FETs

The interest of designers of ICs for space applications in the creation of vertical structure FETs is determined by the fact that such devices can be manufactured with a very short gate length, as well as with very moderate requirements for the lithography process accuracy. The gate length in these structures is determined by thickness of metal film forming the gate.

There are two ways to implement a vertical FET structure. The first class of such devices includes permeable base transistors.

Metal islands were buried in the active layer of the device structure (Figure 3.5(d)), which, similarly to the gate (or permeable base), are used to control the electrons flow intensity between the source and the drain. The most difficult part in the production process of such transistors is the growing of GaAs film over the metal islands. Using the scanning transmission electron microscopy, the defects can be observed that occur in the transition layer between GaAs and metal. These defects have an adverse effect on the flow of current between the metal islands.

There is another type of vertical structure FET. The structure of these transistors (Figure 3.5(d)) resembles that of permeable base transistors, but unlike the latter, no GaAs film needs to be grown on the metal gate islands while manufacturing this type of vertical structure FET. The gate is adjacent to the vertical sides of the channel in the area between the source and drain junctions.

Parameters of such transistors in the microwave range appeared to be very low. Transistor with a characteristic line width of 0.5 microns had a frequency limit of 12 GHz. However, the complexity of manufacturing process makes creation of ICs on their base very expensive.

3.5   Heterojunction Bipolar Transistors

The heterojunction bipolar transistor (BT) structure was first introduced in 1957. It was shown that it was possible to obtain higher gain in a transistor with emitter area formed in semiconductor material with a larger bandgap compared to conventional bipolar transistor, all other conditions being equal.

The most typical heterojunction BT consists of n-type A1GaAs emitter, p-type GaAs base area, and n-type GaAs collector area. Currently, two different types of heterojunction transistor structures—normal and inverted—are designed for digital ICs (Figure 3.7).

In the conventional structure shown in Figure 3.7(a), the emitter area is located at the chip surface, and between the A1GaAs layer and the metal junction to the emitter area, the n-type GaAs layer is formed in order to facilitate the ohmic contact. In order to have the collector area isolated, such devices are manufactured on semi-insulating GaAs substrates. At the same time, to ensure access to the collector layer, windows have to be etched through A1GaAs–GaAs layers while making the ohmic contacts. To make a junction to the base area, ion implantation is used. Bipolar transistors with such structure have been used in the first ultrafast ICs with emitter-coupled logic (ECL) [10].

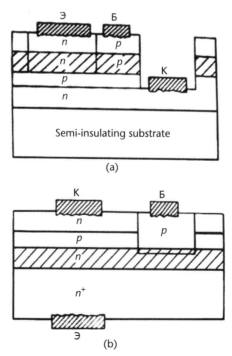

Figure 3.7   Heterojunction bipolar transistors: (a) common structure; (b) inverted structure.

Figure 3.7(b) shows an inverted transistor structure in which the emitter area is buried. The devices having such structure are easy to use as the base for design of ICs with injection logic, since the n^+-type substrate can be used for forming the emitter area. In this case, the n-type A1GaAs layer is grown directly on the substrate. The junction to the base area is formed by ion implantation. The main advantage of the inverted transistor structure is that the n^+ type substrate may serve as a junction to the emitter area. Due to this fact, the window etching operation is eliminated from the workflow to make a junction to the emitter, allowing the use of planar technology in the manufacture of ICs. Gate arrays are created on the basis of inverted bipolar transistor structure in order to design superfast bipolar microcircuits.

3.6   Optoelectronic Devices on GaAs

3.6.1   LEDs

There are two types of light-emitting diodes on GaAs: LED with p-n-junction and GaAs-A1GaAs heterojunction. The device principle is very simple: with the passing of electric current through p-n junction or heterojunction in the forward direction, the charge carriers are recombined and emit the photons (due to transition of electrons from one energy level to another).

The light is emitted within a narrow spectral range, with its spectral characteristics depending on the degree of GaAs doping.

The cost of high-power LEDs used in portable spotlights and car headlights is quite high today. However, such LEDs, compared to other light sources, have high luminous efficiency and long life.

3.6.2   Solar Batteries

A solar battery is an array of photovoltaic cells directly converting solar energy into direct electric current due to photoelectric effect. The main property of material from which the solar cell is manufactured is photoelectric conversion coefficient. For first generation GaAs this value was 25.1%, while for Si it was 24.7%.

Currently, the GaInP structure is used for solar cells production.

3.7   New Devices on GaAs

The manufacturing technology based on $A^{III}B^V$ semiconductor materials allows using a number of methods to improve the product parameters. Current researches in this field are aimed at the development of new materials, improvement of production processes, and the use of special features of carrier transport in semiconductors.

The study of $A^{III}B^V$ semiconductor materials, the structure of which is compatible with InP substrates, has been particularly active at the present time. Such materials include InGaAs and InGaP compounds. The interest in these materials is caused by their ability to change the band gap width, which allows creating optoelectronic devices operating at optimal wavelengths for communication systems.

Furthermore, these materials are promising for the manufacture of generation and amplification devices, since in many cases the maximum charge drift velocity in them is higher than in GaAs.

Development of new manufacturing processes makes it possible to create new types of microelectronic devices. An important place among such devices is occupied by permeable base transistors, the appearance of which was made possible due to the development of new methods of epitaxial film growth. In addition, as the devices become smaller, new technological methods for making submicron structures are becoming increasingly important.

Another promising way to improve the designed device parameters is to design new device structures that use nontraditional carrier transfer processes. The best known approach to the implementation of such properties is making a device smaller, thereby allowing us to take advantage of the effects of exceeding the saturation velocity or ballistic phenomena. Currently, the possibility of creating such transfer conditions also in vertical structure devices is being investigated. Another possibility of changing the carrier transfer process is to use the compression of electron trajectories. This is the principle on which the functioning of some devices with selectively doped heterostructures is based. Here are two examples of such devices.

1. *Devices operating on the basis of quantum dimensional effect:* Improvement of $A^{III}B^V$ growing technology allowed creating fundamentally new types of device structures. For example, combining the molecular beam epitaxy (MBE) methods with classical organometallic vapor phase epitaxy helps to grow heterostructures where the layer width is comparable to de Broglie wavelength of electrons in the conduction band. The motion of electrons in such structures is quantized by discrete states. The use of such structures with the potential wells has made it possible to significantly improve the parameters of industrial semiconductor lasers.

2. *Resonant tunneling diode:* The principle of this diode operation is also based on the tunnel effect due to quantum-mechanical phenomena.

Conventional diodes with an increase in forward voltage monotonically increase the passing current. In the tunnel diode quantum-mechanical tunneling of electrons adds a hump in the current-voltage characteristic, in this case, because of the high degree of doping of p- and n-areas the breakdown voltage is reduced almost to zero.

Resonant tunneling diodes are widely used as generators and high-frequency switches in modern electronic equipment.

The main advantages of $A^{III}B^V$ materials become obvious from the earlier analysis.

Therefore, serious investments were made in recent years into the expansion of research and development of GaAs-based products, as well as into the organization of mass production of new devices.

Already now we can say that microwave range ICs based on GaAs are widely used in the designed military systems and find a relatively large market in the civilian satellite television receiver systems, as well as in cellular communications. Of course, in many respects microwave ICs based on GaAs compare favorably with silicon ICs. Excellent performance of GaAs devices in microwave range and the possibility of manufacturing semi-insulating substrates required for the manufacture of these ICs ensure that GaAs circuits have some technological advantages over silicon ICs. In particular, certain military and commercial tasks exist, the solution of which is possible only with the use of these ICs. A typical example is the use of these devices as active elements of modern radars and phased arrays for ground-based part of MILDS.

We can say that digital integrated circuits on GaAs compete successfully with silicon ICs in the areas that require very high-speed performance.

3.8   Condition and Prospects of Development of Monolithic Microwave Integrated Circuits

3.8.1   Main Spheres of Usage of Monolithic Microwave Integrated Circuits

Monolithic microwave integrated circuits (MMIC) are widely used not only in military but also in civil engineering, especially in mobile telephony, due to the development of high-speed broadband data transmission systems where the reduction of weight and size of products are constantly required [11, 12].

The prototype of the modern MMIC was the idea that was voiced and patented in 1961 (U.S. patent number 2981877) by Robert Noyce, who worked at that time in Fairchild Semiconductor. In our lectures for Belarusian, Chinese, Bulgarian, and Indian students we often say that he created the first microcircuit with a planar structure using silicon as a substrate.

Noyce interconnected the planar diffusion bipolar silicon transistors and resistors using thin aluminum strips lying on passivation silicon oxide. To produce these strips he used the traditional process for his time comprising the metal layer deposition and photolithography followed by chemical etching of metal. Later, MMIC were manufactured on the basis of GaAs semiconductor chips, which to this day occupies a leading position as a material for the production of MMIC (more than 80% of monolithic microcircuits are made on GaAs substrates and ternary compounds on its base: A1GaAs and InGaAs).

Due to the high mobility of electrons, MMICs on GaAs may be used in frequency ranges from 1 to 100 GHz. Historically, the first applications of MMIC were military and civilian radars, satellite communications and navigation systems, communication equipment, and so on. It can be argued that if at the stage of MMIC formation their development was driven by the need to improve the reliability of military equipment, now this driving force is mostly the ever-growing global market requirements concerning reduced product size (e.g., mobile phones, navigation equipment, and so on).

MMICs are often used in the microwave range in applications where small size and high reliability are required. Examples of such systems based on MMIC may be transmitters and receivers of communication systems, phased-array antennas (PAAs), sensors operating at microwave frequencies, and so on.

Here we could also mention the transceiver module (TM), which is part of active PAA. The TM usually includes transmit and receive paths, operating mode switch, and the phase shifter. Very popular are solid TMs using GaAs FETs and solid-state microwave ICs.

Recently, MMICs are widely used in cellular and satellite telephony, as well as in GPS devices. The advances of MMIC technology became widely used in the manufacture of discrete components, which is most related to bipolar heterotransistors made by MMIC technology. These transistors are in great demand among manufacturers of professional communications and telecommunications equipment.

A characteristic feature of MMICs is their low integration compared to digital ICs.

A modern MMIC is a functionally finished device that does not require any additional external setting units or trimmers.

The most typical MMIC are low-noise amplifiers, mixers, power amplifiers, modulators, and so on. Using an MMIC, it is easy to build a higher level device, such as a receiver. In this case, since the receiver will contain only a few components (MMIC does not require any external components) and due to the fact that MMIC has a relatively high MTBF, then the reliability of such a receiver will be very high, which cannot be achieved when making it from discrete components having the same characteristics. There are also devices implemented entirely as a single MMIC. A good example is a single-chip MMIC receiver. Obviously, the scope application of such chips is very limited, especially when taking into account that the MMIC is a finished device, which does not require any external trimmers, and that such a receiver cannot be adapted for use, for example, in a different frequency range. On the other hand, if some external trimmers are foreseen, then using such MMIC will not give any advantage.

Of course, mass production of such microcircuits is out of question, and the main areas of application of such MMICs are space and military equipment, where the device reliability is put first rather than its price. Due to individual production of this type of microcircuit, it is impossible to use the well-established method of average MTBF statistical forecasting in this case. In turn, this leads to another problem associated with forecasting of individual microcircuit reliability.

Within the period between 2010 and 2013, the MMIC sales almost doubled. In addition, there is an obvious tendency of MMIC sales to increase in the commercial rather than military sector. Perhaps the main reason for this is intensively developing market for wireless communications, navigation, and telecommunication systems. In this regard, the majority of manufacturers of semiconductor wafers (Vitesse, Kopin, TriQuent, Conexant, M/A-COM, RF Micro Devices, ATMI) significantly expanded production of gallium arsenide wafers.

3.8.2 Main Materials for MMIC Production

Obviously, as the MMIC technology developed, their design was also improved. Around the time when the first heterojunction bipolar transistor was made by MMIC technology, the interest arose in the use of other materials for production of monolithic ICs. This interest was primarily caused by the need to create a microcircuit operating at higher frequencies and using A_3B_5 semiconductor compounds as raw materials. InGaAs compound was used for collector-base, and indium phosphide InP was used for emitter-base. The use of indium phosphide made it possible to improve the frequency parameters and increase the collector breakdown voltage. Since the forbidden band of InP is wider than that of In0.53Ga0.47As (1.35 and

0.75 eV, respectively), the heterojunction collector breakdown voltage will not be less than 6 V.

At the time of this writing, there is a large variety of combinations of emitter, base, and collector materials, and the selection of MMIC materials in terms of optimal design and manufacturing technologies deserves a separate analysis, so here we will only note that the most widely used are *n-p-n* transistor heterostructures of InA1As-InGaAs-InP and InP-InGaAs-InP types.

The use of such materials in combination with reduced base thickness, which was made possible by the intensive development of molecular beam epitaxy technology, allows the transistor to operate at frequency limits up to 250 GHz and above.

Another popular material for the manufacture of MMIC is gallium nitride (GaN), the characteristics of which will now be considered in more detail. Microwave frequency devices allow us to achieve high values of output power density. For example, Cree Inc. developed a GaN MESFET with 0.55 μm gate length and 0.25 μm gate width, the output power of which in continuous mode at 4 GHz was 8W.

Accordingly, the output power density of that transistor was 33 W/mm, and the source-drain operating voltage was 120V, the maximum current density in the channel reaching 1.2 A/mm.

3.8.3   MMIC Active Elements and Their Reliability

The main active element of MMICs since their introduction until the present time is a metal Schottky-gate field effect transistor (MESFET). However, the increasingly strict demands of the military make it impossible to use them in some applications. This is caused by the fact that MESFET speed is very difficult to improve by reducing the gate length. Therefore, transistors with high electron mobility and pseudomorphic transistors (HEMT/PHEMT) and bipolar heterotransistors (HBT) became popular in the recent years. A diagram showing the frequency, at which the respective devices can be theoretically used is presented in Figure 3.8 [13].

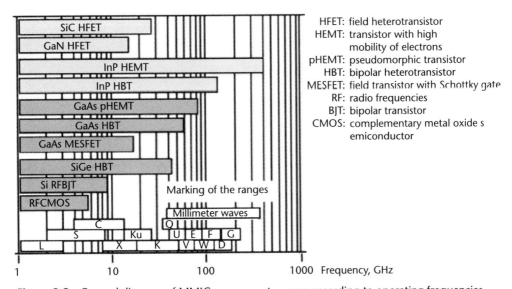

Figure 3.8   General diagram of MMIC components usage according to operating frequencies.

Let us consider the design features of these MMIC active elements. The first gallium arsenide MESFET was made in 1963. This became possible due to the process of controlled growth of high-purity thin films on a gallium arsenide semiconductor developed by GEC Marconi Material Technology.

Basic MESFET structure is shown in Figure 3.9.

The base material is gallium arsenide substrate. The buffer layer is epitaxially grown on a semi-insulating substrate and serves to isolate the substrate from defects in the working part of the transistor. The channel is a thin, lightly doped conductive layer of semiconductor material epitaxially grown on the buffer layer. High-alloy areas, as shown in Figure 3.9, are needed to ensure low ohmic resistance of the transistor contacts.

The equivalent circuit and a typical current-voltage characteristic of another active element widely used in MMIC—MESFET—is shown in Figure 3.10.

To ensure transistor high speed, it is necessary to try to minimize the gate length, which, however, is limited by production capabilities. In addition, it must be remembered that in order for channel current to be controlled efficiently, the channel length L must be greater than its depth a (i.e, $L/a > 1$).

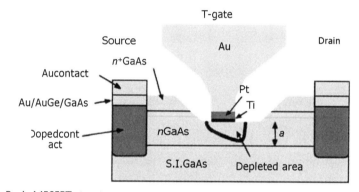

Figure 3.9 Basic MESFET structure.

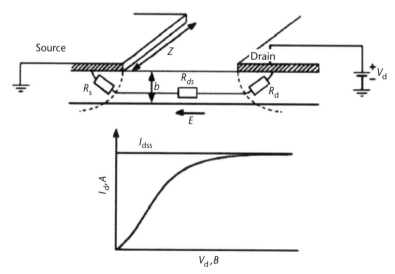

Figure 3.10 Equivalent electric diagram and typical volt-ampere characteristics of MESFET.

Therefore, in most MESFETs the channel depth is 0.05–0.3 μm. This means that for a sufficiently large current, the carrier concentration in the channel must be very high. The small size of transistors leads to a decrease in reliability. This is due to the small cross-section of the gate area, resulting in increased current density. It is usual for the power transistors, in which the main failure mechanism is the electrons migration. Gold is typically used to reduce gate resistance. Since gold creates traps in gallium arsenide, which effectively reduces the carrier concentration and hence the current through the transistor, barrier metal such as platinum should be used. Due to the fact that the channel depth is very small, any diffusion of the gate metal into gallium arsenide leads to significant changes in the current flowing through the channel and reduces the cutoff voltage of the transistor. The small distance between the gate and drain create strong electrical fields that can lead to the generation of electron avalanche.

These hot electrons can then become a trap on GaAs surface or in passivating material, which is usually placed on the surface of the transistor.

These factors of FETs low reliability are mainly technology related. In small-signal devices, degradation of ohmic contacts or mutual diffusion of gate metal and GaAs leads to shift of their main characteristics I_d, g_m and V_p [13].

Although power MESFETs also suffer from parametric degradation, the most common are fatal (sudden) failures. However, recent advances in GaAs devices production technology and ensuring of operation within safe mode limits reduce the number of failures. For power amplifiers, FETs should be designed to maximize the peak output power. For engineers, this means high source-to-drain voltage and high drain current.

Unfortunately, both of these parameters theoretically cannot be maximized simultaneously. Therefore, heterojunction bipolar transistors are used by now for MMIC microwave power amplifiers.

To increase the drain current, a high concentration of carriers or a large gate width is required. However, it must be remembered that the channel depth cannot be substantially increased, as this reduces the frequency range of a device. The carrier concentration cannot be increased either without reducing the gate-to-drain breakdown voltage that needs to be maximized in order to increase the allowable source-to-drain voltage. Therefore, the only alternative is to increase the gate width. However, the long linear elements in the microwave devices design are not elements with uniform potential over the entire length. The basic rule is that the line should be less than a tenth of the wavelength, only in this case it can be considered a homogeneous element. For gallium arsenide in the X-band (8–12 GHz), the maximum usable gate length does not exceed 1 mm.

If higher current is required, MMIC designers use parallel connection of several gates. However, close spacing of parallel gates increases the local temperature of corresponding MMIC area, which also adversely affects the product reliability, since gallium arsenide is a poor heat conductor if compared to silicon.

MESFET Manufacturing Technology

Typical MESFET manufacturing workflow by ion implantation is shown in Figure 3.11.

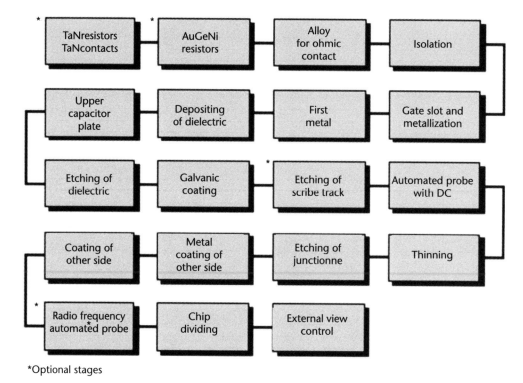

*Optional stages

Figure 3.11   Typical MESFET manufacturing workflow.

Traditionally the first step is production of thin-film resistors. Resistor metal (AuGeNi) is evaporated; then TaN is applied. AuGeNi is usually used for making low-ohmic resistors, while TaN is used for high-ohmic resistors. Mainprocess stages are shown in Figure 3.12.

The second workflow stage is the gate isolation and formation. Due to ion implantation of boron, for example, the GaAs conductive layer is deactivated and the necessary isolation regions are formed. After this step, metal coating is performed and air bridges are formed. At the finishing stages the vias are formed, and the reverse side of the substrate is chemically treated.

3.8.4   Advanced MMIC Design and Technology Solutions

Let us consider the design and technology solutions of the main active elements, which are widely used in modern MMICs.

High Electron Mobility and Pseudomorphic Transistors

As mentioned earlier, in recent years in space and military applications, where low noise and high gain are required, high electron mobility (HEMT) transistors and pseudomorphic (PHEMT) transistors are becoming more widely used.

Both of these transistors are field effect transistors; that is why the basic principles of their operation are quite similar. As we show next, the main difference between HEMT and FET is the layer epitaxial structure.

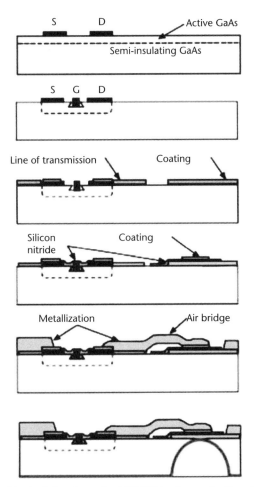

Figure 3.12   Sketches of typical structures for MESFET basic production stages.

Let us consider the special features of a typical HEMT/PHEMT structure.

Epitaxial structure of base high electron mobility transistor (HEMT) is shown in Figure 3.13(a) of pseudomorphic transistor in Figure 3.13(b).

Similar as MESFET, this structure is grown on a semi-insulating GaAs substrate using molecular beam epitaxy (MBE) or more common organometallic vapor phase.

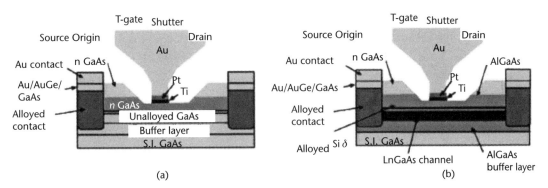

Figure 3.13   Basic structure of (a) HEMT and (b) PHEMT.

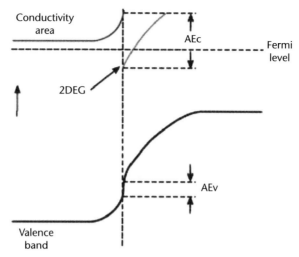

Figure 3.14 Energetic diagrams of HEMT channel.

The buffer layer, usually also gallium arsenide, that is epitaxially grown on the substrate serves to isolate the defects and is designed to create a smooth surface on which then the active layer of the transistor is grown.

The channel energy characteristics corresponding to the standard transistor structure are shown next. In an ideal system, all the conductivity electrons are located in this channel. The most important component in the channel energy structure is two-dimensional electron gas (2DEG in Figure 3.14), resulting from varying width of intervals between energy levels.

The reliability of HEMT and PHEMT is significantly affected by the epitaxial structure parameters, production process, and device geometry. The main known failure mechanisms are [14]:

- Gate sinking due to mutual diffusion of gate metal into the semiconductor, leading to gain reduction;
- Source-to-drain junction degradation as a result of the degradation of the ohmic contacts alloyed area, as well as increased source-drain resistance (RDS);
- Surface damage under the influence of hot electrons;
- Increased sensitivity to atmospheric oxygen, which leads to surface reactions resulting from so-called traps;
- Hydrogen poisoning, which reduces the gain and cut-off voltage;
- High humidity, which can cause gate and drain short-circuiting.

A visual representation of the gate sinking effect on the MESFET and HEMT current-voltage characteristics is shown in Figure 3.15. The arrows show the direction of transistor characteristics shift. As we can see from the figure, such serious CVC shift can lead not only to the output characteristics of the device going outside the tolerance limits, but in some cases to fatal failure of the active element (MESFET, HEMT, and so on).

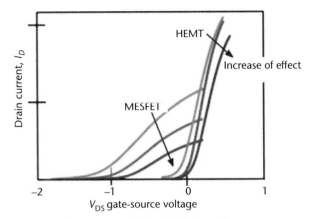

Figure 3.15   Effect of gate sinking on MESFET and HEMT CVCs.

Next we provide a brief review of the main features of HEMT/PHEMT manufacturing technology (Figure 3.16).

The first stage in the production process is careful selection of substrates with the desired characteristics. There are, of course, minor differences between the HEMT and PHEMT manufacturing technologies, but in this case we will not consider them, speaking only about the basic process, which is the same for both devices.

The second stage is formation of the active channel body and insulator implantation; then the ohmic junctions are formed, and finally recessed gate and gate-to-metal area are formed.

After that, source and contacts are etched, air bridges and vias are formed, and the reverse side of the substrate is chemically treated.

Heterojunction Bipolar Transistors

Heterojunction Bipolar Transistors (HBTs) are widely used in digital and analog MICs at operating frequencies above the standard Ku range. Due to their structure, they provide more rapid switching, mainly due to the reduced base resistance and extremely low capacitance between the collector and the substrate. The prices of such transistors are relatively low, due to the less demanding technological process in comparison with, for example, FETs. In addition to high performance, HBTs provide higher maximum permissible voltage as compared to FET. These transistors also have good linearity, they have low phase noise, and they are easy to match.

Figure 3.17 shows a typical vertical HBT structure. The substrate in this case is a gallium arsenide wafer. The epitaxial layers may be grown by various methods, such as molecular beam epitaxy.

Typical volt-ampere characteristics of the HBTs are shown in Figure 3.18 and do not require any special comments.

Let us briefly review the HBT operation principle. In contrast to the previously mentioned active MMIC devices, HBTs have vertical structure. Due to their design they have not only higher frequency than, for example, MESFET, but also they are more convenient to use in a variety of power amplifiers.

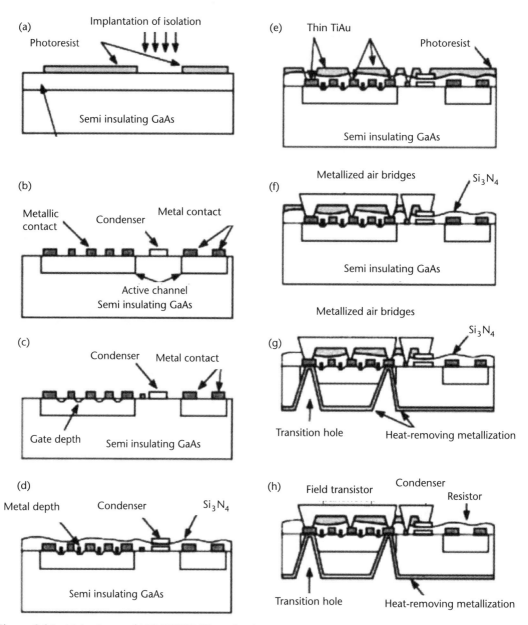

Figure 3.16 Main stages of HEMT/PHEMT production process.

As we can see from Figure 3.19, potential barrier of injected holes (ΔV_p) and electrons (ΔV_n) in emitter-to-base contact differ by spacing width between AlGaAs emitter and GaAs base.

This slight difference significantly affects the I_n/I_p ratio, where I_n is the current of the injected electrons from emitter to base, and I_p is the undesirable current of injected holes from base to emitter.

For gallium arsenide $\Delta Eg \approx 14.6kT$, respectively, $\exp(\Delta Eg/kT) \approx 22{,}106$, so it is possible to perform high base doping and low emitter doping without a significant reduction in current gain. In practice, the doping of the base is usually performed

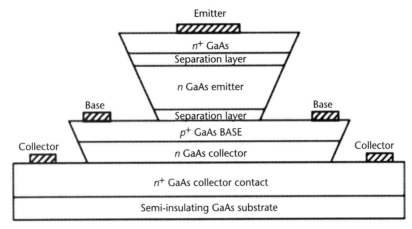

Figure 3.17   Basic structure of heterojunction bipolar transistor.

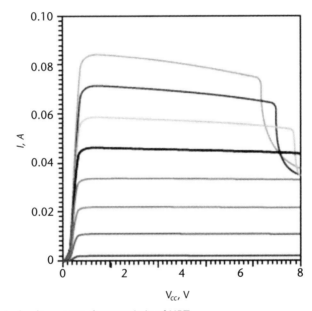

Figure 3.18   Typical volt-ampere characteristic of HBT.

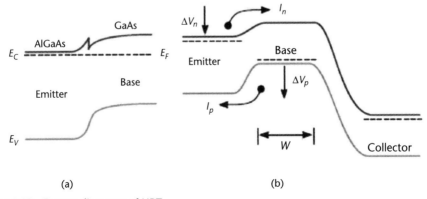

Figure 3.19   Energy diagrams of HBT.

so that the current transfer ratio of the transistor would be approximately 100. Low doping of the emitter reduces the emitter-to-base contact capacity, allowing the transistor to operate at higher frequencies.

As a result of a classical physical aging process, the HBT reliability may deteriorate due to the following factors:

- Reduced current gain and increased base-to-emitter voltage at high emitter currents.
- Increase in contact resistance caused by degradation of junction between the emitter ohmic contacts (metallization) and the emitter semiconductor area. To solve this problem, auxiliary contact layer of InGaAs is commonly used.
- Increasing ofchip defects in emitter-to-base contacts area.
- Current gain drift (reduction) and increasing of base-to-emitter voltage for a particular collector current caused by oxidizing of mesa structure in the emitter-to-base contact area.

Typical HBT manufacturing technology consists of numerous etching stages in order to open the relevant areas and to form the electrical contacts on each layer. Finally, the device is isolated and the necessary conducting paths are formed on it. The basic process stages are shown in Figure 3.20.

3.9   Basic Areas and Peculiarities of GaAs MMIC Application

This section is intended primarily for circuitry engineers, since the famous classic book written more than 40 years ago was titled *The Art of Circuit Design* for a reason, and this branch of applied science is still considered as a kind of engineering art—too many factors are to be taken into account when designing these microcircuits.

Since MMICs are widely used in satellite systems, they are required to be as small and lightweight as possible, highly reliable, and inexpensive [13, 14]. These microcircuits are used in the case where the parasitic reactivity in hybrid integrated circuits reduces the device quality below the maximum permissible level, so the main field of application of MMICs are devices operating in the microwave range. Examples of systems made with MMICs may be transmitters and receivers for communication systems, phased array antennas, for which small size and uniform circuit characteristics are essential, as well as circuits and sensors, ground-based mobile radars operating at high frequencies. The largest market share among the MMICs belongs to microwave receivers and transmitters, a simplified diagram of which is shown in Figure 3.21.

In these circuits the phase shifter (P) can be placed both directly in the local generator (LO) and at the system input/output. The phase shifter is required for the system to be able to perform the function in the way as if each circuit was connected to the PAA radiating element. For other applications the circuit is not modified, except for the phase shifter removal. A specific sample layout of such 30 GHz monolithic receiver is shown in Figure 3.22.

Each functional block of the designed system is usually made on an individual chip, which allows application-specific optimization of material and device

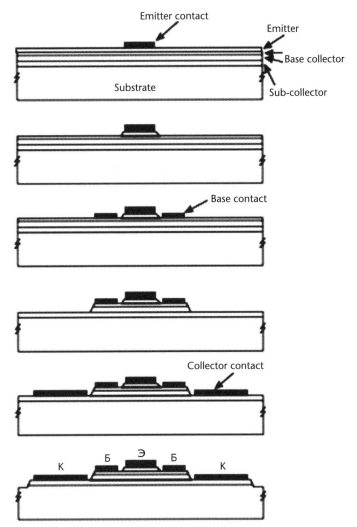

Figure 3.20 Sequence of HBT structure formation.

parameters. Regardless of circuit interconnection level, the reliability of the designed system depends on the reliability of components, which becomes evident when looking at the receiver circuit shown in Figure 3.21(a). The input RF signal has a very low power level and in some cases can be completely blanked out by noises. The low-noise amplifier (LNA) amplifies the received signal, introducing at the same time its own low noise. If LNA gain is high enough, its noise contribution to the system noise is very small, so the noise generated by the subsequent circuits is divided by the LNA gain ratio. This means that the LNA gain and noise ratio determine the noise characteristics of the entire receiver. If the receiver has poor noise parameters, it will not be able to receive a weak signal.

This is the art of being a circuit design engineer—only he or she can make a decision that should be optimal (a trade-off).

The received signal is passed through a narrowband filter and a mixer. A local generator generates a signal of a certain frequency, which is also fed to the mixer. The mixer combines the two signals using a nonlinear device such as MESFET or a

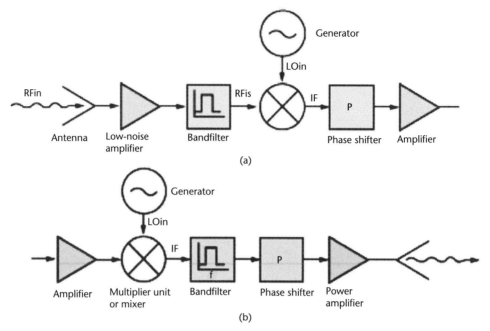

Figure 3.21   Typical structural diagrams of (a) microwave receiver and (b) microwave transmitter.

diode and generates a signal at an intermediate frequency (IF): (fRF-fLG) or (fLG-fRF), as well as intermediate frequency, input radio frequency (RF), and local generator frequency harmonics. To isolate the desired intermediate frequency component they must be filtered out. The conversion efficiency of the mixer usually depends on the generator power. Furthermore, the change in LG frequency leads to the IF shift, which may cause increased attenuation of signal in narrowband filters that are part of the mixer. When the system controls a phased array antenna, the direction and shape of the base signal emitted or received by the antenna is dependent on the phase shift and the power level of each transmitter or receiver. The relative phase of each radiating element is set using the phase shifter. Thus, if the phase shift of the signal passing through the circuit differs from the intended one, the quality of the entire antenna deteriorates. This means that the change in parameters of one of the components may cause the entire system failure.

As it was mentioned earlier, a phase shifter, a local generator, and a mixer are main components of transmitters and receivers of modern military and aerospace electronic systems. The real differences between the two systems are in amplifiers. If the LNA is used as a receiver, it must be able to amplify a weak signal to a level sufficient to operate the mixer, and to generate as low own noise as possible in order to improve the noise immunity of the system. In case of transmitter, the main requirement is the transmitted power and the circuit efficiency. Therefore, the power amplifier must ensure the signal amplification to a desired level.

Low noise amplifiers are commonly used for amplifying radio frequency signals. In almost all military and commercial systems, this is done using MESFET and HEMT transconductivity and HBT current amplification. The most correct operation of the amplifier is at low power levels. Unfortunately, when increasing

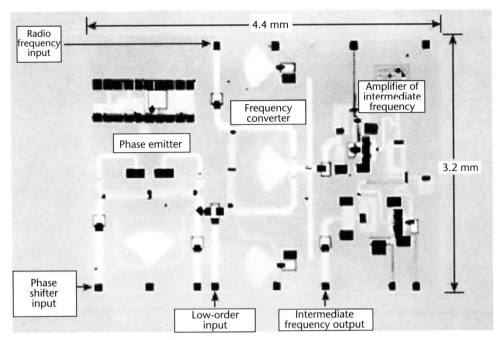

Figure 3.22   30 MHz receiver layout.

the power level, any such power amplifier becomes nonlinear. When operating in the nonlinear region, the output power will always be lower than sum of the input power and the gain ratio of the amplifier in the linear region. Figure 3.23 shows typical characteristics of such standard amplifier.

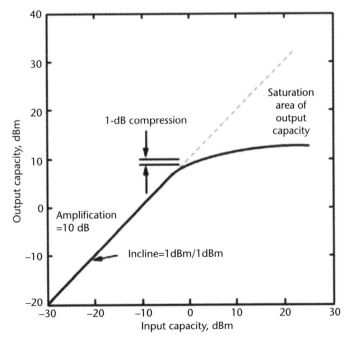

Figure 3.23   Typical transfer characteristic of the amplifier.

Circuitry engineers know that the point at which the output power decreases by 1 dB relative to the linear extrapolated values is called the 1-dB compression point. It should be borne in mind that it is also a criterion for distinguishing between low-power and high-power transistors, as the transistor in a number of applications can be considered simply as an unmatched amplifier.

This difference should be taken into account when studying the mechanisms of failure of space electronics. Shifting point selection by the circuitry engineer working on a military or space project is essential for ensuring the required parameters of the amplifier. Depending on the operating mode of the transistor, output waveform, and, consequently, the efficiency of different classes of amplifiers, we distinguish between class A, B, and C amplifiers. Class A amplifiers are linear, but their efficiency is too low, while Class C amplifiers are nonlinear but have the highest efficiency.

We can also distinguish the following types of MIC circuits: power amplifiers (including LNA), mixers, generators, and so on [3, 14].

First, let us discuss the main features of the power amplifiers design. Power amplifiers have to handle high input and output power. The maximum input signal voltage is limited by the breakdown voltage of the transistor. The flow of electric current through each transistor is limited by the emitter-to-gate resistance. Ohmic losses are converted into heat, resulting in heating and reduced reliability of the device. To increase the device maximum current in high-power transistors, the lay-out designers connect multiple gates or emitters in parallel.

Although such parallel connection increases the total gate width or emitter surface area and reduces the resistance, at the same time it makes the problem of matching the transistor input impedance with the output impedance of the pre-ceding stages more complex. In addition, to ensure heat dissipation the so-called (and useless from the point of view of functional density) free space zone around the transistor is created, which increases the device size. To ensure effective heat dissipation from the transistors of high-power amplifiers, the substrate is usually placed on metal or diamond base. Efficiency of such power amplifiers is their criti-cal parameter. To analyze the amplifiers performance, measurement and analysis of their S-parameters is typically performed.

As is known, the transistors have linearity only at low power levels; as the power increases, their nonlinearity increases greatly as well. In turn, the high-power transistors nonlinearity creates so-called intermodulation distortions, divisible by the input frequency: 2fRF, 3fRF, and so on. The occurrence of these frequencies in matched circuits can lead to distortions, parasitic oscillations, reduced efficiency, and so on, so that they would require special protective measures application.

These intermodulation distortions are usually defined in circuit design compu-tations as a ratio of signal power at the distortions frequency to the useful signal power and are typically indicated in decibels.

In addition to the problems associated with overheating, power amplifiers have failure mechanisms as hot electron traps, electromigration, and diffusion of metal. As we can see, the developers of these microcircuits have enough problems, and consumers (military and NASA officials) are also aware of these problems and of the need to ensure that their projects will be free from such problems.

Low noise amplifiers (LNA) are used to amplify received signals in the receivers and are designed to be used at low power levels. Therefore, the previously mentioned temperature issues, as well as high voltages and currents that affect the amplifier reliability, are not inherent in the LNA.

The most important LNA performance quality criterion is LNA noise. Since HEMT and PHEMT have the lowest noise ratios, they are used in almost all LNA. To reduce the noise ratio, the designers commonly use small gate length, and low gate-to-source parasitic resistance, which is also the art of design.

For spaceborne ECB of 2010–2012, typical values of gate length were 0.1–0.25 μm. Uncertainty factors in these transistors are gate metal sinking and ohmic contacts diffusion, resulting from very short gate length, and, consequently, very thin channels. Again, as the professionals may know, in order to reduce the entire designed system noise rate, it is important to reduce losses in the circuit, especially before the LNA first stage, which includes a transmission line from the antenna to the device. In addition to reducing losses in the circuit, the noise can be reduced when using the amplifier at low temperatures, currents, and voltages. Finally, the LNA noise ratio depends on the degree of circuits matching designed with a view to minimizing the noise and maximizing the gain ratio. From this viewpoint HEMT is an optimal device.

Next, let us consider the *signal mixers* which convert the input signal of one frequency to a signal of other frequency, which is necessary to filter the phase shift and for other data processing operations.

For example, according to SOW, the system must receive data at W-range (74–110 GHz), but filters for W-range have high losses and low Q-factor, which lead to receiver noise parameters degradation. Therefore, the frequency of the received signal must be shifted to a region where high-Q filters with low input loss may be used. Ideally, this should be carried out by developers without reducing the input signal amplitude or adding extra noise.

Modern mixers are made either on diodes or on one of the transistors. Let us view a simple mixer diode shown in Figure 3.24. In this circuit, only two signals

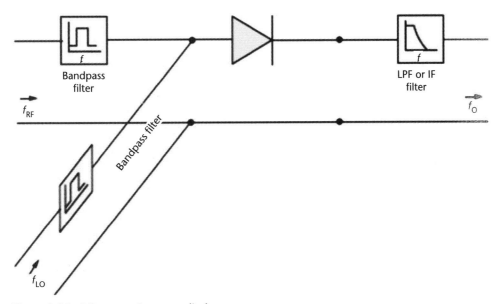

Figure 3.24   Mixer on microwave diode.

are transmitted through the external diode terminals—the local generator and the RF signal.

Typically, the output frequency required by the customer (fRF-fLG) is the so-called intermediate frequency, which is classic for wireless devices and is known to students from the second year of study. The main indicator for the quality of mixers is the IF/RF power ratio, which is called conversion loss and is indicated in decibels. However, several things can make a significant contribution to the increase in conversion losses.

First, it can be a bad impedance matching at the RF and IF ports. (From the customer's point of view, the designer is the one to blame.) Second, this is a real voltage-current characteristic of diode. Depending on the desired parameters mixers may be formed on a single diode or FET, or several, up to eight or more diodes. More sophisticated devices use balanced circuits in order to neutralize the undesirable frequency components and to solve the problem of eliminating the noise created by changes in amplitude of the local generator.

A disadvantage of conventional schemes and designs with multiple mixer diodes or FETs is the need to substantially increase the power output of the local generator, which is quite difficult to achieve at high frequencies. In turn, the mixer reliability issue related to the generation of the undesirable harmonics, which can lead to spurious generation of other IC circuits, signal distortion, occurrence of $1/f$ noise, and other undesirable effects—and all this is on the conscience of a circuitry engineer!

And finally, let us consider the last class of MMIC—*generators*—that generate periodic high-frequency signals and are widely used in modulators, superheterodynes, and phase-locked loop circuits (PLL) of the spacecraft onboard systems.

As is known to the circuit designers, any generator may be obtained from any amplifier by introducing a positive feedback. Students in institutes, universities, and technical colleges are taught that classic generators usually are based on the LNA with a feedback loop, which introduces a delay divisible by 2π. In addition, a very common type of generator is a voltage controlled oscillator (VCO). The generator is designed to ensure the required capacitive and current loads. Critical parameters of the generators are long-term frequency instability, phase noise level, and output impedance. The phase noise of the generator is a short-term instability of the generated RF signal, because this noise cannot be eliminated completely. When using the generators in radars, as well as in digital telecommunications systems, a certain maximum phase noise must be ensured; otherwise, the phase noise can cause a system error, and in case of data transmission—their distortion.

This noise can be generated by a variety of both known and previously unknown physical mechanisms. First, the cause of noise may be the kinetic energy of electrons, which is proportional to the operating temperature. This type of noise is commonly referred to as thermal noise. The thermal noise has a very wide band, so it is also often referred to as white noise. The second type of noise proportional to $1/f$ is the flicker noise, which occurs in active solid-state devices as a result of processes of generation and recombination of major carriers on the semiconductor surface. A typical spectrum of this noise is shown in Figure 3.25.

To minimize the phase noise it is necessary to use high frequency resonators, as well as transistors with low flicker noise. The use of high-Q resonators in MIC

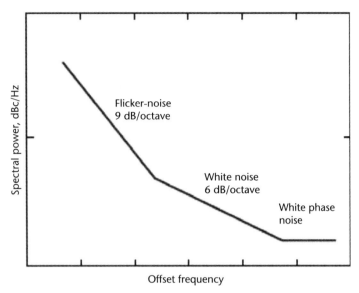

Figure 3.25 Typical noise spectrum of a generator.

is quite difficult because the thin-film elements on GaAs substrates have high conduction losses.

Among all transistors, the HBTs have the lowest flicker noise. Therefore, they are most frequently used in generators, although the temperature effects on them may result in the transistor parameters drift leading to frequency shift or generation disruption. Temperature compensation can be built using varactors or other elements.

Constantly growing demands on the volume of transmitted information with simultaneous reduction of weight and size of space and special purpose devices ensure further development of both MMICs design and technology of their manufacture.

In which direction will integrated technology of RF and microwave devices develop in the near future? The obvious global development trend will be broadband and ultra-broadband microwave devices. As for the design, it is primarily the improvement of structural and technological features of MMIC, further improvement of microcircuit manufacturing on silicon carbide (SiC) and gallium nitride (GaN) substrates, and modernization of serial microcircuit manufacturing technology on sapphire substrates.

Another important task for manufacturing of military and space microcircuits of this class is to create a device with an extremely high power density: more than 1W per 1 mm gate length. Continuous improvement of MMIC design and technology and the emergence of new active microcircuit elements creates the need for appropriate modification of classic reliability models, including new failure mechanisms associated with the use of new materials and technologies. In addition, despite the well-established statistical methods of reliability forecasting applicable to serial microcircuits, the methods for evaluation of individual reliability required for selection of individual MMIC samples used in military and space equipment are not well developed yet. Among these methods we can mention various methods of spectroscopy, with or without backlight, as well as a number of other methods [15, 16].

However, many of the methods of forecasting individual MMIC reliability are destructive, which is inadmissible given the high cost of the chip, while others do not provide for the required accuracy. Such a situation poses a difficult task: to evaluate the reliability of a particular instance of a device only by the results of nondestructive measurement of any of its electro-physical parameters. Concerning the problem of selection of specific informative parameters, we can say that it is basically solved; regarding the availability of reliable mathematical models on their basis, it is impossible to say the same. [17] Therefore, this task remains very relevant at the moment and also needs to be addressed.

3.10 Main Technical Parameters of Overseas-Made Gallium Nitride Microcircuits of Active Phased Array Antenna Transceiver Modules

The use of the active phased array (APAA) radar systems in modern radars and electronic warfare systems is constantly increasing. Several countries have already successfully demonstrated unique capabilities of APAA in aircraft armament systems, such as F/A-18E/F (United States), Rafale (France), and others [17–19]. Similar work is carried out in Russia, where, as it is planned, the APAA will become part of the fifth generation jet fighter radar [20]. Key elements of high frequency range APAA are microwave transceiver modules (MTM) based on the monolithic integrated circuit (MIC). The market forecasts for microwave modules for military and space equipment show that in the coming years, MTM can become one of the most popular and demanded products of microwave technology. It is estimated that MTM sales in 2012 have already exceeded $5.3 billion, and the annual APAA systems sales for the period 2014–2019 are predicted to grow from $6 billion (in 2013) to $13 billion. In addition, this airborne APAA market segment will 100 times exceed ground-based and naval APAA market segment [21]. Today, therefore, providing transceiver modules for APAA radars is a priority task for the global microwave industry.

Until recent time, MTM were based on GaAs microwave circuits. As part of two major gallium nitride research programs conducted in the United States and Europe (WBGSTI and KORRIGAN), intensive development of the next generation MTM based on this material is carried out. In the United States, in addition to demonstration and reliability testing, GaN modules are supplied for more than five years for space and military equipment, including the radars for the well-known strategic missile defense system [22, 23]. To draw level with the United States, the European companies are also taking active measures to develop their own GaN microcircuit technology for MTM [24, 25]. Unfortunately, this activity is not observed in Russia.

In this view, below we will discuss in more detail the GaN MIC components of MTM by European companies.

Of course, the power of MTM on gallium nitride is much higher than that of GaAs devices. Therefore, APAA implemented on the basis of such MTM have either a larger search envelope or longer target tracking range or, all other things being equal, a much smaller aperture [26]. In addition, GaN MTMs are more resistant to high temperatures and have higher efficiency. The MTM consists of several

MICs, including low-noise amplifiers, pre-amplifiers, high-power output amplifiers, switches, and others. According to Thales (France), the MTMs and MICs included in these account for about half the cost of the entire APAA [11, 14]. Let us consider here a relatively old GaN design of European MTM MIC components.

As we have discussed, GaN MIC amplifiers are key components of MTM. The devices previously made by European companies are of major interest. A set of microstrip MICX-band amplifiers for MTM, which includes a pre-amplifier, power amplifier, and low noise amplifier, was developed jointly by military electronics unit of EADS Deutschland GmbH (Ulm) and the Fraunhofer Institute of Applied Physics of the solid body (Freiburg, Germany) [27]. These GaN MICs were made by Fraunhofer Institute experts, while their design and performance measurements were conducted by EADS military electronics unit. AlGa/GaN HEMT structures were fabricated on silicon carbide substrates with a diameter of 75 mm by metal-organic chemical vapor deposition (MOCVD-method). The gate of 0.25 mm length and a field electrode of transistor were formed using electron-beam lithography. SiC substrate, after processing of upper surface, was thinned to 100 μm, and then through-holes were made from its reverse side.

Since the number of publications on this topic in open periodical press has been very limited since 2010, we know from the open sources that the first preamplifier stage is formed by a transistor with a gate width of 8×60 μm, the second stage—transistor with a gate width of 8×125 μm. On the input and output the amplifier is matched to 50 ohms impedance. Maximum output power was achieved in saturation mode during gain compression of 5 dB and has exceeded 38 dB·mW. But even at 1 dB compression this output power of the amplifier was enough for powering of one or two high-power amplifiers in the range of 8.5–14 GHz.

In each of two stages of high-power output amplifier, transistors were installed with equal gate widths of 8×125 μm: two transistors in the first stage and four in the second stage. The output adder of amplifier was optimized for maximum output power in the frequency range from 8.5 to 11 GHz.

When operating such a power amplifier in AB class, its maximum output power was equal to 20W, which corresponds to the transistor specific power of 5.7 W/mm. In this case, the gain ratio in the small signal mode is 18 dB, and the efficiency is 31%. In the 8.5–11 GHz frequency band, the gain ratio in the small signal mode is no less than 15 dB. In the entire frequency range from 8.75 to 11.5 GHz, the measured amplifier output power exceeded 14W.

The low-noise amplifier (LNA) from this series was formed by two stages based on the transistors of 8×30 μm size. Minimum measured noise ratio of a separate low-noise AlGaN/GaN HEMT did not exceed 0.8 dB at 10 GHz and the drain voltage (Uc) was 10V. The minimum LNA noise ratio was 1.45 dB, tand he saturation power was 24 dB·mW with the input power of 16 dB·mW. An important characteristic of any LNA is the maximum possible input power. The amplifier in question was stable up to the input power of 4 W. However, its output power in this case was substantially reduced due to the change of the first stage transistor shift.

MTM module has been assembled on the basis of three designed GaN MIC amplifiers using industrial technology of low-temperature co-fired ceramic (LTCC). The transmitter unit was made on the basis of a pre-amplifier, two parallel-connected power amplifiers, and a circulator. Amplifiers were mounted on CuMo heat sink.

The receiver unit consisted of a low noise amplifier and limiter, and GaAs switch was used for channels switching. Pulse-type power supply circuits of high-power GaN MIC for voltage of 30V; LNA shift circuit and a control switch circuit were also mounted on the multilayer LTCC substrate.

The results showed foreign customers the possibility of implementing GaN APAA transceiver modules with over 20W transmitting output power and receiver channel noise ratio less than 3 dB.

Typical representatives of this class microwave product may also include GaN MICX-band amplifier with 20W output power designed to order of the military department for the advanced APAA radar system [28].

This amplifier was developed jointly by the SELEX Sistemi Integrati and Consorzio OPTEL, as well as the University of Rome and the Polytechnic Institute of Turin (Italy). MIC with microstrip structure was implemented on GaN HEMT by SELEX Sistemi Integrati. The epitaxial structures of GaN/AlGaN/GaN were grown on semi-insulating SiC substrates either by MOCVD or molecular beam epitaxy (MBE) method. The transistors gate of 0.5 μm in length was formed by stepper photolithography. Ohmic contacts were obtained by deposition of Ti/Al/Ni/Au structure on GaN/AlGaN epitaxial layer with subsequent annealing at high temperature. MIC was passivated by SiN film applied by chemical plasma deposition. After active devices manufacturing thin film NiCr resistors, galvanically formed inductances, transmission lines, and, where necessary, air bridges were applied. Then the wafer was thinned to 70 μm, and on its reverse side through holes were formed by plasma enhanced dry etching, the surface of which was electroplated with 10 μm layer of gold.

In terms of circuit design, this amplifier consists only of two transistor stages. The first transistor stage comprises four transistor cells; the second transistor stage has eight cells. The total gate width of transistor stages is 4 and 8 mm, respectively. The size of a single transistor cell gate is 10×100 μm. The amplifier was designed for the frequency range 8.5–10.5 GHz.

Features that are already used in onboard NASA amplifier equipment on plate measured in a pulse mode at a duty cycle of 1% pulse, pulse width 10 ms, 20V drain voltage in a class AB mode. The output power of the amplifier in the frequency band 8–10.5 GHz was 21–28.5W at 12.9–16.5 dB gain and 29–43% efficiency. At points 8.5 and 9 GHz the output reached saturation at 30 with 40% efficiency.

We should also note a number of similar products (known from the periodical press) for X-band—high-power AlGaN/GaN HEMT and MIC L/S band (1–4 GHz) and X band (8–12 GHz) amplifiers [29]. The transistors and MICs have been developed jointly by the Fraunhofer Institute of Applied Physics of the solid body (Freiburg), United Monolithic Semiconductors (Ulm), Germany, and NXP Semiconductors (Nijmegen, the Netherlands).

GaN/AlGaN heterostructures were grown on SiC substrate using the MOCVD method in the reactor allowing us to simultaneously process 12 wafers with a diameter of 75 mm. The length of the MIC transistor gates is 0.25 μm. Furthermore, the structure uses additional optimized field modulation electrodes. The breakdown voltage of HEMT on two electrodes exceeds 100V.

To assess the quality of this GaN HEMT, the characteristics of 21 transistor cells were measured on each of five wafers.

The measurement results are quite convincing. The output power of two-stage X-band amplifier exceeds 11W at a frequency of 8.56 GHz with efficiency of 40%, 17 dB amplification, and 3 dB compression. The output power of single-MIC amplifier at a frequency of 8.24 GHz at 20V is 6W, with efficiency of 55% and linear amplification of 12dB.

We should also mention the results obtained in 2 GHz range for high power HEMTs intended for civil and military mobile communication systems. Thus, with the average transistor voltage of 50V, the average power density was 10 W/mm with efficiency of 61.3% and linear amplification of 24.4 dB. At this frequency, transistors operated stably even at a higher voltage equal to 100V. In this case, the transistor power density reached 25 W/mm with efficiency of ≥60%.

We know from open information sources that gallium nitride amplifiers are also used in the most advanced electronic warfare systems. Let us consider the GaN MIC amplifiers with 2–6 GHz range for electronic warfare systems [30].

The amplifiers were designed by engineers of Polytechnic University of Madrid, INDRA Sistemas (Madrid, Spain), SELEX Sistemi Integrati (Rome), and QinetiQ (Malvern, UK).

It is interesting that, in the course of work, both companies developed two different designs of amplifiers in parallel: microstrip version based on SELEX technology and coplanar design based on QinetiQ technology. Because the GaN HEMT impedance is greater than GaAs HEMT, their matching in a wide frequency band 2–6 GHz selected for designing of amplifiers is simplified. Both amplifiers worked in class AB mode and had two amplifier stages. The required output power of each amplifier stage was achieved by adding the output signals of the four transistors with a gate width of 1 mm.

In general, the frequency characteristics of QinetiQ amplifier corresponded to the estimated values except 4–6 GHz band, where a slight deviation (1–2 dB) of output power (8W) from the estimated value was observed. Adequate conformance was obtained also for the gain ratio for the small signal mode (18 dB) and efficiency (20%) [30].

Another SELEX amplifier was designed based on the estimated output power of 10W at a voltage of 25V and current of 1.3A. The characteristics were measured at voltages of 20 and 25V and currents of 1.1 and 0.9A. The measured gain ratio in the small signal mode across the band of 2–6 Hz exceeded 15 dB. Saturation output power was 10W in continuous mode with efficiency of over 25% at 20V current and 4 GHz frequency. At this frequency in the pulse mode (pulse duration 20 ms; duty ratio 1.1%), the amplifier power was 17W [30].

GaN MIC switches should be mentioned as a separate independent microwave IC class. High operating voltage of GaN transistors gives them a significant advantage when operating as part of the transceiver modules switches. When using GaN transistors the number of series-connected transistor stages can be reduced compared to GaAs switches [31]. In some cases, the circulator can even be eliminated from the module circuits [32].

We should also mention the high-power microwave microstrip GaN HEMT switches for the ranges of 8–12 and 2–18 GHz [14]. The switches were jointly developed by Elettronica SpA, SELEX Sistemi Integrati, CNR-IFN, and Tor Vergata University (Rome, Italy). The switches were built using the single-pole, double-throw

(SPDT) scheme, with the SELEX microstrip GaN HEMT technology. The GaN/AlGaN/GaN epitaxial structures were grown on semi-insulating SiC substrates using the MOCVD method or MBE method. When working with masks of different levels, both classic stepper photolithography and electron beam lithography were used. The gates of 0.25 μm in length were formed by electron beam lithography. Other processes of MIC switches manufacturing were the same as for SELEX MIC amplifiers.

If we talk about the results of studies of these products, the switch insertion loss is less than 1 dB in a specified frequency range 8–11 GHz, with isolation exceeding 37 dB. In this case output matching of the switch is better than 13 dB. While measuring the characteristics, the switch was mounted in a special holder with coaxial outputs. The measurements were conducted in pulse mode (pulse duration—100 ms, duty cycle—25%). Even when applying an 8W signal to the input, the switch retained the linearity of transfer characteristics.

As part of the same work, the broadband switch was created for the range of 2–18 GHz, with insertion loss below 2.2 dB, isolation better than 25 dB, and transmission power exceeding 5W at 1-dB compression.

Another GaN SPDT MIC X-band switch with an operating power of more than 25W [32] was developed by TNO Defense, Security and Safety (Hague, the Netherlands), and QinetiQ (Malvern, UK). The switch ICs were manufactured by QinetiQ GaN MIC coplanar technology. The structure includes a 25 nm undoped Al0.25Ga0.75N layer on 1.9 μm insulating iron-doped GaN layer grown on semi-insulating 4H SiC substrate with a diameter of 50 mm.

Ti/Al/Pt/Au ohmic contacts and the 0.25 μm Ni/Au T-gate were formed using conventional methods. The devices were passivated by $SiN_x/SiO_2/SiN_x$ multilayer structure with a breakdown voltage of 200V, which also served as a dielectric for MIM capacitors. Thin film resistors were made on NiCr film with a surface resistance of 27 ohm/sq. Inductances and coplanar lines were created in the 3 μm galvanic gold layer. The air bridge was formed of an evaporated layer of gold 0.8 μm thick.

The transistor conductivity was 230 mS/mm, current of 1030 mA/mm, maximum current gain frequency of 40 GHz, and gate breakdown voltage over 100V. The switches were designed using the ADS Momentum software, which by now is seriously updated and widely used in microwave devices design.

It should be noted that the designers received a fairly high level of isolation (35 dB in the X-band). Return input and output loss was better than 10 dB. The insertion loss was considerable (3.5 dB) due to relatively low isolation of transistors in shutdown mode. Measurement in the pulse mode showed a linearity of the switch output power versus input power up to 25W, although the homogeneous isolation was maintained across the dynamic range.

In 2008, the European Defense Agency KORRIGAN project aimed at the development of GaN microwave devices and microcircuits was completed [25]. Its main result was the creation of a European infrastructure, which includes independent technology centers for the production of GaN microwave MICs for civil (cellular communication base stations) and military (APAA transceiver modules) applications. European industrial companies and universities participated in the development of MIC GaN components for transceiver modules with the support of ministries of defense of some countries (France, Germany, and Spain).

These results confirm the perspective of GaN MIC application as part of transceiver modules—both in high-power (preliminary and output amplifiers, switches) and in low-power (low noise amplifiers) stages. GaN MIC exhibit good broadband characteristics at high output power levels. Further application of GaN transceiver modules in specific APAA will depend on the readiness of GaN MIC industrial technology and designers of respective radars. So far two groups of companies have worked on the projects of airborne APAA radars for European fighters. One of them is headed by Thales, which developed a 1,000-module array; the second group is a consortium of SELEX Galileo, EADS Electronics, and INDRA, which have successfully designed and put into series production a 1,425-module APAA system [24].

3.11   Brief Comparative Overview of the World Market Situation of MMICs Based on SiGe, GaN, AlGaN/GaN

Today GaAs technology is so perfected that GaAs devices find not only space and military, but also commercial, application. Components based on gallium nitride and indium phosphide are firmly gaining market niches such as high-power high-frequency transistors for WCDMA systems, as well as transistors for centimeter and millimeter waves. But, as we have shown in the previous chapters, the silicon technology does not stand still either: today it is so improved that the transistors based on this material are invading the realm of gallium arsenide devices. In addition, the traditional silicon offsprings—silicon-germanium and silicon carbide—are also gaining strength [25].

Today about 80 companies supply their products to the semiconductor microwave devices market. Stimulated by rapid development of the Bluetooth-standard systems, mobile communications, wireless LANs, and RFID, the manufacturers offer a wide range of microwave devices—from small-signal low-noise and high-power transistors to the single-chip radio receivers, transmitters, and transceivers. Although companies with their own production of developed products continue to thrive, a rapidly growing number of companies deal only with the microcircuits design (fabless companies) and outsource the manufacture from silicon factories (foundries). At the same time, the efforts of many software companies are directed also at the development of various new technologies of RF and microwave devices manufacturing. For example, the Hittite Microwave (www.hittite.com, created by former Raytheon engineers) research programs are aimed at creating high-frequency devices and MMICs on GaAs, InGaP/GaAs, InP, SOI, and SiGe. One of the company's latest developments passed to the silicon production plant is a series of broadband quadrature modulators and amplifier units for 250–800 MHz and 4–7 GHz bands, formed on SiGe HBTs and packaged in a 3×3 mm housing for surface mounting. HMC495LP3 type modulator to a lower frequency range is designed for use in GSM, CDMA, WCDMA, and WLL-Systems, and HMC496LP3 type modulator to a higher frequency is designed for WLAN systems of IEEE 802.11 standard and microwave radio receiving devices.

Centellax Company (www.centellax.com) founded in 2007, has outsourced from a foundry the manufacture of a fractional-N synthesizer created by Centellax engineers. A broadband voltage controlled generator unit, being part of the synthesizer microcircuit, is made on SiGe HBT with a limiting frequency of 300 GHz, while the remaining analog and digital blocks of synthesizer are formed by conventional CMOS technology. This fractional-N synthesizer is capable of generating frequencies up to 30 GHz with up to 25 kHz increments.

The synthesizer output power is +5 dBm; the frequency sweep is approximately 1 ms. The microcircuit is designed for tunable heterodynes used in digital radio stations, chirp radars, instrumentation signal generators, and clock frequency generators with adjustable frequency for optical communication systems.

Modern SiGe devices successfully compete with GaAs microwave devices in wireless communication systems operating at relatively low voltage and power output. As a result, the number of companies producing SiGe devices is growing steadily. In addition to IBM, which developed SiGe technology, and SiGe Semiconductors, which was one of the first to use this technology, SiGe devices are now being created and manufactured by Atmel, Hittite Microwave, IceFyre, Infineon Technologies, Inphi, Intersil, Maxim Integrated Products, Sirenza Microdevices, RF Micro Devices, and many others.

SiGe Semiconductors located in Ottawa (www.sige.com) back in 2008 announced the creation of a series of two-stage power amplifiers (PA) for cell phones IS-95 and CDMA-systems operating in the frequency range of 824–849 MHz. The linearity of SE5103, SE5106, and SE5107 type amplifiers made by SiGe BiCMOS technology is less than −50 dBc at +28 dBm peak output power, with power added efficiency exceeding 41%. Each microcircuit in these series can withstand electrostatic discharges up to 4kV. In addition, they have a mismatch resistance feature. As a result, the amplifier operates reliably at a VSWR of 10:1. Each amplifier includes a digital or analog in-circuit shift control unit, power detector, a 2.8V voltage regulator, and matching and harmonic tuning modules. All this eliminates the number of external components, including the voltage regulator and the output signal detector. Since power attenuation in adjacent channels does not exceed 2 dB and VSWR mismatch is 4:1, the isolation between the power amplifier and the low noise amplifier is not required. According to the designers, due to these features the use of power amplifier in a new series of mobile phones will save up to $3 each.

These amplifier microcircuits are supplied in QFN-package with standard pin-out, so they can directly replace the previously installed devices and thereby ensure savings on equipment upgrading. Each sealed microcircuit has separate outputs for control signals and amplifying cells, which allows us to adjust the value of VCC to 0.8V. The standby mode current consumption is 2 μA. The SE5103 type amplifier includes digital shift control module and is supplied in a package of $4 \times 4 \times 0.9$ mm at a price of $0.8 per piece when buying a batch of 100,000 pieces. SE5106 and SE5107 amplifiers are available in a $3 \times 3 \times 0.9$ mm package with digital and analog control modules, respectively, and cost $0.85 in the same batch.

One of the SiGe Semiconductors devices is a 5 GHz Range Charger family PA designed for PCMCIA-cards, PDAs, and 802.11a/b/g compliant devices. The three-stage SE2534A type amplifier includes power detector, analog shift circuit, and interstage matching modules. The output power of the module is 17.5 dBm at a current of 160 mA and error vector magnitude under 3%, which reduces the error rate in the transmission of data packages and at the same time provides maximum bandwidth and maximum transmission distance. Supplied in industry standard module is a 10-pin LGA-type body of 5 × 5 mm size, compatible with PA connectors used in 802.11a equipment. The cost per unit is $1.93 with a minimum order of 100,000 units.

Infineon developed and patented its own process of forming 70-GHz devices, and then mastered the production of high-speed n-p-n transistors for WLAN systems. Noise rates of BFP640 and BFP650 type transistors are 0.65 dB at 1.8 GHz and 1.3 dB at 6 GHz, respectively, which is comparable with the parameters of GaAs devices.

The most famous manufacturer who has developed the SiGe technology and has been providing services for the manufacture of SiGe devices—IBM—owns a variety of device-specific processes. So, for the manufacture of HBT used in many circuits of wireless communication systems, BiCMOS 7HP 0.18 μm process, is the most advanced, allowing us to form self-aligned emitters, small and deep isolation grooves, and transistors with a frequency limits up to 120 GHz. The designer companies also know very well a foundry called Taiwan Semiconductor Manufacturing Company (TSMC).

Unfortunately, high-power SiGe devices are still rare. We could mention a high-power SiGe transistor designed by engineers of Northrop Grumman (www.es.northropgrumman.com) sector of electronic sensors and systems for air traffic control radar. Gain ratio of WPTB48F2729C type HBT is 7 dB in the frequency range of 2.7–2.9 GHz with 46% efficiency of the collector circuit. In the case with C-class amplifiers, when connected in the common-base circuit, their transistor output power exceeds 180W when 60 ms pulse is applied to the input with a duty ratio of 6%.

But although silicon-germanium is suitable for the manufacture of high-power components, the developers of this class of devices are more attracted by other advanced material—silicon carbide. Strong support for the development of SiC-technology is provided by NASA's Glenn Research Center. A series of high-power MESFETs was released by Rockwell Scientific (www.rockwellscientific.com)—leading designers of SiC RF and microwave devices. The gain of the transistor, designed for a cutoff frequency of 3.6 GHz, is 12 dB at 2 GHz; the minimum output power is 25W. Drain efficiency reaches 40% at a voltage of 50 VDC and current of 1200 mA. The average IM3 is equal −30 dBc. The transistors are designed for CDMA and WCDMA systems.

High-power SiCFETs have been produced since 2009 by Cree Microwave (www.cree.com)—supplier of SiC wafers and manufacturer (foundry) of SiC MMICs. The minimum output power of a CRF24060-101 type MESFET at 2 GHz is 50W at 1 dB compression; the minimum gain of the small signal at the same frequency is 13 dB. Transistor operating frequency is 2.7 GHz. Transistor drain efficiency is 45% at the supply voltage of 48V and current of 250 mA; IM3 equals 31 dBc. Despite the high output power of the device, the minimum noise ratio is low—3.1 dB. These transistors are designed for military broadband communications systems, amplifiers of classes A and AB, as well as for TDMA, EDGE, CDMA, and WCDMA systems.

As we have mentioned before, the attention of high-power RF and microwave device designers is increasingly attracted to *gallium nitride*, which allows achieving greater power density and obtaining higher efficiency devices, as well as ensuring minimal signal distortion as compared to other semiconductor materials. Today, they are developed by more than 100 research institutions, including such major contractors of the U.S. Department of Defense as BAE Systems department of information and military electronics systems (www.baesystems.com) and Northrop Grumman sector of electronic sensors and systems. The development of GaN technology is complicated by high cost of source wafers, the diameter of which is only 50 mm (against 150 and 300 mm diameters of silicon and gallium arsenide wafers, respectively). Nevertheless, at the end of 2003, Triquint Semiconductor (www.triquint. com) and major military contractor Lockheed Martin announced the development of high-power GaNHEMT. The transistor power density is 11.7 W/mm with output power of +34 dBm, the weak signal gain is 9.83 dB, and power added efficiency is over 50%.

A potential market for high-power GaN devices is third generation wireless systems (3G systems). Progress has been made in the area of increasing output power and power density of GaN power amplifiers, and the developers are now close to the implementation of stringent requirements such systems have. According to these requirements, the amplifier output power must be 150W at operating voltage of 48V. This is why the designer's efforts are focused on the creation of a transistor with drain voltage exceeding 50V. And Cree Microwave in 2008 announced the development of GaNFET for PA on silicon carbide substrate with output power density of 32 W/mm and PAE of 55% at a frequency of 4 GHz. At a frequency of 8 GHz, the output power density is 30 W/mm at 50% efficiency. Drain offset voltage is 120V. This work was partially funded by the Office of Naval Research and DARPA.

Gallium nitride based PA for WCDMA-systems was also created by NEC. Its frequency is 2.1 GHz, output power is 150W, total power efficiency is 54%, and drain voltage is 63V.

Release of AlGaN/GaNHFETs for third-generation mobile communication systems was started in 2010 by the company established in 1999 by graduates of North Carolina State University, Nitronex. The special feature of these devices is that they are made by the company's patented technology called SIGANTIC, allowing the growing of high-quality gallium nitride on silicon wafers of 100 mm in diameter. HFET tests with the 72 mm gate periphery (0.7 mm gate length) at the operating current of 2A and drain voltage of 28V showed that gain ratio in small signal mode is equal to 16.3 dB, the drain efficiency is 62%, and the output power in the saturation mode is 138W. This was a record value of power for GaN devices on silicon substrate operating at a voltage of 28V. The results of high-temperature tests allowed assessing the life cycle of such devices as 20 years.

In 2009, Nitronex began to supply prototypes of high-power AlGaN/GaN HFETs with the drain voltage of +28V and output power exceeding 10W (Model N10) and 20W (model N20). The gain ratio of the transistors in 1.8–2.2 GHz frequency range is 11.5 dB with an average efficiency of 25%. In the near future the company intends to release the devices for 36W voltage.

Thus, in connection with the development of wireless mobile communication systems, 3G GaN devices now pose a serious threat to LDMOS widely used in base stations PAs. In addition to ensuring high power, other important advantages of GaN transistors are increased reliability and efficiency, as well as the ability to operate at high temperatures, which will reduce the size of the PA module by eliminating the cooling modules. In addition, replacement of LDMOS devices by GaN makes it possible to use one transistor where previously two were used, which significantly simplifies the device matching process.

However, GaN transistors have been widely used in the apparatus only since the end of the first half of 2006.

Now LDMOS accounts for about 90% of PA transistors market for cellular base systems (the rest is the share of GaAs instruments and a small portion of traditional bipolar transistors). Although theoretically LDMOS is not the best technology to manufacture PAs for 3G communication systems, changing the situation will not be easy. Continuous improvement of these devices will undoubtedly contribute to maintaining their strong position in the market. An example of LDMOS technology improvement are transistors made by Philips (www.philips.com fifth generation), which are second after Motorola's on the market of high-power devices for cellular base stations.

Philips had planned to release the fifth generation of transistors in the market, so that developers could implement hardware based on them. The new technology is transferred to the plant for the production of modern CMOS microcircuits, which has mastered 0.14 μm technology. LDMOS will be made using this next-generation technology.

Significant progress in improving LDMOS transistors was made by Infineon (www.infineon.com) and STMicroelectronics (www.st.com/rf), collectively occupying the third place in the market of components for cellular base stations. Thus, the continuous-mode output power of a new generation LDMOS transistors of GOLD-MOS family type PTFA211001E type at a frequency of 2.1 GHz is 100W at 1 dB compression, gain is 16.4 dB, efficiency is 57%, and supply voltage is 28–30V. In dual channel 3GPP WCDMA system the average transistor output power is 22W, gain ratio is 16.5 dB, efficiency is 30%, power weakening on the adjacent channel is under −42 dBc, and IM3 equals to −37 dBc.

Another serious competitor for LDMOS transistors in PA for the next generation wireless communication systems are GaAs devices. According to the research conducted by PA Consulting Group (UK), LDMOS is advisable to use in devices operating in a stable environment and with limited power range, which is acceptable for GSM systems. The output power in WCDMA base stations depends on traffic, and thus PAs on LDMOS will not always be able to operate in a sweet spot. For such systems, GaAs transistors are more advisable.

Gallium-arsenide transistors (mainly HBTs) today also dominate the mobile phones PA market. In this market there are three leaders, accounting for 80% of sales.

Devices on well-known materials—gallium arsenide and indium phosphide—occupy a strong position in the millimeter wave sector. And here, the manufacturers (foundries) become more important. We can mention Velocium (www.velocium.com)—a former TRW branch, and now a Northrop Grumman Group company. Velocium designs and produces HEMT with a minimum element size of 0.1 μm both on gallium arsenide and indium phosphide. The cutoff frequencies of transistors are 120 and 180 GHz, respectively.

Monolithic GaAs microcircuits also show high rates of development. Among the latest achievements in the field of GaAs devices is the monolithic medium power amplifier PHEMT operating in enhancement mode (E-pHEPT), MGA-425P8 model made by Agilent Technologies. The microcircuit is intended for use as a master device in wireless communication systems, designed for frequencies up to 10 GHz and wireless LAN standard IEEE 802.11a; in the unlicensed national information infrastructures at a frequency of 5 GHz; wireless LAN standard 802.11g/b, operating in the 2.4 GHz band allocated for industrial, scientific, and medical systems (ISM) and wireless local area networks, wireless phones at 2.4 and 5.8 GHz.

At a frequency of 5.25 GHz, supply voltage 3V, and 58 mA current consumption, the output power on this microcircuit with an error vector amplitude of 5% is 13.3 dBm, PAE is 10.3%, output power at 1 dB compression is 20.3 dBm, gain ratio is 16 dB, noise ratio is 1.7 dB, and VSWR at the output resistance of 50 ohm is 2:1. It should be noted that, despite low current consumption, the device meets all linearity requirements of wireless LAN. The required shift is set by an external resistor, and due to the smart shift function, IP3 linearity can be adjusted between 20–35 dBm by adjusting the external resistor. This allows the use of microcircuit in various modules of the same system.

The microcircuit is packaged in a standard industrial leadless eight-pin enclosure DRP-N LPCC measuring $2 \times 2 \times 0.75$ mm. When purchased in lots from 5.5 thousand to 14.5 thousand pieces, the unit price is $1.73.

Recently, designers also have paid attention to microwave devices based on indium phosphide—material with a direct bandgap, high electron mobility, and breakdown voltage. Besides, indium phosphide is the only material on which devices can be produced generating, modulating, amplifying, and receiving light at wavelengths of 1.55 and 1.3 μm (i.e., at the wavelengths used in the single-mode telecom optical fiber systems). Therefore, as the capacity of 10 Gbps OC-192 systems grows, and 40 Gbps OC-178 systems are being developed, the designers of telecommunication equipment, in order to reduce cost, will have to seek new materials, including indium phosphide. The features of this technology are demonstrated by second generation HBT created by Vitesse Semiconductor, which is produced by double diffusion. Its limit frequency exceeds 300 GHz, and the breakdown voltage is 4.5V. The advantage of this new device is that it can be manufactured in wafers with a diameter of 100 mm by four-layer metallization technology, which is close to CMOS technology. The new HBT was used by BAE Systems designers in the digital frequency synthesizer with a record operating frequency of 152 GHz.

It should be noted that all work done by Vitesse together with BAE Systems and the University of Illinois were funded by DARPA.

One of the main problems of InP devices technology is the complexity of integrating optical and electronic elements. By now, they were able to integrate only a short-range pin-detector on a chip with digital devices. Of course, in five years there will be indium phosphide microcircuits with complex pin-detectors, integrated with the current-controlled amplifier, modulation, and control circuits. Here we should mention the approach of Inphi (InP devices designers) and Broadcom (CMOS devices designers), who proposed the use of two chipsets—one on InP for input modules and the other on CMOS—for output modules.

Our national developers are also active in this market sector. As an example of microwave designs of Belarusian companies, we should mention M55326 typical multifunction centimeter wave microwave transceiver module.

These microwave modules are multifunctional transceiver devices operating in centimeter wave range, designed for receiving and double conversion of input signal when used in special-purpose equipment.

The modules are based on advanced gallium-arsenide monolithic and hybrid integrated circuits, which ensures a wide dynamic and temperature range. High technical parameters and high reliability are provided using monolithic circuits and modern thin-film multichip technology, as well as a sealed package. Figure 3.26 shows a general view of M55326 series modules.

Table 3.2 shows the principal electrical characteristics of these modules. The design type is a $70 \times 64 \times 12$ mm one-piece module, weighing less than 150g.

Figure 3.26   M55326 multifunction centimeter wave microwave transceiver modules.

Table 3.2   Electrical Specifications

Input frequencies operating range	5–16 GHz
Output frequencies range	5–16 GHz
Transfer ratio	≥ -10 dB
Transfer ratio variation in input frequencies range	≤ 6 dB
Maximum input signal power at 1 dB output compression	≥ 0.25 mW
Frequency range at heterodyne inputs	9–14.5 GHz
Value of parasitic penetration of signal from input to output:	
in 5–6 GHz frequency range	≤ -30 dB
in 6–16 GHz frequency range	≤ -50 dB
Rejection level of parasitic harmonics in the output signal spectrum from the carrier in the linear range of operation	≥ -30 dB
Current consumption	≤ 0.7 A

3.12 Use of GaAs MMIC Technology in Foreign Aerospace and Military Equipment

GaAs monolithic integrated circuits (MICs) became widely used in low-noise amplifiers of civil direct television receivers, high-power cellular telephony amplifiers and switches, cable networks modems, computer networks, and so on [33]. On the other hand, the current level of MMIC technology allowed returning to the creation of active phased-array antennas (APAA). This opened the way to the mass use of MMIC in space technology and various weapons systems. In this section we will review the experience of foreign companies in the organization of development of MMIC technology, as well as the most typical results of their application in certain types of space and military electronics [15, 16, 34–39].

3.12.1 MIMIC Program and Its Role in the Development of MMIC Technology

Back in early 1980s the designers understood that gallium arsenide, due to its high electron mobility and possibility of obtaining a semi-insulating material, is more preferable for creating microwave devices than silicon. But, the attempt was first made to use GaAs for creating a digital IC rather than MMIC. In 1982, the Defense Advanced Research Projects Agency (DARPA), in order to develop high-speed digital processing ICs on GaAs, initiated an advanced onboard signal processing (AOSP) program. Gallium arsenide was selected due to the possibility of manufacturing GaAs VLSI circuits on semi-insulating substrates. However, GaAs-circuits yield was unsatisfactory due to low quality of material and the fact that gallium arsenide technology was insufficiently tried and tested at that moment, as compared to silicon. Practical experience has shown that the necessary production discipline should be tried and tested on production lines with a capacity of at least 100 wafers per week. A number of companies have established pilot lines for production of both digital and analog microwave circuits.

The main objective of the approved program became the increase in volume of processed GaAs wafers. As soon as this GaAs microcircuit design program was finished, DARPA opened a new microwave and millimeter wave monolithic integrated circuits (MIMIC) program. The volume of financing amounted to $600 million. The purpose of the program was to create microwave and millimeter wave devices with the required electrical, mechanical, and climatic parameters, the price of which would allow its use in existing space and military systems. In particular, the task was set to replace the hybrid-integrated structures with monolithic devices. This replacement promised more advantages in speed, weight, and size, as well as other parameters of electronic equipment. Launched in 1987, MIMIC program lasted until 1995 inclusive. Initially, the share of manual labor in the manufacturing process was significant. The diameter of GaAs wafers was only 50 mm. The question remained open as to which method of forming active layers—ion implantation or epitaxy—should be preferred. As part of a separate technical task, a transition was performed from vapor phase epitaxy to molecular beam epitaxy, and from MESFET to HEMT. When implementing this program, the process control was introduced based on statistical methods. In addition, rapid on-wafer measurements of LF and UHF device parameters were implemented (at frequencies up to 95 GHz), a relation between the parameters of manufactured devices and processes was identified, and statistical models of devices for system design were created. As a result, it became possible for the designers to design the required monolithic circuit at the first dash.

By the end of the first program phase (1988), they succeeded in reducing the MIC cost from 20 to 10 USD/mm^2, and at the end of the second phase (1991)—to 0.1 USD/mm^2. The result of MIMIC program is the development of the United States production of GaAs microwave microcircuits with well-developed infrastructure for the development of materials, substrates, masks, equipment, measuring devices, circuit design, and so on. This industrial capacity, on the one hand, has become the basis of a large MMIC commercial market and, on the other hand, has stimulated serious modernization of radio electronics in aerospace and military systems. Based on the monolithic microwave technology, several systems were upgraded, such as Longbow millimeter wave fire control system for attack helicopter radars; sense and destroy armor munitions (SADARM); 40–60 GHz EHF communication system; X-ROD precision-guided weapons (95 GHz); and others. Following the United States, GaAs MMIC production started in Europe. New companies were established, such as United Monolithic Semiconductors (France), TNO and OMMIC (Netherlands), and Filtronic and Bookham (UK). Similar production facilities were built in Japan and Taiwan.

In the case of military equipment, GaAs MMIC are most widely used in the APAA, which requires a large number of transceiver modules. Important parameters of such modules are output power and efficiency, determining the total power radiated by the array, and heat dissipation capabilities allowed by the design. A common module contains several monolithic circuits including a monolithic output power amplifier chip. The pulse power of a typical modern GaAs MIC of an X-band APAA transceiver module (10 GHz) is 10W, average is 1–3W [40]. The APAAs based on these modules and manufactured by several foreign companies have already begun to be implemented in the latest weapons systems.

3.12.2 MMIC-Based Weapons Systems

The first large-scale applications of MMICs developed under MIMIC program were HARM (antiradar missile) system and COBRA (counter artillery C band APAA radar system) [33]. The APAA transceiver modules of COBRA radar made by Electric Electronics Lab contained six GaAs MICs: preamplifier, two powerful amplifiers with yield addition, a phase shifter, a variable gain amplifier, and a low-noise amplifier. The program provided for the supply of 25,000 such systems, complying with the stringent requirements of military standards. To ensure the desired yield, an extremely careful preselection of chips with the required parameters was very important at the stage of modules assembly. Here, the on-wafer MIC measuring method has played a crucial part, in particular, the pulse method for measuring the high-power amplifier parameters, which was developed during the third phase of MIMIC program. It was the mass production of MICs that contributed to the perfection of the entire process flow—from MIC design to manufacture, testing, soldering, and assembly of modules. Today, the APAA radars are created for various platforms—ground, naval, airborne, and satellite.

Let us consider some examples of airborne and ground-based systems.

Airborne APAA Radars

In this respect there is particular interest to AN/APG-79 radar developed in 2009 by Raytheon. The radar development began in 2000. The combat capabilities of the radar were to be evaluated in 2006, and in 2007 the radar had to be supplied to the army [34]. It is now installed on all U.S. Navy F/A-18E/F Super Hornet fighters (Figure 3.27) [3]. In 2005, Raytheon won a five-year contract for the supply of 190 APG-79 APAA radars to Boeing [35].

APG-79 had a number of important advantages over APG-73 that had been previously used on F/A-18E/F aircraft and had no APAA. It is capable of tracking many more targets and operating simultaneously in various modes: terrain mapping, moving ground targets tracking, air-to-air search mode, and so on. Due to APAA, the reliability of APG-79 has increased four times compared to APG-73. The radar's MTBF exceeds 15,000 hours. The radar is compact and lightweight; its weight is only 43 kg. Operating cost per flight hour is twice as low as that of APG-73. The radar transceiver modules are produced by Raytheon in Raytheon's RF Components subdivision. The area of subdivision's production facilities with Class 100 clean rooms is approximately 2300 m², the GaAs MIC experimental manufacturing facility occupies approximately 840 m².

According to international classification, the company is regarded as IDM, so they have all processes necessary for the creation of modern GaAs MICs on MESFET, PHEMT, and E/D (enhancement/depletion) PHEMT.

A similar APAA radar of APG-81 type (Figure 3.28) is now being developed by Northrop Grumman for the future F-35 fighter [36].

In recent years, much attention is being paid abroad to unmanned reconnaissance aircraft. DARPA has fulfilled two major programs to build such aircraft of type X-45 (Boeing Company) and X-47 (Northrop Grumman). The amount of funding was $1 billion each [37]. The projects use APAA GaAs-based transceiver modules

Figure 3.27   Photo of Raytheon AN/APG-79 radar for F/A-18E/F aircraft.

Figure 3.28   Photo of APG-81 APAA radar for F-35 fighter.

similar to those used in APG-79 radar. The company responsible for development of antenna technology for these programs was Raytheon.

The Joint English-French airborne multirole solid state active array radar (AMSAR) program was started in Europe in 1993. The purpose of the program was the creation of airborne APAA radar for Typhoon aircraft. It was successfully completed in 2010 [38]. The general view of the radar is shown in Figure 3.29; the radar transceiver module is shown in Figure 3.30.

Figure 3.29   AMSAR radar for European Typhoon fighter.

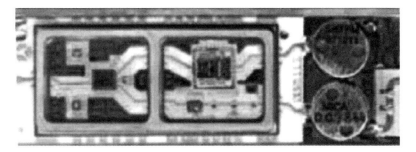

Figure 3.30   AMSAR radar transceiver module.

Monolithic circuits and modules based on GaAs HBTs were developed by UMS. The engineers created a monolithic output power amplifier of APAA module (Figure 3.31) with the following parameters:

- Output power 10W;
- Gain ratio 18dB for 10 GHz;
- Efficiency > 35% for 10 GHz;
- Chip size 4.74 × 4.36 × 0.1 mm.

The level of gallium arsenide technology of Filtronic (UK) allows us to produce more than 200,000 reliable and affordable transceiver modules based on GaAs PHEMT for airborne APAA systems [39].

The GaAs MICs and modules for aircraft APAA radars that we have reviewed are only a part of the scope of work that, as we know from open sources, is being done in this sector. A more detailed list of various types of airborne radars with electronic scanning of various stages of readiness—from development to delivery—as of the end of 2010 is shown in Table 3.3 [16, 38].

Ground-Based APAA Radars

In 1992, Lockheed Martin Missiles and Space received its first contract from the U.S. Department of Defense for the design of mobile missile defense system (Theatre High Altitude Area Defense, THAAD, see Figure 3.32) [34, 40]. THAAD system was designed to destroy ballistic missiles in the higher atmosphere and beyond. The company that became subcontractor for the design of ground-based APAA radar and its constituent solid-state transceiver modules was Raytheon. Multifunctional THAAD radar fulfills the following tasks: ballistic target observation, detection, tracking, recognition, target acquiring for antimissile weapons, and evaluation of results.

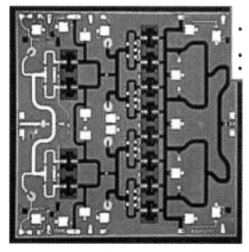

Figure 3.31   Output power amplifier monolithic circuit—AMSAR radar APAA module.

Table 3.3   Airborne APAA Radars

Radar Type	Company	Carrier
AN/APG-81	Northrop Grumman	F-35, fighter
AN/APG-77	Northrop Grumman/Raytheon	F/A-22, fighter
AN/APG-79	Raytheon	F/A-18E/F, fighter
AN/APG-63(V)2 AN/APG-63(V)3	Raytheon	F-15C, fighter
AN/APG-80	Raytheon	F-16E/F, fighter
AN/APQ-181	Raytheon	B-2, bomber
AMSAR	BAE Systems	Typhoon, fighter
	Thales	Rafale, fighter
	EADS	
Seaspray 7000E	BAE Systems	helicopter
Not specified	Mitsubishi Electric	F-2, fighter
AEW&C NORA	Ericsson	JAS-39, fighter

The main specifications of THAAD system are as follows:

- Frequency range: $\times$ (10 GHz);
- APAA aperture: 9.2 m^2;
- Number of APAA transceiver modules: 25,344;
- Target detection range: up to 1000 km;
- Striking range: up to 250 km;
- Striking height: up to 150 km.

Subsequently, the U.S. Department of Defense purchased more than 30 ground-based radars that are part of the system.

Speaking of the prospects of development in this sector, it should be noted that the example of introduction of GaAs MIC in military radio equipment confirms the well-known rule that the time between the development of new technology and its introduction in aerospace and weapons systems is at least 10–15 years. And now, airborne and ground-based systems with most advanced performance characteristics, provided with GaAs MMICs that were designed in the late 1990s, have been made operational in the developed Western countries.

This circumstance may also be of economic importance. The United States are going to execute contracts for sale of F/A-18 E/F American aircraft with APAA APG-79 radars to India [41]. Of course, introduction of APAA technology in specific systems depends on the ratio of technical specifications, cost, weight, size, power consumption, and so on [42]. However, Raytheon already owns the inexpensive missile APAA technology, and Thales demonstrated fully monolithic 95 GHz APAA for homing heads [43]. The introduction of monolithic circuits technology into radiolocation allows the designers of aerospace and military equipment to completely eliminate

Figure 3.32 General view of THAAD mobile anti-missile defense system.

the tube microwave transmitters with all their inherent problems. Apparently, in the future this trend will continue, as MMIC technology continues developing rapidly. In 2004, DARPA announced the launch of a new program—Wide Band-Gap Semiconductor Technology Initiative (WBGSTI). The purpose of the program is to create microwave devices and MICs based on wide band-gap semiconductor compounds (GaN, SiC, and AlN) [44, 45]. The first phase of the program (2005) ended in development of technology for obtaining stable SiC substrates with a diameter of 75 mm and epitaxial growing of AlGaN/GaN HEMT structures. During the second phase (2005–2007), reliable microwave and millimeter-wave GaN transistors with high performance and yield were implemented. The third phase (2008–2009) proved that it was possible to manufacture low-cost reliable GaN MICs and use them in various types of modules, including: radar X-band transceivers, broadband power amplifiers for electronic warfare systems, and millimeter-wave power amplifiers for space communication systems.

The parameters of these modules (Table 3.4) according to power level, efficiency, and bandwidth are far superior to the results obtained for GaAs MICs within MIMIC program and subsequent works. As we already noted, the period between the development of new technology and its introduction into the system is always quite extended.

DARPA had set a task to reduce it during WBGSTI program implementation. To this end, DARPA instructed participants on the third phase of the program to prepare business plans for the accelerated introduction of its products in specific systems [46]. Thus, apparently, GaN devices and MICs created within WBGSTI program will be used in weapons systems sooner than GaAs devices within MIMIC program.

Table 3.4 Target Parameters of GaN Modules Developed Under WBGSTI Program

Module Type	Operating Frequency, GHz	Output Capacity, W	Efficiency, %	Gain Ratio, dB	Main Developer
Radar transceiver	8–12	60	35	18–20	Raytheon
Broadband amplifier for electronic warfare	2–20	100	20	30	Northrop Grumman
Amplifier for Q-band communication systems	> 40	20	30	13	TriQuint

Europe did not stay aside. A large-scale multinational European project KOR-RIGAN was implemented. Its objective was to create modern production of GaN HEMT and microcircuits on their base in Europe [47]. The project involved seven countries: France, Italy, Netherlands, Germany, Spain, Sweden, and the United Kingdom, and the project leader was French company Thales Airborne Systems. Total cost was €40 million EUR. The European unified supply system has been established, which provided space and military industry with reliable, modern GaN devices.

Among domestic research programs in the field of microwave electronics, in addition to the federal target programs, we should mention the work carried out by joint Russian-Belarusian programs Pramen and Mikrosistemotekhnika.

3.13   Microwave Devices Based on Gallium Nitride

3.13.1   Power GaN Transistors

It was impossible to implement theoretic advantages for GaN power transistors over Si power transistors until recently, because of its on-state in normal conditions. Nowadays GaN-cell process make it possible to produce devices with off-state in normal conditions. First practical applications, realized on eGaN®FET technology, are DC/DC-converters for space and commercial applications, D class power amplifiers, inverters, wireless power transfer systems, lidars, envelope tracking systems, and systems with increased radiation resistance and thermal stability.

At the time of printing, an overwhelming part of the market is occupied by silicon semiconductors. However, over the past few years there has been a sharp increase in the power elements fraction, manufactured by gallium nitride (GaN) and silicon carbide (SiC).

The discussion about GaN transistors can raise a lot of question for designers. Why have changes come to the GaN process? Why has such sharp development begun only now? What features are in new manufactured devices? What is the range of application for these developed devices? Does this technology have the ability to grow?

The domination of silicon devices has lasted more three decades—from the emerging point of power MOSFET at the end of the 1970s. There is no competitor for a silicon for long period, as other known semiconductor (germanium, selenium) had considerably poor key properties in practice. Later, semiconductor properties were discovered for new materials—gallium nitride, gallium arsenide, silicon carbide, and so forth.

It should be noted, that GaN is not so new, though. Its special characteristics was discovered as early as at 1975 by T. Mimura, and detailed research was carried out at 1994 by M. Cann.

Even at that time of this study, GaN was shown to be far better future-proof material than silicon (Table 3.5) [48].

High critical strength for GaN gives potential possibility to implement devices at higher voltage. Large bandgap provides highest stability for properties with temperature change or radiation exposure, which are extremely important for space and military electronics and for hardwire in harsh environment.

Table 3.5 Semiconductor Material Properties

Parameter	Material		
	GaN	Si	SiC
Bandgap, eV	3.4	1.12	3.2
Critical strength, MV/cm	3.3	0.3	3.5
Drift saturation velocity for electrons, $\times 10^7$ cm/s	2.5	1	2
Mobility, $cm^2/(V\bullet s)$	990–2000	1500	650
Dielectric constant	9.5	11.4	9.7

High electron mobility and high drift velocity define significantly less on-state resistance and high power density, compared to silicon.

It is quite meaningful to study resistivity dependence on breakdown voltage for different semiconductor materials (Figure 3.33) [49]. This characteristic is linear practically for each material, however GaN showed significantly less resistance at same breakdown voltage. Consequently, it results in GaN-transistor power density by an order of magnitude greater, which leads to considerable minimization of overall size, compared to conventional silicon elements.

Let's take a look at the easiest conventional GaN-transistor structure (Figure 3.34) [49]. AlN protective layer is growing on silicon substrate. GaN/AlGaN heterostructure is formed on it. Then protective dielectric layer and electrodes is created.

GaN and AlGaN are inherently polar materials. Spontaneous polarization occurs at the interface in growth process to form surface charges. In addition, GaN has expressive piezoelectric property. Therefore, it additionally is polarized under mechanical stress and deformation. So GaN and AlGaN have lattice disregistry, and inevitable strains arise at this interface [48].

Therefore, the polarization processes cause charge formation in the form of two-dimensional plane (so named electron gas, 2DEG) [48].

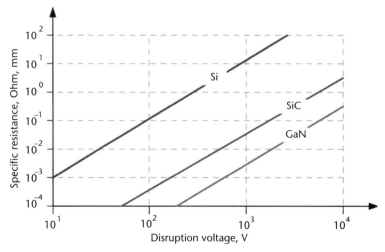

Figure 3.33 Resistivity versus semiconductor breakdown voltage.

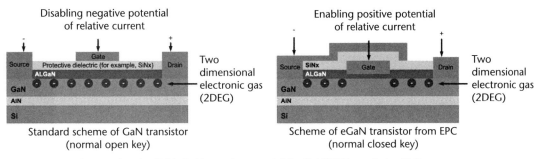

Figure 3.34   Conventional cell (a) GaN-transistor and (b) eGaN®FET, made by EPC.

If voltage is applied to source and drain electrode in this base cell, current will flow, even though gate voltage is equal to zero. And so this device is named after normal-opened among designers. To stop current flow, it needs to put negative voltage on the gate relative to the source (Figure 3.34).

For obvious reasons, this transistor is quite uncomfortable in use. First, to avoid the circuit burning effect, it needs to provide a safe off-state to the transistor before the main voltage supply will be applied. It's also necessary to have an additional negative voltage source.

Simple cell problems are not over at this point, since the earlier mentioned structure has an extremely oversimplified type. It's necessary to use substrate for effective heat removing from GaN layer, but all conventional materials (Si, SiC, sapphire) have lattice disregistry over GaN. So process engineers must introduce additional matching layers to decrease strain. Appending additional layers between other dissimilar materials in the cell by analogy is necessary. Structure is provided to be complex in the issue [48].

These additional problems determine the requirement in additional studies to reveal optimal materials, best layer thickness, and so on.

Here, it is practical to consider principal technological decisions designed by Efficient Power Conversion (EPC), one of the world leaders in this field.

Engineers from this company offered advanced transistor cell structure, which was named eGaN®FET (Enhancement Mode) (Figure 3.34). Basic changes are applied to the formation of the gate area and under the gate region [49].

AlGaN depletion area is formed under the gate in this cell, which provides gap formation in the electronic gas region (2DEG).

The transistor proved to be a normal-off device. It needs to apply positive voltage on the gate to the source, to form a conductive channel between the drain and the source in this case. In the issue eGaN operation is absolutely identical to usual N-MOSFET operation.

Certainly, the presented structure is as much as possible simplified here, and practical cell has a much more complex form.

Comparison results to this new transistors and its silicon anchestors are presented in the Table 3.6, which lists advantages and limitations for representatives out of each technology [50].

Here the open channel resistance of the channel Rds open, mOhm is one of the most important characteristics, determining the losses on the switch. The type

Table 3.6 Comparison of the Power Switches Characteristics for the Voltage of 100V

Parameter	Type Silicon MOSFET 100V	eGaN EPC2022 100V
Rds open, mOhm	Units-tens	2.4
Resistance variation of Rds open at the temperature variation of 125°C/25°C	2.2	1.4
Energy for switch-overs	High	Low
Restoration time of the reverse diode	Great	Unavailable
Threshold voltage Ugs threshold, V	2–4	0.7–2.5
Variation of Ugs threshold at the Temperature variation of 125°C/25°C	0.66	1
Maximum voltage Ugs, V	±20	−1.5
Operating temperature, °C	150	150
Input resistance of the gate R_g, ohm	Several	0.3
Input current Igs	Several nA	1 mA

value of resistance is close for the series samples of the both transistors, but here it is worthwhile making several remarks.

First, GaN has more stable characteristics. For it the value Rds within the temperature range of 25–125°C varies 1.4 times approximately. Resistance of MOSFET varies over 2.2 times.

Second, the resistance dependence on the maximum operating voltage for GaN is considerably weaker than of MOSFET. This is related to the fact that the length increase of the channel Drain-Source does not have such a critical effect on the resistance value. For instance, the resistance for 30V of transistor EPC constitutes 1.3 mOhm, and 200V of the transistor EPC2034 is just 10 mOhm.

Capacity C, pF, determines the transistor's performance. The proposed in the work flat structure eGaN FET has the minimum values of the capacities CGD and CDS. This makes it possible for the devices to commutate the voltages of hundreds of volts with the gigahertz frequency and to reduce the dimensions of the voltage converters.

The threshold voltage Ugs threshold, V, constitutes for GaN only 0.7–2.5V. It is important to note that the maximum voltage value at the gate of GaN-transistor constitutes, as a rule, only +6 V/−4V.

The input gate resistance R_g, ohm, determines the recharge rate of the input capacity. For GaN the given resistance is quite small, which results in the high performance and better protection dv/dt. At the same time the values of input currents increase, and this means the control power increases.

The design engineers usually use in such systems the reverse diode. However, the type circuit of the GaN-transistor does not have the reverse diode as such. However, there is the reverse conductivity mechanism, which performs its function. Meanwhile, interestingly, with the reverse direction of current there does not occur a build-up of the minor carriers, opposite to MOSFET. And this means that the time losses for restoration of the reverse diode's resistance are missing.

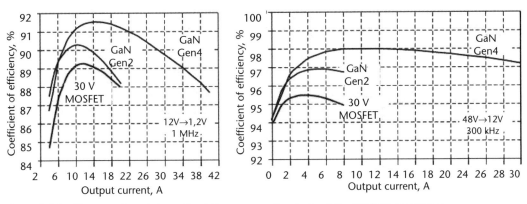

Figure 3.35   The experimental results of the comparative tests of the DC/DC converters.

Thus, the eGaN FET-transistor has the excellent electric characteristics, but still is inferior as compared with MOSFET with regard to the control convenience. Therefore, the given type of the power switches will not always have an advantage as compared with MOSFET. Nonetheless, it is possible to indicate the prospective applications for eGaN FET: DC/DC-converters for the various space applications, power amplifiers of the D class, inverters, space systems of the wireless power transfer, lidars, supply tuning systems of the high frequency amplifiers (envelope tracking), and any systems with the enhanced temperature and radiation resistance.

As a demonstration of the advantages of using eGaN FET, one can refer to the test results of the DC/DC-converters of 12V/1.2V with the operating frequency of 1 MHz and 48V/12V with the operating frequency of 300 kHz (Figure 3.35). From these schedules it is obvious that when increasing the operating frequencies and output currents the advantage of eGaN FET as compared with MOSFET increases. The best results were demonstrated by the GaN transistors of the so called fourth generation: if for the first converter the maximum coefficient of efficiency constituted over 91%, then for the second one it is already over 98%. This is significantly higher of the analogous indicators of a DC/DC on the MOSFET-switches.

It is worthwhile noting that from the moment of emergence of this class of devices there were created four generations of power components. From 2015 there were in mass production representatives of the two last generations: Gen2 and Gen4.

Here there's some sense to provide a concise review of the range of eGaN FET of one of the major producers—EPC. The line of the power components, made by the company EPC, constitutes three major groups:

- Discrete power eGaN FET;
- eGaN FET for the high frequency applications;
- Integrated subassemblies of eGaN FET.

All these elements are made in the LGA-packages (Figure 3.36.)

A large group of the discrete eGaN FET comprises the representatives of two generations of transistors with the operating voltages of 30–450 V (Table 3.7).

All these represented power switches have the small resistance. The record Rds constitutes 1.3 mOhm for EPC2023—a 30V switch. Meanwhile, the resistance

Table 3.7 General Purpose Discrete eGaN FET

Identification	Uds max., V	Ugs max., V	Rds (on) max, mOhm, at Ugs = 5 V	Qg type, nK	Qgs type, nK	Qgd type, nK	I_d, A	I_d pulse, A	Tj max., °C	LGA-package, mm
EPC2023	30	6	1.3	20	5.8	1.9	60	590	150	6.1 × 2.3
EPC2024	40	6	1.5	19	6.4	2	60	550	150	6.1 × 2.3
EPC2030	40	6	2.4	18	5.2	3.4	31	495	150	4.6 × 2.6
EPC2015C	40	6	4	8.7	3	1.4	36	235	150	4.1 × 1.6
EPC2015	40	6	4	10.5	3	2.2	33	150	150	4.1 × 1.6
EPC2014C	40	6	16	2	0.7	0.3	10	60	150	1.7 × 1.1
EPC2014	40	6	16	2.5	0.67	0.48	10	40	150	1.7 × 1.1
EPC2020	60	6	2	16	5	2	60	470	150	6.1 × 2.3
EPC2031	60	6	2.6	17	5.2	3.2	31	450	150	4.6 × 2.6
EPC2035	60	6	45	0.88	0.25	0.16	1	24	150	0.9 × 0.9
EPC2021	80	6	2.5	15	3.8	2.1	60	420	150	6.1 × 2.3
EPC2029	80	6	3.2	13	4	2.5	31	360	150	4.6 × 2.6
EPC2022	100	6	3.2	13	3.7	2	60	360	150	6.1 × 2.3
EPC2032	100	6	4	14	4.2	3.1	31	340	150	4.6 × 2.6
EPC2001C	100	6	7	7.5	2.4	1.2	36	150	150	4.1 × 1.6

EPC2001	100	6	7	8	2.3	2.2	25	100	125	4.1 × 1.6
EPC2016C	100	6	16	3.4	1.1	0.55	18	75	150	2.1 × 1.6
EPC2016	100	6	16	3.8	0.99	0.7	11	50	125	2.1 × 1.6
EPC2007C	100	6	30	1.6	0.6	0.3	6	40	150	1.7 × 1.1
EPC2007	100	6	30	2.1	0.52	0.61	6	25	125	1.7 × 1.1
EPC2036	100	6	65	0.7	0.17	0.14	1	18	150	0.9 × 0.9
EPC2033	150	6	7	10	3.5	1.7	31	260	150	4.6 × 2.6
EPC2018	150	6	25	5	1.3	1.7	12	60	125	3.6 × 1.6
EPC2034	200	6	10	8.5	2.6	1.4	31	140	150	4.6 × 2.6
EPC2010C	200	6	25	3.7	1.3	0.7	22	90	150	3.6 × 1.6
EPC2010	200	6	25	5	1.3	1.7	12	60	125	3.6 × 1.6
EPC2019	200	6	50	1.8	0.6	0.35	8.5	42	150	2.7 × 0.95
EPC2012C	200	6	100	1	0.3	0.2	5	22	150	1.7 × 0.9
EPC2012	200	6	100	1.5	0.33	0.57	3	15	125	1.7 × 0.9
EPC2025	300	6	150	1.85	0.61	0.3	4	20	150	1.95 × 1.95
EPC2027	450	6	400	1.7	0.6	0.25	4	12	150	1.95 × 1.95

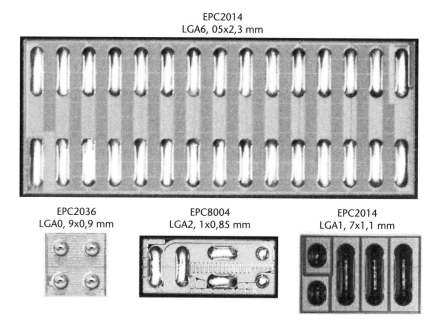

Figure 3.36 Examples of the package versions of discrete eGaN FET.

dependence on the operating voltage inside the given segment turns out to be not as strong as in MOSFET.

The maximum mean-square currents of the given group of elements in the majority of cases constitute tens of amperes, and the pulse currents—tens and hundreds. Despite the such impressive power, all transistors are made in the miniature packages, the largest of which is LGA 6.1 × 2.3 mm, and the most compact of which is LGA 0.9 × 0.9 mm (Figure 3.36).

The group of the devices eGaN FET EPC800x is intended for operation in the subgigahertz range. It should be stressed that specifically in the high frequency applications the GaN transistors from the very start gained the leading position, because they ensure operation in the systems with the strict switching modes of the power switches and the operating frequencies from tens to hundreds MHz.

Table 3.8 represents the characteristics of the family EPC800x with the operating voltages of 40–100V. The package version for all representatives of this group is similar—LGA 2.1 × 0.85 mm (Figure 3.36).

The basic applications for the devices of the series EPC800x became the supply tuning systems of the high frequency amplifiers (envelope tracking) and power wireless transfer systems, including in the devices of space application.

One more category of the devices is a group of the integrated subassemblies eGaN FET, which contains six components with the operating voltages of 30–100V (Table 3.9). All these subassemblies represent the classic semi-bridge circuits (Figure 3.37).

It is necessary to note an important factor that the integrated subassemblies eGan FET may have a symmetrical and nonsymmetrical structure (Figure 3.38) [51].

In the symmetrical configuration the chip sizes of transistors of the both arms are equal. Appropriately, resistance of the upper and lower switches is equal. Such circuit is used for the D class amplifiers, electric motor drives, wherein the both transistors have an equal load range.

Table 3.8   eGaN FET for High Frequency Applications

Identification	Uds max., V	Ugs max., V	Rds on max., mOhm at Ugs = 5 V	Qg type, nK	Qgs type, nK	Qgd type, nK	I_d, A	I_d pulse, A	Tj max., °C	LGA-package, mm
EPC8004	40	6	110	0.37	0.12	0.047	2.7	7.5	150	2.1 × 0.85
EPC8007	40	6	160	0.302	0.097	0.025	2.7	6	150	2.1 × 0.85
EPC8008	40	6	325	0.177	0.067	0.012	2.7	2,9	150	2.1 × 0.85
EPC8009	65	6	130	0.37	0.12	0.055	2.7	7.5	150	2.1 × 0.85
EPC8005	65	6	275	0.218	0.077	0.018	2.7	3.8	150	2.1 × 0.85
EPC8002	65	6	530	0.141	0.059	0.0094	2*	2	150	2.1 × 0.85
EPC8010	100	6	160	0.36	0.13	0.06	2.7	7.5	150	2.1 × 0.85
EPC8003	100	6	300	0.315	0.11	0.034	2.7	5	125	2.1 × 0.85

In the asymmetric configuration the chip size of the upper transistor is approximately four times smaller than that of the lower one. Their resistances also turn out to be different. Such asymmetry is used, for instance, in the DC/DC-converters, operating with the small pulse durations, with the great differences between the input and output voltage. In such cases the lower switch always turns out to be loaded substantially greater.

Application of the integrated subassemblies instead of the discrete components makes it possible for the hardware designers to create the more compact solutions; meanwhile, such a subassembly possesses the minimum values of the parasitic inductances, which provides the additional advantages at the higher frequencies (Figure 3.39) [51].

Of course, for control of such eGaN FET, possessing a number of peculiarities, it is necessary to use the special drivers. These drivers of eGaN FET should not only

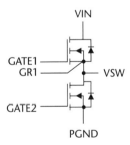

Figure 3.37   Equivalent electric circuit of eGaN FET subassemblies.

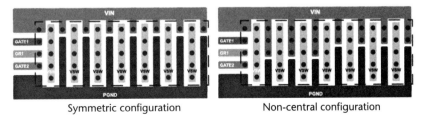

Symmetric configuration   Non-central configuration

Figure 3.38   Type configuration of eGaN FET subassemblies of EPC production.

Table 3.9 Integrated Subassemblies eGaN FET of General Purpose

Identification	Type	U_{ds} max., V	U_{gs} max., V	Rds on max., mOhm, at U_{gs} = 5 V	Qg type, nK	Qgs type, nK	Qgd type, nK	I_d, A	I_d pulse, A	TJ max., °C	LGA package, mm
EPC2100	Noncentral	30	6	8; 2	3.5; 15	1.4; 4.6	0.57; 2.6	9.5; 38	100; 400	150	6.1 × 2.3
EPC2101	Noncentral	60	6	11.5; 2.7	2.7; 12	1; 3.7	0.50; 2.5	9.5; 38	80; 350	150	6.1 × 2.3
EPC2102	Symmetrical	60	6	4.4	6.8	2.3	1,4	23	215	150	6.1 × 2.3
EPC2105	Noncentral	80	6	14.5; 3.5	2.5; 10	1; 3.2	0.50; 2	9.5; 38	75; 320	150	6.1 × 2.3
EPC2103	Symmetrical	80	6	5.5	6.5	2	1.3	23	195	150	6.1 × 2.3
EPC2104	Symmetrical	100	6	6.3	7	2	1.2	23	165	150	6.1 × 2.3

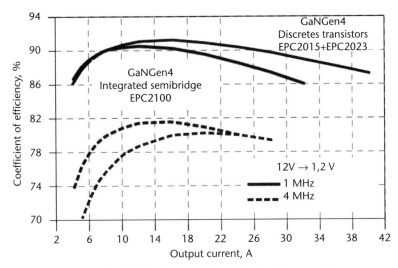

Figure 3.39   Advantage of eGaN FET subassemblies at the high frequencies.

form the appropriate control currents and voltages, but also have the additional functions. This is related to the applied controllers.

First, they should have the increased operating frequencies. Second, they should differ with the minimum own losses. Third, the transistors should be turned on and off with the safe levels of *dv/dt* and *di/dt*.

Table 3.10 lists the recommended control integrated circuits by the company EPC. Here most optimum is application of the drivers LM5113 and LM5114 from Texas Instruments.

Table 3.10   Recommended Drivers and Controllers of eGaN FET

Identification	Functional	Producer	Description
LM5113	Driver	Texas Instruments	5A, 100V semi-bridge driver eGaN FETs
LM5114	Driver	Texas Instruments	7,6A single channel driver of the lower switch
UCC27611	Driver	Texas Instruments	4A/ 6A high speed 5 V single channel driver
ADP1851	Controller	Analog Devices	Step-down controller 2.75V/20V
ISL6420	Controller	Intersil	Synchronous step-down controller 4,5V/16V
LM27403	Controller	Texas Instruments	Synchronous step-down controller 3V/20V
LTC3833	Controller	Linear Technologies	Step-down controller 4.5V/38V
LTC3891	Controller	Linear Technologies	Synchronous step-down controller with the low consumption 60V
MAX15026B	Controller	Maxim	Synchronous step-down controller 4.5V/28V
MCP19118/19	Controller	Microchip	Step-down pulse PDM-controller 4.5V/40V
SC419	Controller	Semtech	Step-down controller with the integrated diode 3V/28V
TP253219A	Controller	Texas Instruments	Synchronous step-down controller 4.5V/25V
TPS40490	Controller	Texas Instruments	Step-down pulse PDM-controller 4.5V/60V
UCC24610	Controller	Texas Instruments	Synchronous step-down controller for the secondary winding

Of course, creation of the power converters and high frequency devices requires high skill and enhanced attention [52]. For minimization of errors at the first stages during the design development, it is logical to use the ready solutions of debugging means and design from EPC.

Thus, the company EPC provides the design engineers with complete informational support: documentation on the components, models (PSPICE, TSPICE, LTSPICE), libraries of seats for Altium Designer, and heat models. However, a great variety of intricacies may result in shortcomings or even errors in design. In order to avoid this and also to promptly master the new devices, it is necessary to use the ready solutions:

- Debugging circuit boards;
- Demonstration kits;
- Completed modules DrGaNPLUS.

The debugging printed circuits eGaN FET (Tables 3.11 and 3.12) represent the finished printed circuit boards with the semi-bridge circuit, driver, required additional passive components, and logic. Practically for each eGaN FET there is its own debugging circuit board. This pertains to both the discrete transistors and integrated subassemblies (Table 3.11) and also eGaN FET for the high frequency applications (Table 3.12).

Note that EPC produces a whole group of demonstration kits step-down converter, kits of wireless power transfer, and amplifiers of the D class (Table 3.13).

Here it is necessary to especially note the modules DrGaNPLUS, which represent the miniature circuit boards, intended for the direct embedding into the finished devices. Their devices constitute only 11×12 mm (Figure 3.40). Operating voltage constitutes 30V (EPC9201) or 80V (EPC92013) (Table 3.14).

3.13.2 Microwave Amplifiers Based on GaN Technology

Let's consider the following solid-state power amplifiers manufactured under GaN-HEMT GaN- technologies used in microwave amplifiers for the aerospace and defense industries.

As shown earlier, the solid-state microwave power amplifiers and broadband amplifiers are especially important elements of space and radar systems for various applications, so they are subject to particularly high requirements. Microwave power amplifiers for radar based on power transistors and modules are the devices that determine the most important parameters of the system, such as power radiation and power consumption, operating frequency band width, size and weight, durability, reliability, and cost.

The sphere of wide-area semiconductor materials (silicon carbide SiC and gallium nitride GaN) and devices based on them has become the breakthrough trend in the development of high-power microwave components [53, 54]. Therefore, the leading companies of the world in the production of components used in solid-state power amplifiers are primarily developing technologies based on gallium nitride GaN [53, 55].

Table 3.11   Base List of Debugging Circuit Boards eGaN FET of General Purpose

Identification	Description	Uds max., V	I_d max., A	Type of Transistor
EPC9036	Debugging circuit on the basis of the integrated semi-bridge	30	25	EPC2100
EPC9031	Semi-bridge circuit with a driver	30	40	EPC2023
EPC9018	Debugging circuit on the basis of the integrated semi-bridge	30	35	EPC2015/ EPC2023
EPC9016	Semi-bridge circuit with a driver for applications with the minimum fill-up coefficient	40	25	EPC2015
EPC9032	Semi-bridge circuit with a driver	40	35	EPC2024
EPC9005C	Semi-bridge circuit with a driver	40	7	EPC2014C
EPC9005	Semi-bridge circuit with a driver	40	7	EPC2014
EPC9001	Semi-bridge circuit with a driver	40	15	EPC2015
EPC9037	Debugging circuit on the basis of the integrated semi-bridge	60	22	EPC2101
EPC9038	Debugging circuit on the basis of the integrated semi-bridge	60	20	EPC2102
EPC9033	Semi-bridge circuit with a driver	60	30	EPC2020
EPC9049	Semi-bridge circuit with a driver	60	4	EPC2035
EPC9046	Semi-bridge circuit with a driver	80	22	EPC2029
EPC9034	Semi-bridge circuit with a driver	80	27	EPC2021
EPC9041	Debugging circuit on the basis of the integrated semi-bridge	80	20	EPC2105
EPC9039	Debugging circuit on the basis of the integrated semi-bridge	80	17	EPC2103
EPC9019	Semi-bridge circuit with a driver for applications with the minimum fill-up coefficient	80	20	EPC2001/ EPC2021
EPC9040	Debugging circuit on the basis of the integrated semi-bridge	100	15	EPC2104
EPC9035	Semi-bridge circuit with a driver	100	25	EPC2022
EPC9006	Semi-bridge circuit with a driver	100	5	EPC2007
EPC9010C	Semi-bridge circuit with a driver	100	7	EPC2016C
EPC9010	Semi-bridge circuit with a driver	100	7	EPC2016
EPC9050	Semi-bridge circuit with a driver	100	2.5	EPC2036
EPC9002	Semi-bridge circuit with a driver	100	10	EPC2001
EPC9047	Semi-bridge circuit with a driver	150	12	EPC2033
EPC9014	Semi-bridge circuit with a driver	200	4	EPC2019
EPC9017	Debugging circuit on the basis of the integrated semi-bridge	100	20	EPC2001
EPC9013	Debugging circuit on the basis of the integrated semi-bridge	100	35	EPC2001
EPC9004C	Semi-bridge circuit with a driver	200	3	EPC2012C
EPC9004	Semi-bridge circuit with a driver	200	3	EPC2012
EPC9014	Semi-bridge circuit with a driver	200	4	EPC2019
EPC9003C	Semi-bridge circuit with a driver	200	5	EPC2010C
EPC9003	Semi-bridge circuit with a driver	200	5	EPC2010
EPC9042	Semi-bridge circuit with a driver	300	3	EPC2025
EPC9044	Semi-bridge circuit with a driver	400	1.5	EPC2027

Table 3.12 List of Debugging Circuit Boards eGaN®FET for Radio Frequency Applications

Identification	Description	Uds max., V	I_d max., A	Type of Transistor
EPC9024	Semi-bridge circuit with a driver	40	4.4	EPC8004
EPC9027	Semi-bridge circuit with a driver	40	3.5	EPC8007
EPC9028	Semi-bridge circuit with a driver	40	2.2	EPC8008
EPC9022	Semi-bridge circuit with a driver	65	1.6	EPC8002
EPC9025	Semi-bridge circuit with a driver	65	2.2	EPC8005
EPC9029	Semi-bridge circuit with a driver	65	3.5	EPC8009
EPC9023	Semi-bridge circuit with a driver	100	2.2	EPC8003
EPC9030	Semi-bridge circuit with a driver	100	3.2	EPC8010

Table 3.13 Basic Demonstration Kits eGaN FET

Identification	Description	U_{in}, V	U^{out}, V	I_d max., A	Transistor Type
EPC9101	Step-down converter 19 V/1.2 V, 1 MHz	8–19	1.2	18	EPC2015/EPC2014
EPC9102	Step-down converter 48 V/12 V	36–60	12	17	EPC2001
EPC9105	Step-down converter 48 V/2 V, 1.2 MHz	36–60	12	30	EPC2001/EPC2015
EPC9106	150 W/8 Ohm audio amplifier of D class	–	–	–	EPC2016
EPC9107	Step-down converter 28 V/3.3 V	9–28	3.3	15	EPC2015
EPC9111	Demonstration kit of the wireless energy transfer, complying with the requirements A4WP	8–32	$U_{вх}$	10	EPC2014
EPC9112	Demonstration kit of the wireless energy transfer, complying with the requirements A4WP	8–32	$U_{вх}$	6	EPC2007
EPC9115	Step-down converter 48 V/12 V	48–60	12	42	EPC2020/EPC2021
EPC9118	Step-down converter 48 V/5 V, 400 kHz	30–60	5	20	EPC2001/EPC2021
EPC9506	Demonstration circuit board of the wireless amplifier of D class	8–32	$U_{вх}$	10	EPC2014
EPC9507	Demonstration circuit board of the wireless amplifier of D class	8–32	$U_{вх}$	6	EPC2007
EPC9508	Demonstration circuit board of the wireless amplifier of D class	7–36	$U_{вх}$	3	EPC8009/EPC2007

The development of GaN devices using GaN-on-SiC HEMT-technology is going on in the following areas: increase in the maximum capacity, power density, and maximum voltage; increase in the upper band of microwave frequencies (advanced developments are aimed at C, X, and Ku wavelength bands); reduction in the cost to the level of one dollar per watt; increase in reliability and radiation resistance (see Table 3.15); and downsizing the products [53, 56].

The history of the development of GaN technology counts for more than 30 years. It should be noted that in 1993 there was a first LED GaN-LED, and in 1997 the first sample of GaN-transistor and samples of amplifiers was based on it. Military and government programs assisted in development and introduction of

Figure 3.40   External appearance of the finished modules DrGaNPLUS.

GaN technology into practice. It's a well-known American WBGSTI program, and later, European MARCOS, TIGER, KORRIGAN, and Japanese NEDO [53, 57].

In 2001 the first commercial GaN-transistor was produced [58]. Then all world's leading electronics companies, previously associated only with the production of GaAs-components, began to make their own investments in new GaN technology primarily for the manufacture of amplifiers and MMICs devices. In 2006–2007, the first commercial mass GaN-products were produced: high-power universal packaged transistors in L and S bands (2–4 GHz) with the power output of 5 to 50 watts (later up to 120–180 watts).

The pioneers entering the commercial market were Eudyna (Sumitomo) [57], Nitronex [59], Cree [56], and RFHIC [57], later joined by Toshiba, RFMD, M/A-COM Tech, TriQuint [60], OKI [61], Microsemi [53, 62], NXP, and others.

It is known that the larger the band gap of a semiconductor material is, the higher the permissible operating temperature and the more the operating range of devices created on the basis of respective semiconductor materials is shifted to

Table 3.14   Basic Technical Characteristics of Modules DrGaNPLUS

Identification	Description	Uds max., V	I_d max., A	Transistor Type
EPC9201	Semi-bridge circuit on the discrete transistors	30	20	EPC2015/EPC2023
EPC9203	Semi-bridge circuit on the discrete transistors	80	40	EPC2021

Table 3.15 The Electronic Properties of Semiconductor Materials Si, GaAs, InP, 3-4-6H-SiC, GaN, C (Diamond)

Material Properties	Units	Si	GaAs (AlGaAs/ InGaAs)	InP (InAlAs/ InGaAs)	3C-SiC*	4H-SiC*	6H-SiC*	GaN (AlGaN/ GaN)	C (Diamond)
Width of the band gap, E_g	eV at 300K	1.12	1.42	1.34	2.4	3.26	3	3.39	5.47
Electron mobility, I_n	300 K, cm^2/V*s	1500	8500	4600	1000	950	500	2000	2800
Mobility of holes, I_p	300 K, cm^2/V*s	600	400	150	40	120	80	200	2100
Speed of the electrons drift at saturation, v_{sat} * 10^7	cm/s	1.0	2.1	2.3	2.5	2.0	2.0	2.7	1.5–2.0
Critical electric field, E_c	mV/cm	0.025	0.4	0.5	2.0	2.2	2.5	5.0	20.0
Thermal conductivity, K	Wt/cm*K at 300 K	1.5	0.55	0.7	3.0–4.0	3.0–4.0	3.0–4.0	1.3	24.0
Dielectric constant, ε	—	11.68	12.8	12.5	9.7	10	10	9.5	5.7
C FoM**	—	1	8.5	21	—	250	—	660	75000

* 3C, 4H, 6H—proposed crystal structures of SiC materials.

** Combined figure of merit (C FoM)—quality factor of the material with respect to silicon for power and frequency.

shorter wavelengths. For example, the maximum operating temperature for the GaN devices reaches 350–400°C; for devices based on C (diamond) the operating temperature reaches 500–600°C and above. The band gap correlates well with the melting point of the material. Both of these values increase with the increase in the binding energy of atoms in the crystal lattice, so wideband semiconductor materials have high melting point, which brings certain difficulties for creating pure and structurally perfect single crystals of semiconductor materials.

Of course, such physical parameters as the carrier mobility largely determines the frequency response of any semiconductor device. Therefore, to create microwave devices one should use the semiconductor materials with high values of the carrier mobility and capable of operating at high temperatures and high levels of radiation, which is very important for the space industry.

Thus, the larger the band gap is, the more stable operation of the microwave transistor (see Table 3.15) at high temperatures and high levels of radiation, and the higher the concentration of electrons, and the higher the current density in the transistor channel section, which leads to its high coefficient of amplification. The higher the maximum critical electric field strength of the semiconductor material used is, the higher the maximum drain voltage of the microwave transistor (50–100V) and its breakdown voltage (100 to 300), which increases the reliability and lifetime of the product. Microwave transistors made under GaN technology have high power densities—up to 10 watts or more per 1 mm gate width, which is much greater than the specific output power of microwave transistors produced using GaAs-technology [57, 63]. Still important but gradually solved problem of GaN technology is ensuring heat removal from the active structure of the crystal and growing GaN epitaxial structures [53, 57].

It is known that earlier solid-state amplifiers for radar with active phased array radars (APAR) or active electronically scanned array (AESA) were manufactured on the basis of GaAs technology. However, it had a lower power density (0.5–1.5 W/mm; see Table 3.16) compared to high power density of GaN-HEMTs-technology (4.0–8.0 W/mm microwave transistor gate width) [53, 63].

The high density of microwave power transistors manufactured using GaN technology can significantly reduce the size and weight of the solid-state radar amplifier, which is very important for the air and space applications requiring weight and dimensions minimization of radars, including APAR [53, 64, 65]. For example, instead of five GaAs LDMOS amplifiers, one can use only one GaN amplifier, which provides much better technical characteristics of the product as a whole [53, 58].

Table 3.16   Comparative Characteristics of the Basic Parameters of GaAs- and GaN Materials

Material Properties	Units	GaAsi	GaN
Output power density, $\rho(P_{out})$	W/mm	0.5–1.5	4.0–8.0
Operating voltage, V_{ds}	W/mm	5–20	28–48
Reverse voltage, V_{br}	W/mm	20–40	>100
Maximum current density, $\rho(I_{max})$	A/mm	~0.5	~1.0
Thermal conductivity factor, K	W/m*K	47	390(z)/490(SiC)

Experts in physics know that the thermal conductivity factor of GaN material is 8–10 times higher than that of GaAs, which allows for better and quick removal of heat from the MMIC chip and higher power density (see Table. 3.16 and [53, 58]).

A solution to the problem of the production of space communication systems and radars on the basis of new GaN-products for frequency bands C, X, and Ku is important.

Solid-state amplifiers and other products based on GaN technology open large prospects not only for the development of new devices, but also for upgrade of critical frequency bands of 4.1 GHz, 2–6 GHz, 4–12 GHz, 6–18 GHz, and 2–20 GHz already used in manufacturing, which will be competitive microwave tube devices (transistors, teratrons, magnetrons, klystrons, and so on) in output power, efficiency, overall dimensions, reliability, and price.

It is obvious that the GaN technology takes its rightful place in the market of military and space systems in problems of replacement of GaAs MMIC products in transceiver modules APAR of centimeter and millimeter ranges [53, 57]. In this connection it is of great interest for developers of microwave systems to consider the technical parameters of microwave amplifiers based on GaN technology of the Sumitomo company, one of the recognized leaders in this field.

The Sumitomo company was established in 1897 and initially focused on manufacture of copper products. In the beginning of the 1990s Sumitomo was one of the first companies to introduce GaN technology. In 2004, the division of semiconductor products of Sumitomo was renamed to Eudina Devices, and in 2009 Eudyna Devices completely changed its name to Sumitomo Electric Device Innovations. At the time of publication of this book, Sumitomo Electric Device Innovations (hereinafter Sumitomo) was the leader in the development, design, and mass production of microwave products for space communication systems, radio astronomy, general radars, and special-purpose radars for various industries and special applications.

For developers of space purpose systems, it is advisable to examine and use the information about the products of this company (Tables 3.17 and 3.18).

The product line includes:

- GaN-HEMT transistors: for base stations, radar and general application;
- Monolithic integrated circuits MMIC: Ku and V bands, low-noise, Ka-band high-power, C and V bands;
- High-power GaN amplifiers on pallets and MMICs;
- Converters of Ku and Ka bands;
- Multiplexers of Ku and V bands;
- Generators with a wide dynamic range;
- GaAs FET transistors: high-power, power and low-power, on-chip.

Microwave transistors for space, air, ground, and surface radars (APAR), wireless communication systems, and identification systems are manufactured using different technologies. Thus, the microwave transistors of L and S wavelength bands are manufactured by Sumitomo based on GaN technology (GaN-HEMTs, GaN-on-Si, GaN-on-SiC), which is currently the most effective and promising. The key products of Sumitomo company developed under GaN-HEMT technology at the end of 2016 are shown in Figure 3.41.

Table 3.17 Main Applications of Sumitomo Products Based on GaN Technology

Products	Technology	Series	Application/Pulse Duration	Wavelength Band	Operating Frequency Band, GHz	Output Power, W/dBm	Voltage (V)/Impedance (Ohm)/Efficiency (%)
Transistors	GaN-HEMTs	EGN13	Pulse radars/3ms/10%, 1.5ms/25%	VHF-L	1.2–1.4	170	50/50/
Transistors	GaN-HEMTs	EGN	Pulse radars/5ms/10%, 750µs/25%	VHF-S	2.7–3.5	120–600	50/50/
Transistors	GaN-HEMTs	SGN	Pulse radars/300 µs/10%	VHF-S	2.9–3.5	150–600	50/50/
Transistors	GaN-HEMTs	EGNB	Continuous wave radars (CW)	VHF-S	3.5	/40	50/50/
Transistors	GaN-HEMTs	EGN31	Continuous wave radars (CW)	VHF-S	3.1	/45.5	50/50/
Amplifiers on pallets, high-power	GaN-HEMTs	SMC	Pulse radars/300 µs /10%	VHF-S	2.9–3.5	150–600	50/50/50
Amplifiers on pallets, high-power	GaN-HEMTs	EMC	Pulse radars/ 5ms/10%, 750µs/25%	VHF-S	3.1–3.5	100	50/50/50
Amplifiers New	GaN-HEMTs	SGNC	Wireless communications and base stations	UHF-S	0.9–2.6	300/66	50//
Amplifiers	GaN-HEMTs	EGNC-EGN3	Wireless communications and base stations	UHF-S	0.9–3.5	70–270/47–53.5	50//
Amplifiers	GaN-HEMTs	EGNB	General application	UHF-S	0.9–3.5	10–90/41–53	—
Amplifiers	GaN-HEMTs	SGNE	General application	UHF-S	0.9–3.5	10–90/40.5–51	—

Table 3.18   Main Applications of Sumitomo Products, Including MMIC Based on GaAs Technology

Products	Technology	Series	Application/Pulse Duration	Wavelength Band	Operating Frequency Band, GHz	Output Power, W/dBm	Voltage (V)/ Impedance (Ohm)/ Efficiency (%)
Amplifiers, converters	GaAs WLCSP* MMICs	SMM/ EMM	Communication systems	C-E	12.7–30.0	/26–33	6
Amplifiers	GaAs MMICs	SMM/ EMM	VSAT and communication systems transceivers	S-C-Ka	3.4–30.0	/30.0–33.5	6/14–29
Amplifiers	GaAs MMICs	EMM/ SMM/ FMM	VSAT and communication systems transceivers	S-C-V	3.4–64.0	/26–34	3–7/17–29
Amplifiers	GaAs MMICs	EMM/ SMM/ FMM	VSAT and communication systems transceivers	Ku-V	12.0–64.0	/7–20	/13.5–23
Converters	GaAs MMICs	SMM/ FMM	Satellite radio communication	Ku-Ka	12.0–32.0	/5.0	/10–12
High-power amplifiers	GaAs FETs	FLM/ ELM	Radio communication	L-C	2.0–15.3	/39	10/9.5–11.5
High-power amplifiers	GaAs FETs	FLU/FLL/ FLC/FLX/ FLK	Mobile and cellular communication, WCDMA, LTE, and WiMAX	L, S, C, X, Ku	2.0–15.0	40–80/	10/6.0–13.0
Low-power amplifiers	GaAs FETs	PSU/FSX	Average power amplifiers and generators	C	8.0–14.5	/15–24	8–10/10/18
Amplifiers, mixers, converters	GaAs HEMTs	FHC/FHX	DB S-converters, mobile and cellular communication, radio astronomy, and other	S-C	4.0–12.0	—	2.0/10.0–15.5

* WLCSP—wafer level chip scale package

Application of the finished amplified submodule-pallets with a minimum impedance of 50 ohms is important. They are used in aircraft, wireless communication
systems, and radars with S and L frequency bands (Table 3.17).

In many cases, two or more discrete elements are mounted on a pallet to achieve
gain that cannot be obtained using a single element. The pallet is made under special
technology [53, 57], which allows you to create a compact device. Small size of the
pallet is perfect for space and air applications.

Besides, Sumitomo is actively developing the products based on GaN-on-SiC
technology (gallium nitride on silicon carbide) [53, 55]. Products based on this
technology are widely used in the aerospace industry.

The following parameters are important for microwave amplifiers: frequency
band (MHz), supply voltage (V), output power (W), values of the input/output
impedance, gain (dB), efficiency (%), and fill factor (%) (see Table 3.17).

High-power microwave transistors operate in a wide band of frequencies and
provide the maximum power in the load.

The popularity of GaN transistors is related to their real benefits at the system
level, such as ease and low cost of schematic implementation of high-power amplifiers; ease of obtaining the gain broad bands; overlap by one high-power amplifier
of several subbands of space communication station; lower power consumption of
the product (radar) and related costs; and reducing the complexity and cost of cooling systems. For example, an amplifier with the power of 10W of L-band has an
efficiency of more than 70–80%, and the amplifiers with the power of 100–600W
have 50% efficiency.

Developers of space-purpose devices very often face the question of how to
find a compromise between GaN- and GaAs-technologies or plan the use of new

New products of Sumitomo Electric Europe, Ltd

SMC2935L3012R* Pallet	**SMC2933L6012R Pallet**	**SGN2933-600D-R**
GaN-HEMT amplifier on a pallet, Pout > 300 W in the frequency band 2.9-3.5 GHz (S-band), amplification (Gain) = 12.8 dB, Vds = 50 V, Ids = 1.5 A, P/D = 300 µs/10%, impedance 50 Ohm, Eff = 48 %	GaN-HEMT amplifier on a pallet, Pout > 600 W in the frequency band 2.9-3.5 GHz (S-band), amplification (Gain) = 12.8 dB, Vds = 50 V, Ids = 3.0 A, P/D = 300 µs /10%, impedance 50 Ohm, Eff = 50 %	GaN-HEMT amplifier on a pallet, Pout > 600 W in the frequency band 2.9-3.3 GHz (L-band), amplification (Gain) = 12.8 dB, Vds = 50 V, Ids = 3.0 A, P/D = 300 µs /10%, impedance 50 Ohm, Eff = 50 %

Figure 3.41   The main products of Sumitomo Electric Europe produced using GaN technology.

technologies on diamond for advanced space systems. To answer this question and to compare the possibilities of application of GaN GaAs transistors and MMICs in the circuits of broad-band power amplifiers, as well as the possibilities for the optimum technical solution, one can apply the method of migration from one SiC material to another one and conduct a simple qualitative analysis of their specific parameters (referred to 1 mm width of the transistor gate). For this purpose, it is recommended to use well-known estimates for the class A amplifier with a maximum output power P_{max} and the transistor load optimal resistance R_{opt} [53, 57].

For example, developers of GaN devices show great interest in broadband communications systems. This primarily relates to the use of GaN transistors in base stations of WCDMA UMTS and WiMAX standards. It is known that in 2014 within the total output of GaN devices the application market in commercial and TV broadband communication systems made up 63% versus 26% in the space and military systems [53, 57].

At the time of writing, the centimeter and millimeter wavelength band became the scene for serious battle and compromise of two industrial production technologies of high-power solid-state monolithic integrated circuits MMICs, discrete microwave products, and amplifiers on pallets almost equal in frequency and amplifying characteristics. One of them (GaN) provides significant benefits to the maximum electrodynamic and minimum weight and dimensional parameters and minimum cost of power sources, and the other (GaAs) still has minimum cost and maximum level of development for mass production [53, 57].

One should understand that a natural change of generations of technology should provide a jump to a higher level to improve the parameters of solid-state amplifiers not only for space, but also for commercial purposes. It is still possible to have time to implement many projects on the basis of GaN technology. But time goes forward, and there are already new technological solutions (i.e., the technological solutions on diamond—diamond FETs).

It should be noted that the artificial synthesis of diamond single crystals has been studied in the world very actively since the 1990s. Since 1990, Sumitomo Electric Industries has been applying to the Bureau of Patents and Trademarks of the United States and has received several key patents for the development of field-effect transistors on diamond.

Diamond, due to its quite unique characteristics, is considered as an ideal material for a new generation of nanoscale electronic components [53, 66–68]. It was found that if treated in hydrogen plasma, the diamond surface becomes p-type conductive. Such properties of hydrogenated diamond made it possible to create the first field-effect transistors based on diamond. For example, back in 2009 the diamond transistor gate width of 50 nm was achieved, which is 1000 times smaller than the thickness of a human hair, and 2 times less than the dimensions of the previous world record-holder of Japanese company NTT [53, 67, 68].

Traditional materials GaN and GaAs have their strengths and weaknesses, while diamond is nearly universal. Diamond transistors (diamond FETs) can be widely used not only in medical terahertz scanners, in car safety systems, and so on, but also in devices for space application [53, 69].

It should be noted that the high radiation resistance of the diamond allows us to consider it as a promising material for onboard UV detectors for space research.

CVD diamond can also be the material for high-energy particle detector (alpha particles, gamma rays, and neutrons), used on spacecraft for investigation of deep space.

None of the existing semiconductor materials can compete with diamond in the set of parameters shown in Table 3.15 and Figure 3.42 [53, 69]. The starting material for the creation of such transistors on diamond is made by the CVD method in hydrogen-vapor phase. To form the structure of such transistors on the surface of the diamond film, the electron-beam lithography technology is used. Due to the large size, the diamond films are promising for creation of position-sensitive and UV and X-ray microstrip detectors mounted on orbital and interplanetary spacecraft, as well as on the devices designed to move along the surface of the planets, moons, and asteroids.

3.13.3 Microwave Devices Based on RFHIC's Gallium Nitride Technology

Ga-N-based devices of Sumitomo, a Japanese company, were reviewed in the previous section, and in this section the reader is invited to become acquainted with the Ga-N devices of RFHIC Corp., a South Korean company being recognized as a leader in this field.

It is known that performance improvement of structures based on GaN, applied for microwave transistors implementation, became feasible due to the development of new methods. One of the approaches is technology of gate region recession by plasma chemical etching generally combined with the process of slit etching in dielectric [70]. This results in improvement of numerous parameters, such as transistor transconductance increase due to the gate-channel distance reduced, source

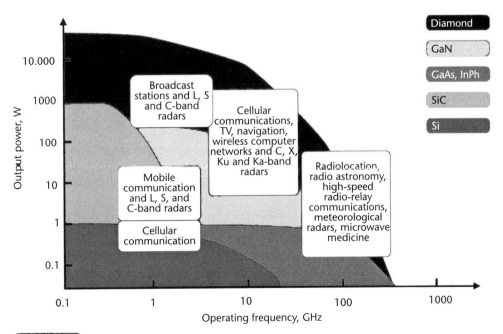

Diamond FETs Transistors on diamond will replace high-poewr tube devices of microwave band

Figure 3.42   Generalized diagram of dependence of the output power and operating frequency of the microwave systems and devices on the type of semiconductor material.

and drain resistance decrease due to the lack of the transistor gate-source and gate-drain regions depletion, and transient processes reduction or even elimination during transistor switching-on due to the decrease of traps influence in the gate-drain region as their location surface can be shifted to a safe distance. In connection with this, process engineers offer to grow up dielectric passivation layer directly after all the heterostructure layers are grown up [70].

Furthermore, in recent years investigations have been directed at searching new passivating materials for transistor heterostructures based on GaN and its solid solutions. Application of new materials can both increase the transistor pulsed current and its transconductance by more than twice and significantly reduce switching times due to the compensation of surface states [70, 71].

As shown in the previous section, researchers and developers from Cree, TriQuint, Northrop Grumman, and other companies achieved high frequency characteristics of transistor structures, which became the basis for the development and implementation of efficient ICs of power amplifiers functioning in various bands. Mass and dimensional parameters of these ICs exceed gallium arsenide (GaAs) based integrated circuits more than 10 times [53, 70].

Mentioned manufacturers have already mastered mass production of power amplifiers on the basis of GaN heterostructures with frequencies up to 100 GHz, and QuinStar Technology together with HRL are in the process of developing transmit-receive modules for 94-GHz ranged radars with an output power of more than 5W [53, 70].

Thus, key production and technological problems that hindered advancement of transistors and GaN-based monolithic ICs to a new commercial level have been resolved [53, 72]. New solutions in the field of GaN technology are offered by the world's leading microwave producers, in particular, RFHIC Corp.

RFHIC Corp. was founded in 1999 and gained its leading position in the telecommunications/cable TV equipment market due to its innovative approach to the technology of components making. To provide its customers with high-quality products at optimal conditions, a full production cycle was implemented in the RFHIC-owned fab, including design and development of the product and its assembly (i.e., die/chip attaching to a *substrate*, wire bonding, packaging and sealing). RFHIC produces a wide mix of products based on GaN for microwave electronics ranging from amplifiers for cable TV to power amplifiers for radars and space applications. Key features of the company's products are considered next.

First, RFHIC's wireless communication systems is worth noting (see Figure 3.43). These amplifiers are designed for application in advanced networks used with different data formats, including LTE, CDMA, WCDMA, and WiMAX. They can be divided into groups based on power and voltage operating range.

The first group includes amplifiers with 5–10W power rating and 28V operating voltage. These amplifiers are designed to operate in the 2–3 GHz band and have a gain of 40 dB. The next group includes amplifiers with a rated power of 28W and operating voltage ranging within 48–50V. These amplifiers are designed to operate in the band of 0.8–3 GHz and have a gain of 44.5 dB. The third group includes amplifiers with a rated power of 56W and operating voltage in the range of 48–50V. They are also designed to operate in the band of 0.8–3 GHz and have a gain of 47.5 dB. The fourth group includes amplifiers with a rated power of 80W

Figure 3.43 RFHIC's amplifier for wireless communication systems (external view).

and operating voltage in the range of 48–50V. These amplifiers are designed to operate within 0.8–3 GHz band at 55 dB gain.

Furthermore, amplifiers for wireless space communication systems include hybrid amplifiers with 1–5W power and 28V operating voltage and amplifiers designed under the Doherty scheme, with 7W power and 31V operating voltage. These amplifiers are designed to operate in the frequency band of 0.8–3 GHz at the gain of 27–38 dB and in the frequency band of 1.5–3 GHz at the gain of 14–16 dB, respectively.

Basic models and characteristics of RFHIC's amplifiers for wireless communications are shown in Tables 3.19–3.21.

Table 3.19 Basic Models and Characteristics of RFHIC's GaN-Based Amplifiers for Wireless Communications

Model	Frequency, MHz	Gain, dBm	Output Power, dBm	Efficiency, %	Dimensions, mm
5–10W amplifiers (28V)					
RTP21005-11	2110–2170	45	37	40	$100 \times 50 \times 20$
RTP21010-11	2110–2170	45	40	40	
RTP26010-N1	2570–2690	40	40	38	
28W amplifiers (48–50V)					
RTP08028-20	869–894	44.5	44.5	42	$125 \times 90 \times 20$
RTP18028-20	1805–1880			42	
RTP21028-20	2110–2170			42	
RTP26028-20	2496–2690			40	
56W amplifiers (48–50 V)					
RTP08056-20	869–894	47.5	47.5	42	$150 \times 90 \times 20$
RTP18056-20	1805–1880				
RTP21056-20	2110–2170				
RTP26056-20	2620–2690				
80W amplifiers (48–50V)					
RTP08080-20	869–894	55	49	42	$170 \times 100 \times 20$
RTP18080-20	1805–1880				
RTP21080-20	2110–2170				
RTP26080-20	2620–2690				

Table 3.20 Basic Models and Characteristics of RFHIC's 1–5W (28V) GaN-Based Amplifiers for Wireless Communications

Model	Frequency, MHz	Gain, dBm	Output power, dBm	Efficiency, %	Operating voltage, V
HT0808-15A	869–894	37	33	26	28
HT1818-15A	1805–1880	33	33	25	28
HT1818-15M	1805–1880	34	33	27	5/28
HT2121-15A	2110–2170	33	33	24.5	28
HT2121-15M	2110–2170	32	33	26	5/28
HT2626-15A	2610–2690	30	34	25	28
HT2626-15M	2610–2690	30	33	25	5/28
HT2627-15A	2650–2750	27	33	31	28
HT0808-30A	869–894	38	37	28	28
HT0909-30A	925–960	36	37	27	28
HT1818-30A	1805–1880	38	37	26	28
HT1919-30A	1930–1995	38	37	26	28
HT2121-30A	2110–2170	36	37	26	28
HT2626-30A	2610–2690	32	37	28	28

Since these amplifiers are designed on the basis of GaN, they can operate at high values of temperature and voltage, which is particularly important for space applications.

Amplifiers for pulsed radar systems (see Figure 3.44) on the basis of GaN have wide frequency ranges covering almost the entire frequency spectrum band from 135 MHz to 10 GHz and high power level.

Amplifiers for radar systems shall be conventionally divided into several groups as well. The first one includes hybrid amplifiers. They cover quite a wide range of operating frequencies from 400–450 MHz to 9.3–9.5 GHz. Their gains can take on values in the range of 11–33 dB. The efficiency of amplifiers in this group can reach 40 to 70%.

Next group includes low noise amplifiers. These amplifiers cover the range of operating frequencies from 1.2–1.4 GHz to 9.3–9.5 GHz and the gain takes values

Table 3.21 Characteristics of RFHIC's GaN-Based Hydride Doherty Amplifiers of 7W (31V) for Wireless Communications

Model	Frequency, MHz	Capacity (bandwidth), MHz	Gain, dB
RTH18007-10	1805–1880	75	16
RTH20007-10	1930–1995	65	16
RTH21007-10	2110–2170	60	15
RTH26007-20	2620–2690	70	14

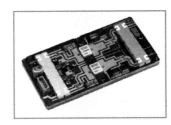

Figure 3.44 RFHIC's amplifier for pulse radar systems (external view).

in the range of 10–18 dB. The efficiency of amplifiers in this group can reach 18 to 30%.

Pallet amplifiers shall be attributed to the third group. They cover the frequency range from 135–460 MHz to 9.0–10.0 GHz and the gain takes values ranging 8–39 dB. The efficiency of amplifiers in this group can reach 5 to 20%.

High-power amplifiers can be isolated in a separate group. The value of output power for this type of amplifiers can reach 200W to 2.6 kW; operating frequency range overlaps practically the entire band from 1.02–1.03 GHz to 9.30–9.50 GHz. The value of the gain is in the range of 20–63 dB, while the efficiency is 20–45%.

Basic models and characteristics of RFHIC's GaN-based amplifiers for radar systems are shown in Tables 3.22–3.26. This category of products includes GaN-based transformer-rectifiers. Their main characteristics and models are also shown in Tables 3.22–3.26.

Table 3.22 Characteristics of RFHIC's GaN-Based Hydride Amplifiers for Radar Systems

Model	Frequency, MHz	Gain, dB	Output power, W	Efficiency, %	Operating voltage, V
HR2731-50A	2700–3100	26	50	50	50
HR2933-50A	2700–3500	25	50	50	50
HR2933-70A	2900–3300	24	80	50	50
HR5459-25B	5400–5900	20	25	40	50
RRC94030-10	9300–9500	17	25	40	50
HR9395-08A	9300–9500	11	8	40	50
HR9395-30A	9300–9500	9	30	40	50
RNP04006-A1	400–450	33	4	70	24

Table 3.23 Characteristics of RFHIC's GaN-Based Low-Noise Amplifiers for Radar Systems

Model	Frequency, MHz	Gain, dB	P1dB, dBm	Input power, dBm	Operating voltage, V
CL1302D-L	1200–1400	18	20	30	5
CL3102D-L	2700–3500	11.5	20	30	5
CL5602	5400–5900	15	23	18	5
CL9402	9300–9500	10	22	18	5

Table 3.24   Characteristics of RFHIC's GaN-Based Pallet Amplifiers for Radar Systems

Model	Frequency MHz	Gain, dB	Output power, W	Efficiency, %	Operating voltage, V
RRP03250-10	135–460	31	300	20	50
RRP10350-10	1030–1090	28	350	10	50
RRP10800-10	1030–1090	27	800	10	50
RRP1214500-14	1200–1400	14	560	20	50
RRP2731080-39	2700–3100	39	100	20	50
RRP2731200-08	2700–3100	8	250	20	38
RRP2731330-09	2700–3100	9	400	20	50
RRP2731160-35	2700–3500	35	180	20	50
RRP2735200-30	2700–3500	32	230	20	50
RRP5257550-35	5250–5750	35	600	5	50
RRP5657500-35	5600–5700	35	550	10	50
RRP9095080-08	9000–9500	8	80	10	50
RRP9095150-18	9000–9500	18	150	10	50
RRP090100120-15	9000–10000	15	120	10	50
RRP3842075-30	3860–4140	29	75	10	50

Table 3.25   Characteristics of RFHIC's GaN-Based High-Power Amplifiers for Radar Systems

Model	Frequency MHz	Gain, dB	Output power, W	Efficiency, %	Operating voltage, V
RRP102600-10	1026–1034	20	2600	40	25.5
RRP131K0-10	1200–1400	53	1000	45	12.5
RRP291K0-10	2700–3100	60	1000	35	12.5
RRM291K5-10	2700–3100	62	1500	30	12.5
RRP311K0-10	2900–3300	60	1000	35	12.5
RRM27312K0-62	2700–3100	63	2000	30	12.5
RRM52571K0-50	5250–5750	50	1000	25	12.5
RRM9395200-56	9300–9500	56	200	20	50

Table 3.26   Characteristics of RFHIC's GaN-Based Transformer-Rectifiers for Radar Systems

Model	Frequency, GHz	Output Power, W	Gain, dB	Duty Cycle, %	Pulse Width, μs
RFCR91-XTRM-015SP-200A	9.0–9.2	15	42	10	100
RFCR93-XTRM-015SP-500A	9.0–9.5	15	42		100
RFMR57-CTRM-020SP-500A	5.4–5.9	20	40		50
RFMR31-STRM-200-400B	2.9–3.3	200	53		500
RFMR13-LTRM-250-100B	1.3–1.4	250	54		2000
RFMR13-LTRM-250-200B	1.2–1.4	250	52		2000

Another type of RFHIC products is a high-power amplifier for space and commercial applications (see Figure 3.45). Amplifiers of this type are also developed based on GaN, thus making it possible to use the devices type in severe operating conditions, particularly at elevated ambient temperature. These amplifiers cover sufficiently wide band of frequencies from about 20–512 MHz up to 2.0–6.0 GHz. Gain values are within the range from 17–60 dB, and operating voltage values are in the range of 24–33V. The basic models of the group and their characteristics are shown in Table 3.27.

It should be noted that RFHIC Corp. goes beyond the supply of serial products and provides customized product manufacturing services in addition. This advantage can be used by developers of microwave components for space vehicles.

3.13.4 Microwave Devices Based on the Technology of Heterogeneous Integration

Unlike digital ICs, which were historically developing exclusively based on silicon technology, from the beginning microwave devices were developing based on the technologies for manufacturing complex semiconductors—first, gallium arsenide (GaAs) technology and later based on gallium nitride (GaN), silicon carbide (SiC), and other technologies.

Overseas, those technologies were developing mostly due to funding of the famous DARPA in different programs known in the open press as MIMIC [53], WBGS-RF, and devoted mostly to the development of GaAs and GaN microwave devices for military applications.

Starting from 2014, in frames of those programs there is actively developing a new technology, the so called heterogenous integration for microwave ICs [53] built on the integration in a single semiconductor chip of different technologies, both silicon ones and those of complex semiconductors. The final target of one of the sections of this program is creation in the near term of an integrated technology that will allow optimization of the operating conditions of each device on the basis of complex semiconductors, such that by means of certain digital tuning those might be capable to perform the required specialty operations (functions). When the assigned target is achieved, this technology may sufficiently affect the development of modern microvawe electronics.

Figure 3.45   RFHIC's multipurpose high-power amplifier (external view).

Table 3.27   Basic Models and Characteristics of RFHIC's GaN-Based High-Power Amplifiers for Radar Systems

Model	Frequency, MHz	P3dB [P1dB], dBm	Gain, dB	Operating voltage, V	Current, A
RWS02520-10	20–512	43	40	28	2.5
RWS02540-10	20–512	46	41	28	3.2
RWS05020-10	20–1000	43	36	28	2.3
RWP03040-10	20–520	46	42	28	3.8
RWP03040-50	20–500	46	39	28	4
RWP03160-10	20–500	52	43	28	11
RWP05020-10	20–1000	43	40	28	2.3
RWP05040-10	20–1000	46	38	28	3.5
RWP06040-10	450–880	45	40	28	3
RWP06040-60	500–1000	46	42	28	4.5
RWP15040-10	500–2500	47	38	32	5.0
RWP15080-10	700–2700	50	53	32	10.0
RWP17050-10	700–2700	47	37	32	4.5
RUP15010-10	500–2500	40	17	28	1
RUP15020-10	500–2500	43	15	28	2
RUP15020-11	500–2500	43	50	30	3.5
RUP15030-10	500–2500	45	13	28	4
RUP15050-10	500–2500	47	13	28	5.5
RUP15050-11	500–2500	47	60	30	7
RUP15050-12	500–2500	47	60	30	7
RWP15020-50	1000–2000	43	29	28	3.6
RNP19040-50	1800–1900	47,5	33	28	3.7
RWP25020-50	2000–3000	44	25	28	2.8
RNP21040-50	2100–2170	47,5	33	28	3.9
RFC042	400–800	—	23	24	0.4
RFC092	800–1000	[30]	23	24	0.4
RFC1G22-24	20–1000	[30]	22	24	0.4
RFC1G21H4-24	20–1000	36	21	24	0.55
RFC1G21H4-24-S	20–1000	36	21	24	0.55
RFW2500H10-28	20–2500	36	17	28	0.7
RWM03060-10	20–520	49	55	28	7
RWM03125-10	20–520	51	55	28	9
RWM03125-20	20–520	51	55	28	9
RUM15040-10	500–2500	46 (Psat)	56	28	5.5
RUM15040-20	500–2500	46 (Psat)	56	28	5.5
RWP15100-R0	500–2500	50	42	33	10
RUM43010-10	2500–6000	40 (Psat)	29	28	2.2
RUM43020-10	2000–6000	43 (Psat)	35	28	4

As is shown in [53], the microwave IC technologies based on the GaAs and GaN materials resulted from the execution of MIMIC and WBGS-RF programs and became the basis for making completely new radio-electronic weapons, particularly AFAR radar system [73–75]. One of the strategic tasks of the DARPA program is, based on the heterogenous integration technology, to cardinally improve the performance of typical microwave ICs and modules and to further enhance the effectiveness of modern radio-electronic control systems for armament and military hardware.

Each type of complex semiconductors has its own peculiarities used for designing microwave devices. For example, an InP allows having transistors with maximum frequency over 1 THz and GaN provides higher breakdown voltages and higher output power. So far, all complex semiconductors are inferior to silicon in terms of IC integration (number of transistors per circuit). For now, the most promising in this respect is the InP yields to silicon by more than five orders (not including GaN, which provides lower integration levels). Simultaneously, silicon CMOS technology, according to the Moore's law, continuously increases IC speed and integration level while providing a higher percentage yield, which is so far unachievable with complex semiconductors. Moreover, silicon ICs allow placement on one chip of built-in correction and optimization circuits for parameters and units operating simultaneously with high-frequency and digital types of signals. The perspectives for further development of integrated microwave electronics lie in the area of technology integration for different types of complex semiconductors and silicon (i.e., their heterogenous integration). With a view of development of the heterogenous integration technology for microwave ICs in frames of the base DARPA program, a series of specialty projects [76–79] have been performed with the major results received in the course of performance, as is considered next.

3.13.4.1 COSMOS Project

The compound semiconductor materials on silicon (COSMOS) project was realized during the period from 2007 to 2013 [76]. Usually, issues for integration of complex semiconductor chips and silicon are solved by means of assembly of several IC types in specialty multichip modules (microassemblies), though parameters of the module simultaneously deteriorate. This is stipulated by different mismatching effects due to big lengths (areas) of semiconductors connecting heterogenous types of chips and devices. Therefore, in the COSMOS subprogram, an integration process was transferred to the level of transistors implemented, as minimum, on one type of complex semiconductor and which are manufactured along with transistors on wafer on the upgraded CMOS process. The length of interconnections between different transistor types and the distance between the conductors to different transistors did not exceed several micron (Table 3.28). In the COSMOS subprogram, InP was taken as a material for complex semiconductor, and standard ICs were taken as pilot sample devices—differential amplifier (DA), digital-to-analog converter (DAC), and analog-to-digital converter (ADC). As known from [53], in frames of COSMOS project, three technological heterointegration methods have been studied.

That was actually stipulated by selection of three companies, each one proposing their own technology. Those three companies include Northvop Crummen, Raytheon, and HRL Laboratories. Therefore, those process solutions look as follows:

Table 3.28 Major Technical Parameters of Microwave ICs Obtained in COSMOS Project [53]

Technical Parameters	IC Type		
	DA	DAC	ADC
Percentage yield for heterogenous connections, X (defined on a separate test wafer with a big quantity of heteroconnections switched in parallel)	>99	>99.9	>99.9
IC reliability, % (defined as part of the remained good IC after 100 thermocycles ranging from −55 to 85°C and dwell time at each extreme temperature not less than 10 min)	>50	>95	>95
Minimal length of heteroconnections between CS- and SI transistors, um	<5	<5	<5
Minimal distance between power supply lines of CS- and Si transistors, um	<25	<5	<5
Conductance ratio of the microwave transistors tested before and after heterointegration process, %	>80	>90	>90
Conductance ratio of the Si CMOS-transiistors tested before and after heterointegration process, %	>80	>90	>90
Microwave IC yield, %	>25	>50	>50

- Heteroepitaxial growth of InP HBT transistors coplanarly with silicon CMOS structure on a multilayer silicon platform (Raytheon);
- Micrometric assembly, when finished microchips (chiplets) with InP transistors are mounted on a fully processed silicon substrate (a group of companies headed by Northrop Grumman);
- Only producing epitaxial InP heterostructures by printing method on a processed silicon substrate, followed by separate processes of making InP transistors and ICs (HRL Laboratories).

One mandatory requirement for all participants of the project was to present a specific manufactured and operating integrated device of enhanced complexity. In the first stage (2007–2009), all methods of heterogenous integration were studied theoretically and experimentally, and simultaneously, the first differential amplifier IC with record parameters was made. That IC contained ten heterogenous connections, five InP HBT transistors, and four CMOS transistors. The major condition of the second stage (2009–2011) was enhancement of the percentage yield. In conclusion of the second stage, there was manufactured a 13-bit DAC containing over 500 heteroconnections, including 400 InP HBT transistors and 3200 CMOS transistors. In the third stage of the program (2011–2013), it was shown, on the example of ADC, that the developed integration process may be used for manufacturing analog-digital LCI. An experimental ADC circuit counted more than 1,000 InP HBT transistors, 18,000 CMOS transistors and over 1,800 heterogenous conections linking InP-microchip with a master silicon chip.

A combination of the submicron technology of silicon ICs and high-speed InP transistors has allowed us to apply several calibration and self-healing methods that earlier were impossible when only using a single InP technology. In its turn, InP HBT transistors provided higher operating speed and breakdown voltages and also a good transistor matching compared with classical CMOS process. Therefore, the super wideband 130-nm CMOS ADC developed during implementation of the project turned out to be, in the effective number of bits, comparable with ADC on the basis of a purely silicon 32-nm CMOS process. Furthermore, signal-to-noise and distortion ratio of the ADC exceeded 30 dB in the frequency ranges from 2.75 to 8.75 GHz and 14.25 to 20.25 GHz, which corresponds to the best ratings received at that moment on the basis of silicon.

During execution of COSMOS project, each participant used his own design and process solutions. Particularly, Raytheon developed the process of integration of GaN HEMT and Si CMOS transistors on the modified standard silicon-on-insulator wafer, which included the base Si (111) substrate and Si (100) patterned layer (Figure 3.46). Through the window used in the layers of silicon dioxide, on the Si (111) surface of the base substrate, a GaN HEMT-transistor epitaxial layer was grown. Based on this solution, a power GHz-range GaN-amplifier circuit was created, which included GaN HEMT transistors integrated with a silicon transistor offset control circuit (Figure 3.47). An output power and the efficiency of the

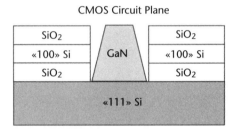

Figure 3.46 Design outline of the multilayer silicon substrate for monolithic integration of GaN HEMT and CMOS IC.

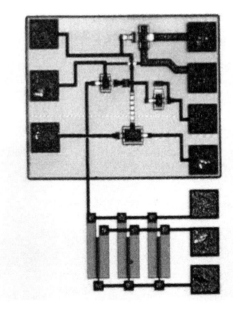

Figure 3.47 Layout outline of the integrated CMOS power amplifier on GaN HEMT.

amplifier on silicon substrate turned out to be close to parameters of silicon amplifier on silicon carbide (SiC) substrate.

In the COSMOS project there was also used the method of multiproject development (multiproject wafer), where on one common wafer several different IC designs are produced, which allows us to sufficiently reduce the costs for mask making for experimental sample ICs. In particular, one of the participants of the project, U.S. Air Force Research Laboratory, has created a four-stage low-noise mm-range amplifier with a noise ratio of not more than 7.2 dB in the frequency band from 75 GHz to 100 GHz and gain ratio of not more than 20 dB. Moreover, amplifier power dissipation only amounted 19 mW.

3.13.4.2 NEXT Project

As, when building complex ICs, the capabilities of GaN as one of the major materials used for heterogenous integration are quite limited [77], in 2009–2012 DARPA financed the development of another new GaN technology on the Nitride Electronic NeXt Generation (NEXT) project, which was targeted to the enhancement of speed and integration level of GaN HEMT transistor ICs at the expense of structural and circuit design solutions and, uppermost, at the expense of reduction of geometrical transistor dimensions and the use of new epitaxial structures. As a result of the project execution, on sample ICs there were increased transistor cutoff frequencies corresponding to the gain ratio 1 (unity-gain cutoff frequency, f_T), and maximum oscillation frequencies (f_{max}). As is known, with simple physical reduction of sizes there is simultaneously decreased the maximum allowable transistor breakdown voltage. For evaluation of optimal correlation of those parameters, designers are using so called Johnson coefficient, which is a simple product of the transistor frequency cutoff and the breakdown voltage. In frames of the NEXT project they managed to raise the Johnson coefficient to the value of 5 THz V (Table 3.29).

Naturally, in the course of the research, to extend the circuit GaN capabilities, it was required to make certain modifications to the operating conditions of GaN

Table 3.29 Major Results of Realization of the NEXT Project on the Improvement of Characteristics of CaN HEMT Transistors

Characteristic	Value
Cutoff frequency in the depletion (D) mode, GHz	500
Maximum oscillation frequency in D mode, GHz	550
Johnson coefficient in D-Mode, THz × V	5
Cutoff frequency in the enhancement (E) mode, GHz	400
Maximum oscillation frequency in E-mode, GHz	450
Johnson coefficient in E-mode, THz × V	5
Transistor yield, %	95
RMSD threshold voltage, mV	30
RMSD cutoff frequency, GHz	30
RMSD maximum oscillation frequency, GHz	5
Operation time, h	>1000

HEMT transistors. As is known, GaN HEMT transistors usually operate in the depletion mode (D mode); however, for operations with digital signals, it is necessary to use so called enhancement mode (E mode) transistors, as the joint use of both transistor types sufficiently simplifies building the microwave IC and allows creation on wafer different logic circuits with immediate links with a possibility of using over 1,000 such base elements.

Thus, the results obtained in frames of the COSMOS project have confirmed the possibility of significant increases in the IC parameter values with heterogenous integration, which was experimentally proved on the example of only one type of complex conductor (InP).

3.13.4.3 DAHI Project

An effective progress of the COSMOS program of the United States DARPA agency [78–80] is the Diverse Accessible Heterogeneous Integration Program (DAHI), which major task was the development of the strategy for distribution of the methods of heterogenous integration to the whole wide spectrum of microwave devices, including GaN, MEMS devices, and temperature control circuitry. The final target of the DAHI project is setting up the industrial production of microwave ICs, both for military and commercial applications.

As a result of analysis of the studies on the DAHI project, from all the heterointegration methods investigated in the COSMOS program, the experts have selected a single method of assembly of an integrated unit from heterogenous, completely processed microchips. Really, this method does not require any introduction of sufficient changes, in either the base silicon technology or the technology of integrated devices on the basis of complex semiconductors. All other investigated methods refer to sufficient changes in the technology before or during the heterointegration process and require significant financial and time costs. It should be said in all fairness that those technologies are quite effectively used in a row of countries, especially in the areas of military and space applications, from what is just briefly publicly announced in the press.

Particularly, in one of the jobs performed on the DAHI project another two technologies of complex semiconductors on one silicon wafer have been implemented (Figure 3.48). In composition of the manufactured oscillator-amplifier chain of the mm-range, an oscillator circuit on 0.25-μm InP HBT transistors and an amplifier circuit on a 0.2-μm GaN HEMT have been designed, the gain ratio of GaN-amplifier being only 15 dB.

As those results obtained in the process of creating the technology of heterogenous integration establish a base for considerable improvement of characteristics for radioelectronic weapons systems, the results of further research in this area unfortunately are not announced in the open press. Anyway, further evolution of this technology will depend on the development of the respective methodology of designing IC. Permanent increase in the requirements for up-to-date microwave systems of military and commercial applications undoubtedly presuppose further complication of the structure of custom ASICs. With relatively low consumption volumes characteristic of the military, the cost of designing such microwave ICs may constitute an inadmissibly big part of the total cost of designing the system

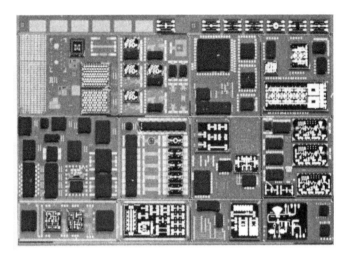

Figure 3.48   Fragment of multiproject—wafers with several ICs of different design.

(device). In such a case, designers have to make a difficult choice between the required IC parameters, the design lead time, and the total cost of such product. Naturally, from experience, sufficiently saving cost and reducing design lead time for radioelectronic devices is possible due to the use of programmed gate arrays. However, the practice of designing such reprogrammable ICs is not yet sufficiently developed in microwave engineering. For practical introduction of the methods of heterogenous integration, it is necessary to develop a completely new approach that provides an economic effectiveness of the design. One of the ways is making, on the basis of semiconductor chips manufactured with the use of different complex semiconductors, standardized reusable blocks with properties of intelligence (reusable IP design blocks). Such blocks combined by means of assembly on the common standard substrate in form of an interposer might become multifunctional and meet the requirements of different military and space systems depending on their assigned tasks (Figure 3.49).

Realization of this approach will require creation of a uniform system capable of controlling the technological processes of making IP blocks from different complex semiconductor materials. Moreover, one indispensable condition is the development of the design flow, including simulation and verification, and which could interface with known commercial design tools. And of course, for a wide commercialization of this heterogenous integration, it is necessary to have a mechanism for exchanging the flows of technological processes between different companies by means of standard interfaces, which is sufficiently difficult to arrange.

3.13.4.4 SMART-LEES Project

It is commonly known that the possibility to make new A^mB^v materials-based and silicon-based ICs is of interest not only for the military. In early 2012, a special group was established by the governments of Singapore and the United States working within the framework of the SMART-LEES project upon a similar task in the commercial area [81]. Key organizations as authoritative and competent participants

Option A

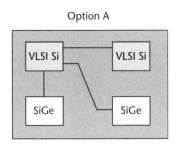

Option B

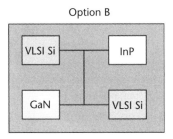

Figure 3.49 Assembly options for IP-blocks in form of an interposer on the common standardized substrate.

of the program are Massachusetts Institute of Technology (United States), National University of Singapore (Singapore), and the Nanyang Technological University (Singapore). Moreover, partners for silicon manufacturing are well known companies such as Global Foundries (Singapore), Tower Semiconductor-Panasonic (Japan), Aixtron (Germany), IQE (UK), EVG (Austria), and Samco (Japan). Program initiators acted on the assumption of the obvious fact that introducing new types of devices into the basic process of silicon IC manufacture is quite difficult both for technological and economic reasons. It was therefore decided to design ICs with devices on $A^{III}B^V$ materials as a separate process first, subsequently introducing them into CMOS process used in a production line (the so-called modular processing method). The following order of interaction has been implemented in such event. The designers acquire prefabricated 200 mm wafers from the specialized silicon fab and make a design of silicon ICs under spec requirements at their own fabs, designated for 200 mm wafers as well, using standard CMOS design method. At this point, wafers were coated with epitaxial layer of compound semiconductors and, thus, the basic engineering substrate was made with all the required dielectric layers and contact pads. After this, this wafer was returned to the silicon fab for the finishing processes of interconnecting silicon components and $A^{III}B^V$ devices inside the chip.

In 2012 the designers team started its work from coating a silicon substrate with GaN, GaAs and InP, and starting from 2014 InGaAs and GaN were used in production, which is the basis for InGaAs HEMT, GaN HEMT transistors, and GaN LED making. Such companies as IQE (UK), Lumileds (The Netherlands), and Analog Devices (United States) joined the project in 2015. Due to significant financial support of this program, the first test chips were supplied in 2016. Another important goal of the program is making a design kit for IC end-to-end design on CMOS and $A^m B^v$ devices and new type ICs launch to the commercial market.

Currently, there are two main technological trends of heterogeneous integration: micrometric assembly approach of ready CS-devices on a silicon substrate (DARPA, the United States) and CS-devices manufacturing method directly on the Si-substrate, followed by finished processing at a standard silicon fab. Although both technologies are dual use, each has priority fields of application: military and space applications in the United States and commercial application in Singapore.

European countries carry out researches in the technology of compound semiconductors and silicon as well. So in the UK, the development of compound semiconductors is one of the most important directions defining industrial and economic leadership perspective of the country [82]. Thus, in 2015 a scientific and industrial cluster was established on the basis of the well-known Institute of Compound Semiconductors at the University of Cardiff and the largest manufacturer of semiconductor wafers, IQE, which combines scientific researches with production [82, 83]. In addition, commercial microwave devices and power devices are planned to be designed on the basis of GaN /Si 150 mm epitaxial wafers and military perspective photonic devices on GaAs- and InP-wafers of the same diameter. Major semiconductor clusters in Europe use silicon technologies and the cluster to be established will be the first one specializing in compound semiconductors. It is hoped that its establishment will lay the foundations of high-tech industries reverse movement from competitors of East Asia to Europe.

References

[1]   Watson, H.A. (rd), *Microwave Semiconductor Devices and Their Applications*, 1972, USSR, translated from English, Watson, H. A. (ed.), *Microwave Semiconductor Devices and their Circuit Applications*, New York: McGraw-Hill, 1969.

[2]   Klampitt, L., *Powerful Electric Vacuum Microwave Devices*, USSR, 1974.

[3]   Howes, M., and D. V. Morgan, *Semiconductors in Microwave Circuits*, USSR, 1979.

[4]   http://vlad-gluh.livejournal/com/420211.html.

[5]   http://blog/i.ua/user/2663242/940249.

[6]   http://armsdata.net/nuclear/116.html.

[7]   http://rnns/ru/14834-udarnaja-sila-poslednee-preduprezhdenie.html.

[8]   http://ss-op.ru/reviews/view/55.

[9]   Hurtov, V. A., *Solid-State Electronics*, Russia, 2005.

[10]   Belous, A. I., A. V. Silin, *Bipolar Microcircuits in Interfaces of Automatic Control Systems*, M.: Radio i sviaz, 1990.

[11]   http://www.twirpx/com/file/400874.

[12]   http://dxdt/ru/2007/11/26/836.

[13]   Scarpulla. J. R., "Reliability and Qualification Challenges for RF Devices," The Aerospace Corporation, Los Angeles, 2004.

[14]   Kayali, S., G. Ponchak, and R. Shaw, "GaAs MMIC Reliability Assurance Guideline for Space Applications," NASA Lewis Research Center, 1996.

[15]   http://www.chipnevs.ru/html.cgi/arhiv _i99_04/stat-56.

[16]   http://www.kit-e/ru/articies/svch/2005_9_174.php.

[17]   http://www.pvsm/ru/radiosvyaz/20389.

[18]   Brookner, E., "Phased-Array Radars: Past, Astounding Breakthroughs and Future Trends," *Microwave Journal*, Vol. 51. No. 1, 2008, p. 30.

[19]   Thales, M. Y., "Components and Technologies for T/R Modules," *Proceedings of the*

Third European Microwave Integrated Circuits Conference, Amsterdam, Oct. 2008, p. 270.

[20] www.RG.RU. No. 227, C. 14, 2008.

[21] "Growing Market for Active Electronically Scanned Arrays (AESA)," *Microwave Journal*, Vol. 52, No. 7, 2009, p. 47.

[22] "Raytheon Demonstrates Gallium Nitride Advantages in Radar Components," *Microwave Journal*, Vol. 51. No. 6, 2008, p. 48.

[23] *Microwave Journal*, Vol. 51. No. 10, 2008, p. 62.

[24] Mumford, R., "Microwaves in Europe: Historical Milestones and Industry Update," *Microwave Journal*, Vol. 51, No. 10, 2008, p. 88.

[25] Vikulov I., N. Kichaeva, "GaN Technology: New Stage of Microwave Microchips Development," Electronics, NTB, No. 4, 2007, pp. 80–85.

[26] Russel, M. E., "Future of the RF Technology and Radars," *Proceedings of the IEEE Radar Conference*, 2007, p. 11.

[27] Schuh, P., et al., "GaN MMIC Based T/R-Module Front-End for X-band Applications," *Proceedings of the Third European Microwave Integrated Circuits Conference*, Amsterdam, Oct. 2008, pp. 274–277.

[28] Costrini, C., et al., "A 20 Watt Micro-Strip X-Band AlGaN/GaN HPA MMIC for Advanced Radar Applications," *Proceedings of the 38th European Microwave Conference*, Amsterdam, Oct. 2008, p. 1433.

[29] Quay, R., et al., "Efficient AlGaN/GaN HEMT Power Amplifiers," *Proceedings of the Third European Microwave Integrated Circuits Conference*, Amsterdam, Oct. 2008, p. 87.

[30] Gonzalez-Garrido, M., et al., "2–6 GHz GaN MMIC Power Amplifiers for Electronic Warfare Applications," *Proceedings of the Third European Microwave Integrated Circuits Conference*, Amsterdam, Oct. 2008, pp. 83–86.

[31] Alleva, V., et al., "High Power Microstrip GaN-HEMT Switches for Microwave Applications," *Proceedings of the Third European Microwave Integrated Circuits Conference*, Amsterdam, Oct. 2008, pp. 194–197.

[32] Jansen, J., et al., "X-band GaN SPDT MMIC with over 25 W Linear Power Handling," *Proceedings of the Third European Microwave Integrated Circuits Conference*, Amsterdam, Oct. 2008, pp. 190–193.

[33] *Microwave Journal*, Vol. 49, No. 6, 2006, p. 22.

[34] www.raytheon.com.

[35] *Compound Semiconductor*, July 2005.

[36] www.northropgrumman.com.

[37] *Compound Semiconductor*, March 2005.

[38] en.wikipedia.org.

[39] *Proceedings of the 35th European Microwave Conference*, 2005, pp. 809–812.

[40] *Aerospace Defense*, No. 2 (21), 2005.

[41] *Aviation Week and Space Technology*, No. 8, 2005.

[42] *Military Microwaves Supplement*, June 2006.

[43] *Microwave Journal*, Vol. 49, No. 1, 2005, p. 24.

[44] *Electronics EEXPRESS*, March 2006.

[45] *GaAsMantech Digest*, 2004, 2005.

[46] *Compound Semiconductor*, May 2005.

[47] *Proceedings of the 13th GaAs Symposium*, 2005, pp. 361–363.

[48] Fedorov, Y., "The Wide-Bandgap (Al, Ga, In)N Heterostructures and Devices for Millimeter Wavelength Band Based on Them," *Electronics: Science, Technology, Business*, No. 2, 2011.

[49]   Lidow, A., and J. Strydom, "WP001 Gallium Nitride (GaN) Technology Overview," *EPC*, 2012.

[50]   Lidow, A., "AN001 Is It the End of the Road for Silicon in Power Conversion?" *EPC*, 2011.

[51]   Lidow, A., D. Reusch, and J. Strydom, "AN018 GaN Integration for Higher DC-DC Efficiency and Power Density," *EPC*, 2015.

[52]   Mehta, N., "SNVA723: Design Considerations for LM5113 Advanced GaN FET Driver During High-Frequency Operation," Application Report, Texas Instruments, 2014.

[53]   Vikulov, I., "Heterogene Integration—New Stage of Integrated Microwave Electronics Development," *Electronics NTV*, No. 1, 2016, pp. 104–112.

[54]   Farmikown, et al., "The Technology of High-Power Microwave LDMOS Transistors for Radar Transmitters of L-Band and Air Applications," *Components and Technologies*, No. 10, 2007.

[55]   SUMITOMO Research & Development, 2012.

[56]   "Rad Effects in Emerging GaN FETs," NASA, July 11, 2012.

[57]   Wireless Device Products 2012–2013, Sumitomo Electric Europe, Ltd.

[58]   "GaN Technology for Radars," CS MANTECH Conference, Boston, MA, April 23–26, 2012.

[60]   Triquint, "Wideband Power Amplifier MMICs Utilizing GaN on SiC.," IEEE, 2010.

[61]   "Gallium Nitride High Electron Mobility Transistor (GaN-HEMT) Technology for High Gain and Highly Efficient Power Amplifiers," *Oki Technical Review*, Issue 211, Vol.74, No. 3, October 2007.

[62]   "Gallium Nitride (GaN) Versus Silicon Carbide (SiC)," *Microsemi PPG*, Apr. 2010.

[63]   Curtis, D., "Development of Class C Multi-Stage Amplifiers for Pulse Radar Applications," *Advanced Electronics*, No. 1, 2007.

[64]   Kopp, C., "Evolution of AESA Radar Technology," Monash University, August 14, 2012.

[65]   Heinz-Peter Feldle, "Current Status of Airborne Active Phased Array (AESA)," IRS, 2009.

[66]   "Model of Diamond Microwave Transistor," *Technology and Design in Radioelectronics*, No. 6, 2011.

[67]   Kasu, M., "Diamond Field-Effect Transistors as Microwave Power Amplifiers," 2010.

[68]   "Diamond Semiconductors Operate at Highest Frequency Ever—A Step Closer to Diamond Devices for Communication Satellites, Broadcasting Stations, and Radars," Vol. 1 No. 7, Oct. 2003.

[69]   "SELEX_GaAs-GaN Enabling Technologies for Microwave," Sapienza Universita di Roma. Oct. 4–5, 2012.

[70]   Fedorov, Y., "The Wide-Bandgap (Al, Ga, In)N Heterostructures and Devices for Millimeter Wavelength Band Based on Them," *Electronics: Science, Technology, Business*, No. 2, 2011, pp 92–107.

[71]   Fedorov Y. V., D. D. Gnatyuk, P. P. Galiev, M. Y. Shcherbakova, Y. N. Sveshnikov, et al., "EHF-Band Power Amplifiers on AlGaN/AlN/GaN/Sapphire Heterostructures," *Proceedings of the IX Scientific & Technological Conference on Solid-State Electronics, Complex Functional Modules for Radar Equipment*, Zvenigorod, December 1–3, 2010, pp. 44–46.

[72]   Kishchinsky, A. A., "GaN-Based Solid-State Microwave Amplifiers—The Status and Prospects of Development," *Proceedings of the 19th Crimean Conference on Microwave Equipment and Telecommunication Technologies*, Sevastopol, Weber, 2009.

[73]   Vikulov I., and N. Kichaeva, "Technology of GaAs Microwave Monolithic Circuits in Foreign Military Engineering," *Electronics: NTV*, No. 2, 2007, pp. 56–61.

[74]   Vikulov, I., "Monolithic Microwave Integrated Circuits: Process Base of AFAR," *Electronics: NTV*, No. 7, 2012, pp. 60–73.

[75]   Vikulov, I., "Microwave Electronics Today: Trends and Challenges," *Electronics: NTV*, No. 3, 2015., pp. 64–72.

[76] Rosker, Mark J., "The DARPA Compound Semiconductors on Silicon (COSMOS) Program," CS MANTECH Conference, 2008.

[77] Rosker, Mark J., et al., "DARPA's GaN Technology Trust," *IEEE Microwave Symposium Technical Digest*, 2010, p. 1214.

[78] Green, D. S., et al., "The DARPA Diverse Accessible Heterogeneous Integration (DAHI) Program: Status and Future Directions," CS MANTECH Conference, 2014.

[79] Green, D. S., et al., "Compound Semiconductor Technology for Modern RF Modules: Status and Future Directions," CS MANTECH Conference, 2015.

[80] Green, D. S., et al., "Heterogeneous Integration for Revolutionary Microwave Circuits at DARPA," *Microwave Journal*, Vol. 58, No. 6. 2015, p. 22.

[81] "Delivering the Future," http:// www.compoundsemiconductor.net/article/ 98066-delivering-the-future.html.

[82] "IQE: Semiconductor Cluster Ambition," http://www.compoundsemiconductor.net/article/98247-iqe-semiconductor-cluster-ambition.html.

[83] "CSC Formally Launched as First Compound Semiconductor Cluster," http://www.semiconductor- today.com/news_items/2015/nov/csc_201115.shtml.

Selected Bibliography

http://epc-co.com/epc/Markets.aspx.

Vance, A., "The Semiconductor Revolutionary," *Bloomberg Business*, 2015.

Lidow, A., and D. Reusch, "AN017 Fourth Generation eGaN® FETs Widen the Performance Gap with the Aging MOSFET?" *EPC*, 2014.

http://epc-co.com.

Kischinsky, A. A., "Solid-State Microwave Power Amplifiers on Gallium Nitride—The State and Prospects of Development," 2010.

Kischinsky, A. A., "Broadband Microwave Power Transistor Amplifiers—A Generational Change, 2010.

NITRONEX Corp.2008 GaN Essentials™ AN-012 Thermal.

Turkin, A. N., "Gallium Nitride as One of the Most Promising Materials in Contemporary Optoelectronics," *Components and Technologies*, No. 5, 2011, pp. 6–10.

Yunovich, A. E., "LEDs Based on Heterostructures of GaN and Its Solid Solutions," *Lighting Engineering*, No. 5/6, 1996, pp. 2–7.

Zolina, K. G., V. E. Kudryashov, A. N. Turkin, and A. E. Yunovich, "The Luminescence Spectra of Blue and Green LEDs Based on InGaN /AlGaN/GaN Multilayer Heterostructures with Quantum Wells," *Physics and Technology of Semiconductors*, Vol. 31, No. 9. 1997, pp. 1055–1061.

Turkin A. N., and A. E. Yunovich, "Noble Award Winners 2014 in Physics: I. Akasaki, X. Amano, S. Nakamura," *Nature*, No. 1, 2015, pp 75–81.

Vikulov I., and N. Kichaeva, "GaN Technology: A New Stage in the Development of Microwave ICs, *Electronics: NTV*, No. 4, 2007, pp. 80–85.

Vikulov, I., "WBGS-RF Program. Phase II: Results and Intentions," *Electronics: NTV*, No. 8, 2009, pp. 62–65.

CHAPTER 4

Microelectronic Element Base of Rocket and Space Technology

4.1 Classification of Modern Microprocessors

The rocket and space industry is often a high-priority area of science and engineering development, and it widely applies to all advances of modern microelectronics. This comprises both the electronic component base (ECB), which is used for construction of all onboard and ground control systems, and SC devices that have not applied microelectronic advances before, such as sensors, gyroscopes, photodetectors (PDs), thermal imagers, television and video cameras, IR visualizers, and night vision instruments. In particular, one of the applications that was previously impossible is a subminiature thruster based on MEMS technology for micro- and nanosatellites and other small SC, which was thoroughly examined at the end of Chapter 1.

Figure 4.1 shows a simplified structure of the main fields where ECB is used for RST.

Here we can see quite a wide range of microelectronic products, including various memory devices, microprocessors, logic and interface microcircuits, and power supply microcircuits. But the main microelectronic components that determine functional characteristics, performance, and power consumption of the designed applications are microprocessors (processors), the features and technical characteristics of which should be examined more carefully.

A microprocessor (MP) is a software-programmable device designed for processing digital information and control of processing on a single or several (set of) integrated circuits with high level of electronic element integration (LSIC, VLSIC) [1]. A MP serves as a processor in digital systems of various applications, such as data processing systems (computers), object and process management systems, data measuring systems, and other types of systems used for industry, special (military), and consumer equipment, communication equipment, and other applications. Microprocessors determine the intelligence of modern RST ground and onboard hardware.

There are many approaches to classification of microprocessors according to various criteria and parameters, but three widely accepted fields of microprocessor development are shown in Figure 4.2.

There are two classes of applied electronics: digital and analog. Due to the size constraints of this book, we will examine digital MPs only.

General-purpose microprocessors are used to build computers: personal computers, work stations, computer complexes, and onboard control systems of parallel

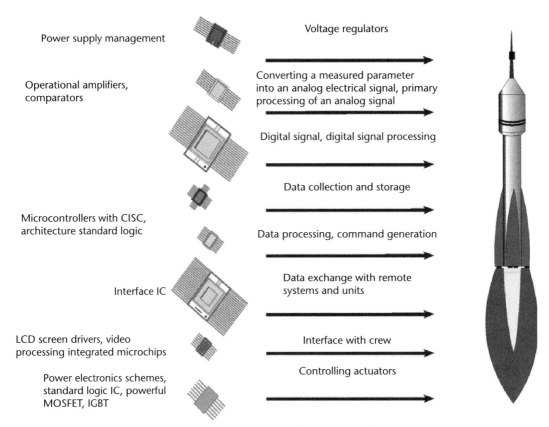

Figure 4.1   Structure of the main ECB fields of rocket and space technology.

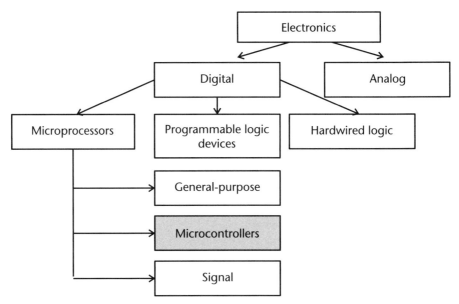

Figure 4.2   Classification of microprocessors according to application fields [1].

supercomputers. They feature cutting-edge solutions in terms of technology, system engineering, and electronics. A key characteristic of such microprocessors is the presence of embedded well-developed devices for efficient implementation of floating point operations on 32- and 64-bit and longer operands.

Microcontrollers (MCs) are the most widespread products of microprocessor technology. They are used for all embedded control systems, including both household devices and special (military and space) equipment. Unlike general-purpose microprocessors, the key parameters of MCs are size, cost, and energy consumption. Microcontrollers (also previously called single-chip microcomputers) allow implementing wide specifications of control systems for various objects and processes at minimum costs by integrating a high-performance processor, memory, and a set of peripherals on a single chip. The use of microcontrollers for onboard control systems ensures exceptionally high effectiveness at such low costs that there is in fact no alternative element base for construction of quality and inexpensive systems. For many customized applications onboard systems can be comprised of a single MC.

Signal processors are designed for real-time processing of digital flows created during digitizing of signals received from analog sensors. They are used for performing tasks that used to be performed by analog electronics. Signal processors have specific requirements [2]. They shall have maximum performance and be easily interfaced with analog/digital and digital/analog converters. They shall also have big processing word width and a small (minimum) set of mathematical operations, including the multiply-accumulate operation and hardware looping. Such parameters as cost, size, and power consumption are also important; however, higher values of these parameters than with microcontrollers should be accepted.

Each microprocessor is also classified by such parameters as *architecture*. Architecture means description (i.e., model of system composition and interaction of system components, according to the standard ISO 15704). Microprocessor architecture means description of organization and interaction of microprocessor parts (components) as determined by properties of components, principles of their design, connection, and development. Architecture includes description of logical (program), functional, and physical components of MP organization during data processing. Description of logical (program) components covers a set of commands, addressing modes, supported types of branches and junctions, types of access, and microprocessor register names. Description of functional and hardware MP components includes description of quantity and data path width of components, ways of interaction between the components during exchange of data, instructions and control signals, description of clock signal exchanges, timing diagrams of MP component operation, and description of layout (i.e., physical arrangement of a component on the chip).

In other words, MP architecture means its representation from the point of view of a programmer. As a first approximation, MP architecture is a function of its instruction set. Choice of architecture affects performance, time for development, and cost of the device.

Architecture of modern microprocessors and microcontrollers manufactured by different companies have much in common, which leads to unification of their architectures and which is of special importance to RST, as it utilizes the smallest amount of ECB compared with other fields. Table 4.1 shows microprocessors that are now present on the world market.

Table 4.1   Main Architectures of Modern Microcontrollers

MP Capacity	MP Architecture			
8-bit	MCS-51	AVR	CPU08	PIC
16-bit	C166	MAXQ	MSP-430	F2MC-16LX
32-bit	68000	AVR32	ARM	MIPS

OJSC Scientific-Research Institute of Electronic Technology (OJSC NIIET), Voronezh	OJSC INTEGRAL, Minsk	ICC Milandr, Zelenograd	Scientific Research Institute of System Analysis of Russian Academy of Sciences (NIISI RAN), Moscow

MCS-51:
series 1830–   1830BE31
             1830BE51
             1830BE32
series 1882–   1882BE53
           → 1882BM1T
under development
(R&D work,
code Slozhnost-2
(complexity-2))
	MCS-51:
	series 1880– 1880BE31
	             1880BE51
	             1880BE81У
	       → 1880BM1У
	under development
	(R&D work, code Dvina 51AC)
	IN80C31/51, IN87C51
		PIC:
		series 1886– 1886BE1У
		             1886BE2У
		             1886BE5У
		
		ARM Cortex-M3
		series 1986– 1986BE91T
		             1986BE92У
		             1986BE93У
		             1986BE94T
			MIPS:
			family KOMDIV-32
			analog of MIPS R3000
			           1B812
			           1890BM1T
			           1890BM2T
			           5890BE1T
			           1900BM2T
			           1907BM1T

Mega AVR (mid-range)
series 1887–   1887BE4У
        { 1887BE7T
     → { 1887BE8T
under development
(R&D work,
code Slozhnost-5
(complexity-5))
	ClassicAVR (low-end)
	IN90S2313 }
	IN90S2323 } not
	IN90S2333 } produced
	series 1881– 1881BГ4T
	       →
	under development
	(R&D work Dvina-135
		ARM Cortex-M0
		series 1986– K1986BE21
		
		ARM Cortex-A9
		is planned to be developed
			Family KOMDIV-64
			analog of MIPS RM7000
			           1890BM 3T
			           1890BM5Ф
			           1890BM6Я
			           1890BM7Я

C166: series 1887– 1887BE3T → 1887BE6T under development (R&D work, code Obrabotka-7 (processing-5))	Based on **licensed** core C 166SV1.2 of Infineon Technologies Company	Based on **licensed** Cortex cores of ARM Company	Based on **licensed** core of MIPS Technology Inc.
ARM Cortex-M4F: 32-bit MP for electric drive control is under development			OJSC Angstrem Zelenograd ↳ Analog MIPS-1 series Л1876BM1, Л1876BM2

This table also shows products of MC class manufactured by main Russian and Belarusian companies that are related to microelectronics and use these microcontroller architectures in their designs.

Before proceeding with detailed description of characteristics and features of national MPs and MCs, we shall briefly mention foreign microprocessors that have been used for onboard radio-electronic equipment (REE) of most prominent space missions.

4.2   Processors of Electronic Control Systems of Foreign Spacecraft

4.2.1   Onboard Processors of Foreign Spacecraft

Hundreds of part types of various integrated and hybrid microcircuits as well as a wide spectrum of discrete semiconductor devices of varying capacity and frequencies including the range of super high frequencies (SHF) are used for construction of any spacecraft.

Developers of SC electronic control systems are mainly interested in choice of central processor unit, since control logic, data storage and processing schemes, various interfaces, and so on are designed in accordance with it.

In 1981 one of the first intelligent American SC, space shuttle (Figure 4.3), used two types of central processing units at the same time: Intel 8086 and RCA 1802. The latter was used for information displays (as display controller).

In two years processor 8086 was replaced with its modifications of increased processing width—80386, 80386S. New onboard control systems used computers APA-101S with processing power of 1.2 MIPS and an ancient ferrite core memory, which was most resistant to cosmic rays. SC *Galileo* (1989) also used this couple—80386 and RGA 1802 (Figure 4.4).

In 1990 the onboard control system of *Hubble Space Telescope* (Figure 4.5) had an 8-bit processor DF-224 (390 processors were used). Due to many faults, REE developers had to replace it with Intel 80486 (processors successfully operated until mission end in 2010).

Figure 4.3   Space shuttle.

Figure 4.4 Spaceship *Galileo.*

It should be noted that at the time all processors weren't specifically developed for space application, but they underwent rigorous selection and various in-process tests.

At last in 1996, spacecraft *Pathfinder* (Figure 4.6) used processor BAE RAD 6000 that was specially designed for space applications under contract of IBM and British Aerospace Electronics. This processor proved its high operational reliability as part of SC onboard equipment and afterward was widely used for other spacecraft.

A widely known SC *Sojourner* (Figure 4.7) launched also in 1996 used the same processor type and another SC of this series had a new processor, FAST (Table 4.2).

Since 1998 the main electronic systems of the International Space Station (Figure 4.8) widely used upgraded versions of Intel processors: 80386SX-20W and coprocessor 80C387. Starting from 2004 space projects *Spirit* (FIgure 4.9) and *Opportunity* rather widely applied the version of BAE RAD 6000 with an operation frequency of 25MHz.

Figure 4.5 *Hubble Space Telescope.*

Figure 4.6   Spacecraft *Pathfinder*.

Table 4.2 also shows other types of central processing unit microcircuits that are used for less known SC and projects. We shall examine in more detail the main technical characteristics of processor microcircuits that are used for onboard control systems of foreign SC and manufactured by most prominent companies that cooperate with NASA and European Space Agency.

4.2.2 Aitech Defense Systems

One of the latest designs of this company, the single-board computer S950, is worth special mention. The S950 with 3U CompactPCI form factor features radiation tolerance and power saving operation modes [3]. Optimized for space equipment the S950 consumes 13.5W in normal mode, 10W in low-power mode, and no more than 8W in sleeping mode. The S950 has microprocessor PowerPC 750FX with clock frequency of 733 MHz.

Figure 4.7   Spacecraft *Sojourner* (on Mars).

Table 4.2   Processors Used on Foreign SC

Project Name	Processor Type	Project Name	Processor Type
Cassini	1750A	Galileo AACS	ATAC (bit slice) and 1802
Cluster (ESA)	1750A	SPOT-4	F9450
MSTI-1,2	1750A	EO-1/WARP	Mongoose V
Rosetta (ESA)	1750A	IceSat Glas	Mongoose V
EOS Terra	1750A (2)	MAP	Mongoose V, UTMC 69R000
EOS Aqua	1750A (4) & 8051 (2)	CGRO	NSSC-1
EOS Aura	1750A (4) & 8051 (2)	Topex/Poseidon	NSSC-1
Clementine	1750A, 32 bit RISC	UARS	NSSC-1
MSTI-3	1750A, R-3000	EUVE	NSSC-1, 1750A
Pluto Express	32 bit RISC	HST	NSSC-1/386, DF-224->486
Sampex	80386, 80387	Coriolis	RAD6000
SMEX	80386, 80387	Deep Space-1	RAD6000
SWAS	80386, 80387	Gravity Probe B	RAD6000
TRACE	80386, 80387	HESSI	RAD6000
WIRE	80386, 80387	MARS 98	RAD6000
FUSE	80386, 80387, 68000	SIRTF	RAD6000
Surrey MicroSat	80386EX (2)	SMEX-Lite	RAD6000
UoSat-12	80386EX (3)	Swift	RAD6000
FAST	8085 (2)	Triana	RAD6000
HealthSat-II	80c186 (2), 80c188	MightySat-II	TMS320C40 (4)
PoSat-1	80c186, TMS320C25, TMS320C30		

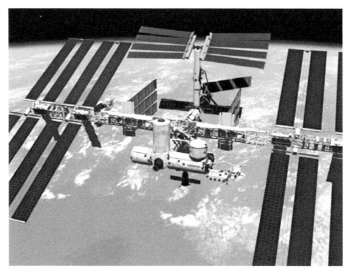

Figure 4.8   International Space Station.

Figure 4.9 Spacecraft *Spirit*.

The board incorporates 128 MB of synchronous dynamic random access memory (SDRAM) arranged in a triple voting architecture and 1 MB of dual-redundant boot flash. L1 cache of 32 Kb feature parity checks on tags and data arrays and L2 cache has an ECC (error correcting code) check on data arrays. ECC and memory controller are implemented on radiation-tolerant field-programmed gate array (FPGA). In addition, 64 MB flash memory, two standard serial ports, four independent 32-bit counters/timers and a hardware watchdog timer are provided for custom use. The computer has a slot for a 32-bit PCI mezzanine card (PMC) and 32-bit interface CompactPCI (33.3 MHz). Since S950 features protection from single event upsets (SEU), it is well-suited for spacecraft where it can perform navigation, data control, and collection tasks.

4.2.3 Microsemi

One of the major foreign developers and suppliers of ECB for space application is the Microsemi corporation, which includes such renowned companies as Actel (Microsemi SoC Product Group), White Electronic Designs (Microsemi Power and Microelectronics Group), Zarlink, ASIC Advantage, Arxan Defence Systems, Endwave, Spectrum Microwave, Maxim Integrated Products, and Symmetricom. In recent years the company has become a world leader in the field of production and package supply of components for onboard electronics of special-purpose complex systems.

Today the range of products includes renowned highly reliable programmable logic devices (PLDs), in particular PLDs with low power consumption, as well as random access and nonvolatile memory chips, 32-bit processors of the PowerPC family, hybrid microcircuits for special applications, and a wide range of radiation-tolerant small-scale integration electronic components: switches, controllers, transistors and diodes, and radio frequency components. Apart from that Microsemi produces highly reliable serial advanced technology attachment (SATA) flash storage and NAND-FLESH BGA (a type of flesh-memory on the principle of information change in cells—conventional designation) microcircuits with memory of up to 32 GB and support of parallel advanced technology attachment (PATA) interfaces. Special mention should go to the system-on-a-chip branch of Microsemi Corporation (Microsemi SoC Products Group) that was founded as Actel Corporation in 1985 and has long specialized in the field of development and production of processors, programmable logic devices for aeronautic, space, and military applications. Apart from radiation-tolerant highly reliable microcircuits, the branch produces reprogrammable arrays with enhanced power consumption and reliability characteristics, including PLDs with analog modules. More than 4500 companies around the world, including more than 300 companies in Russia, use production of this branch, according to the company's website [4]. Today production of Microsemi SoC Products Group is used for the special purpose hardware of space equipment, airborne onboard equipment, and control systems of nuclear power plants.

Microsemi SoC advertises the following advantages of PLD over standard SRAM FPGAs:

- High radiation tolerance;
- Triple modular redundancy for each logic gate;
- Reliability parameters proven in hundreds of NASA projects;
- Readiness to operate on power enabling;
- Low power consumption;
- Complete protection of configuration against copying;
- Simplified and size-reduced system solution.

Microsemi components are designed for application onboard spacecraft of various classes from microsatellites to space stations. Several hundred space missions now utilize Microsemi radiation tolerant PLDs for various units:

- Star trackers and gyroscopes;
- Onboard computers;
- Telemetry systems;
- Digital filters;
- Digital transmitter/receiver;
- Memory controllers and storage;
- Control units of terminators;
- Payload scientific instruments.

High radiation tolerance and reliability allow applying Microsemi PLDs for important control units in nuclear power plants that are exposed to severe neutron irradiation and high temperatures. PLDs with a large number of logic gates (up to 4 million) allow developing complex systems-on-chip that perform a large number of computing and control tasks, replacing massive equipment of previous generation.

Active application of PLD in NASA designs is explained by the availability of a considerable number of ready-made IP cores that allow creating solutions for specific tasks. For example, ready-made solutions for interfaces PCI, MKO, Ethernet, USB, and SpaceWire as well as a wide range of processor cores (ARM, 51, 186, LEON, and other) allow creating not only single-chip computing and process controllers for a wide spectrum of terminators and intelligent sensors, but various computers and many others.

Microsemi PLDs can be found virtually in any modern civil and military launch vehicle as well as in majority of spacecraft. Since 1996 Microsemi PLDs have been used for Russian satellites of GLONASS system, *Ekspress-AM*, *Koronas*, *Luch-5*, *Gonets-M*, *Spektr-UF*, *Resurs-P*, *Meteor-M*, *Elektro-L*, *GEO-IK-2*, *Soyuz*, and *Progres* spacecraft and for large numbers of high reliability ground equipment.

4.2.4   BAE Systems

BAE Systems has developed and manufactured products for military and space applications since 1980s. The company's own engineering and manufacturing center with certified radiation tolerant and submicron technologies is located in Virginia. In a ten-year period, the company developed a wide range of microprocessors for space applications, the main of which are shown in Table 4.3. At the moment more than 500 onboard computers for various spacecraft have been built on the basis of BAE Systems microprocessors.

Microprocessor RAD750 with PowerPC architecture, which is a radiation-tolerant version of commercial microprocessor PowerPC750F long produced by IBM, features most advanced characteristics. The RAD750 is fully compatible with its commercial analog both in terms of program compatibility and pin assignment.

To ensure resistance to outer space exposure factors, the initial product was modified though its functionality remained unchanged. The main alterations include:

- Modification of memory cell circuit diagram and layout
- Modification of circuit diagram and layout of reading-amplifiers, decoders, and other elements that form memory block, phase locked loop (PLL) block, and others;
- Replacement of all blocks consisting of dynamic logic elements with functionally similar blocks based on completely static electronics;
- Replacement of all triggers and registers with fault-resistant analogs;
- Wide introduction of error detection and correction schemes.

Until 2007 the microprocessor was manufactured using 0.25-μm CMOS technology and had operation frequency of up to 166 MHz. It features tolerance to

Table 4.3   Main Microprocessors of BAE Systems

	GVSC 1750	RAD6000	RAD750
Architecture	MIL-STD-1750A	RS\6000 POWER	PowerPC
Year of development	1991	1996	2001
Technology	Radiation-tolerant 1.0 μm CMOS process	Radiation-tolerant 0.5 μm CMOS process	Radiation-tolerant 0.25 μm CMOS process
Frequency, MHz	20	33	166

cumulative exposure of no less than 200 Krad. Absence of latch-up is guaranteed and single event upset linear energy transfer (LET) threshold is no less than 45 MeV • cm^2/mg. At the time predicted error rate at geostationary orbit constituted $1.6 • 10^{-10}$/bit/day.

In 2008 the company presented an improved model of the RAD750 manufactured in radiation-tolerant 0.25 μm CMOS process. Microprocessor frequency was improved up to 200 MHz, radiation tolerance increased to withstand cumulative dose of up to 1 Mrad, absence of latch-up is guaranteed, and tolerance to single event upsets is comparable with that of previous 0.25 μm microprocessor version. In 2010 the microprocessor underwent standard certification for compliance with requirements for space equipment.

4.2.5 Honeywell

Honeywell manufactures a wide range of radiation-tolerant components. The baseline technology is silicon-on-insulator with design rules of down to 0.15 μm, which allows almost complete elimination of latch-up and reduction of single event upset probability. Among the company's products are quite an old 16-bit microprocessor 16750A and a more updated microprocessor with PowerPC architecture named HXRHPPC. This microprocessor is identical in terms of functional characteristics and pin assignment to the commercial microprocessor PowerPC603e produced by the company Freescale.

To ensure resistance to exposure factors, the initial project was improved, but its functionality remained unchanged. In general, the improvements are approximately the same as those made by the company BAE Systems to improve resistance of microprocessor RAD750.

Microprocessor HXRHPPC is manufactured in 0.35 SOI process and has a clock frequency of up to 80 MHz. The HXRHPPC tolerates total cumulative dose of more than 300 Krad and absence of latch-up is guaranteed. Predicted error rate at geostationary orbit is $1.5 • 10^{-5}$/chip/day [4].

4.2.6 Microprocessors with SPARC Architecture

In 1990 the European Space Agency (ESA) started development of 32-bit microprocessor for space applications and in 1997 the ERC32 microprocessor with SPARC v7 architecture was produced [5]. To further develop the product, in 1998 ESA started new program on an improved microprocessor version named LEON. It was developed by Swedish company Gaisler Research. This project aimed at meeting all requirements for resistance to exposure factors in the presence of the following initial constraints:

- The use of commercial manufacturing process;
- A completely synthesized project, and custom units were discarded to ensure transferability to various production processes and plants;
- Construction in accordance with SoC principles based on standard on-chip interface to ensure possibility of quick upgrading and scaling;
- Program compatibility with one of the widespread processor architectures.

The famous architecture SPARC V8 was chosen as a basis, which ensures full program compatibility with ERC32. Furthermore, openness of architecture allowed avoiding possible legal complications. AMBA bus was chosen as a standard on-chip interface.

At the moment two companies are working at development of this microprocessor family: Atmel, which produces microprocessors based on the LEON2FT core (its characteristics will be examined in detail in the next chapter) and equally well-known company Aeroflex, which produces microprocessors based on the LEON3FT core.

4.2.7   Microprocessors of Atmel

Atmel has long produced microprocessors based on the LEON2FT core. The latest design goes under the number AT697E.

Microprocessor contains integer core that is compatible with SPARC V8, floating point unit, statistic RAM and SDRAM controllers, PCI controller, serial ports, and other peripherals.

A feature of this microprocessor is that instead of common means for detection and correction of single event upsets that protect a regular pattern of cache memory and register files, it uses triple redundant latches with a voting scheme, which provides means for protecting irregular control logic latches from single event upsets. A potential fault of a latch is corrected by voting logic, which ensures faultless operation of the whole system.

Microprocessor AT697E is produced using 0.18 CMOS technology and has a clock frequency of up to 100 MHz. The processor tolerates total cumulative doze of more than 60 Krad. Absence of latch-up effect is guaranteed up to LET level of no less than 70 MeV • cm^2/mg. Predicted error rate at geostationary orbit constitutes 1 • 10^{-5}/chip/day.

In 2009 the company introduced an improved microprocessor version under the number AT697F with an increased tolerance to cumulative dose up to 300 Krad and corrected errors of previous version.

4.2.8   Aeroflex

Today the company's microprocessors successfully function as a part of onboard control systems of more than a hundred of operating special-purpose SC of the United States, the European Space Agency, Japan, and other countries (Figure 4.10).

The company Aeroflex Colorado Springs develops and promotes a microprocessor family based on the LEON3FT core, which has a deeper pipeline as compared with LEON2FT (seven instead of five stages) and features multiprocessor support. The company started developing this field in 2008 after the acquisition of the main developer of this series, Swedish company Gaisler Research, which was renamed as Aeroflex Gaisler after the merger. Currently a microprocessor family LEON3FT-RTAX based on Actel radiation-tolerant PLDs RTAX2000S is produced.

Table 4.4 shows the characteristics of eight versions of a LEON3FT-RTAX microprocessor that differ in number of peripheral controllers. All modifications of LEON3FT-RTAX microprocessor have an operating frequency of up to 25 MHz and a tolerance to cumulative dose of up to 300 Krad; absence of latch-up effect

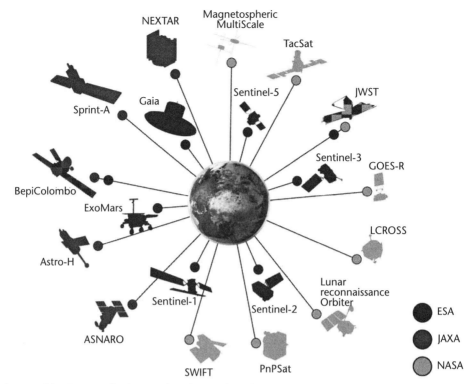

Figure 4.10 Spacecraft of general and special application that use microprocessors based on Leon 2FT core.

Table 4.4 Functional Structure of Microprocessor LEON3FT-RTAX Versions

Number of Modification	1	2	3	4	5	6	7	8
Integer core LEON3FT	+							
Division-multiplication unit					+			+
Power control unit	+							
Floating point unit	+						+	
Embedded memory, Kb	4		4	ESA				
1553 RT controller	+			JAXA				
1553 BC/RT/MT controller			+	NASA				
SpaceWire controller, number of channels		2		3	2		2	
CAN 2.0B controller, number of channels	1							
PCI controller						+		
Ethernet MAC controller								
Statistical RAM controller	+							
SDRAM controller				+				
Package	CQFP352			CQFP624			CQFP352	

is guaranteed up to LET level of no less than 104 MeV • cm^2/mg and single event upset LET threshold is no less than 37 MeV • cm^2/mg.

In May 2009 the company announced microprocessor UT699 that is unlike LEON3FT-RTAX implemented not on PLD, but on custom microcircuit. The microprocessor contains an integer core compatible with SPARC V8, a floating point unit, a PCI controller, four channels of SpaceWire interface, statistical RAM and SDRAM controllers, an Ethernet MAC controller, a CAN 2.0 controller, serial ports, and other peripherals.

Microprocessor UT699 has operating frequency of up to 66 MHz. It tolerates cumulative dose of no less than 300 Krad; absence of latch-up effects is guaranteed up to LET level of no less than 10E8 MeV • cm^2/mg.

This family of processors will be examined in more detail in Chapter 6, Volume 2.

4.3   Domestic Microprocessors and Microcontrollers

Construction of modern onboard control systems for military, airborne, and space equipment requires microcontrollers with a powerful high performance processor core together with equally powerful analog-digital and digital-analog units, extended (often unique) peripherals, improved means for in-system programming and debugging, together with small size and low power consumption.

We shall examine production of microprocessors (MPs) and microcontrollers (MCs) in Russia and Belarus. First, we would like to remind readers that microelectronic companies divide into designing companies and manufacturing companies:

- *Foundry companies* (silicon foundries): Features include the absence of own designers, a limited set of basic technologies, large-scale production of wafers, wafer manufacturing in short time, and available design kits;
- *Fabless companies* (no fabrication): Features include an absence of own wafer (device) fabrication, highly qualified staff, modern hardware and software (CAD), designing in short time, and low costs;
- *Combined companies*: Features include a company that has some wafer fabrication with a limited set of technologies and a design center; interaction within the company between the design center and the fab is the same as in types of companies mentioned earlier. Our joint stock company (JSC) integral belongs to this type.

Analysis results of MP and MC production at national enterprises that are based on information from companies' websites and publications in open access are presented next.

4.3.1   OJSC Angstrem (Zelenograd)

The company is one of the major Russian suppliers of radiation-tolerant electronic components for military-industrial complex enterprises. Customer orders for series fabrication are fulfilled either at company's own plant (1.0–0.6 μm) or at foreign

Table 4.5 Technology Level of Angstrem OJSC Chip Fabrication

Type of wafers	100-mm Diameter Silicon-on-sapphire, silicon carbide	150-mm Diameter Silicon-on-sapphire, silicon-on-insulator	200-mm Diameter Silicon, silicon-on-insulator
Design rules (microns)	1.2	0.6–0.8	0.25–0.35
Number of metal layers	2	2	4
Capacity, wafers/year	48,000	96,000	48,000
Current technologies	SOS	CMOS, EEPROM, CMOS RS, SOS, BiCMOS, double-diffused MOS, IGBT, MRS	
Developed and/or designed technologies	Silicon carbide	MEMS, IR-vision, BCD	CMOS RS, SOI, BCD, FRAM, SHF LD-MOS, Trench DMOS, Trench IGBT

partners' facilities (0.5–0.13 μm), as Angstrem is a combined company (fabless plus foundry). The main technological capabilities of the company are shown in Table 4.5.

It is planned to start fabrication of wafers with design rules of 0.11–0.13 μm (9-layer copper metallization, 14,000 wafers per month) on former AMD equipment in 2018. It is also planned to start fabrication of chips with layout rules of 90 nm on this equipment: an agreement on licensing technology of integrated circuit fabrication has been concluded with IBM.

Table 4.6 Microprocessors and Microcontrollers Produced by the OJSC Angstrem

	Composition	Features, Application
8-bit microcontroller with Tesej (Theseus) architecture	КР1878ВЕ1 (proprietary design)	Set of instructions: 52 instructions Clock frequency: 32 kHZ–8 MHz Cycles per instruction: 2 cycles
16-bit LSI-11/23-compatible set	H1836BM3 H1836BM4	16-bit microprocessor Floating point unit for H1836BMP3
16-bit LSI/2-compatible microprocessor	1806BM2 H1806BM4	6-bit LSI-11/2-compatible microprocessor
32-bit VAX-11/750-compatible microprocessor set	1839BM1Ф 1839BM2Ф 1839ВТ2Ф 1839ВВ1Ф H1839РЕ1 H1839ВЖ2	32-bit central processing unit Floating point unit SRAM controller for 1839BM1Ф 32-BUS and Q-BUS adapters ROM 16Kx32 for Л1839BM1 Majority element (two out of three)
32-bit RISC microprocessor se with MIPS-1 architecture. Rules: 1.2 μm	Л1876ВМ1 Л1876ВМ2 Л1876ВГ1 Л1876ВГ2	32-bit RISC microprocessor Floating point unit SVGA graphic controller VME bus interface controller

Apart from a license for the use of technology, IBM provided Angstrem with design rules needed for organization of chip contract manufacturing, thus extending the range of foundry service for Russian and foreign customers both in commercial and industrial spheres.

Angstrem already offers design centers (fabless enterprises) a technology library for chip design (process design kit, PDK) with the use of 0.13 μm technology based on Virtuoso IC 6.1.3 platform in Virtuoso multimode simulation (MMSIM) of the company Cadence Design Systems.

All microcontrollers shown in Table 3.6 are rather outdated (designs of 1980–1990s). Series Л1876 is worth special mention. A feature of series is that it was manufactured based on architecture of MIPS-1 core of the company MIPS Technology under official license and with all the production documentation in 1990s. It was developed at Scientific Research Institute of System Analysis of Russian Academy of Sciences (NIISI RAN) and adapted to the Angstrem 1.2 μm process.

4.3.2   OJSC NIIME (Scientific Research Institute of Molecular Electronics) and Mikron (Zelenograd)

NIIME and Mikron is the major manufacturer of integrated circuits in Russia and CIS. Until 2010 Mikron worked with the rules of 0.8 μm and Ø 150 mm wafers (since 1997); and rules of more than 1.0–1.5 μm on Ø 100 mm wafers on equipment that remained since Soviet times. While there were a many varieties and specifications of ECB, there were hardly any microprocessors and microcontrollers produced (except for 8-bit microprocessor sets of the 1802 series based on Schottky transistor-transistor logic (STTL).

But in 2010 a new line was acquired from the company ST Microelectronics and introduced to manufacture Ø 200 mm wafers with rules of 0.18 μm (CMOS + EEPROM, 4–6 metals). Technology for manufacturing ICs with layout rules of 90 nm (6–9 metals) was mastered. Not only was a license for manufacturing and sales of products based on this technology obtained from the company ST Microelectronics, but also design rules that can be used by developers of Mikron or outsourced fabless companies under contract.

Mikron manufactures and will manufacture products based on these technologies as a foundry for large market segments, including microprocessor and microcontroller segments. These are commercial and industrial microcontrollers designed for smart cards, electronic ID microcontrollers with high level of protection, including passport and visa documents (MIK51xx microcontroller line) and others. The first representative of this line, an 8-bit MIK5016XC2 microcontroller manufactured with the rules of 0.18 μm, has the following characteristics:

1. Memory:
   - EEPROM 72Kb;
   - ROM from 160 Kb to 256 Kb;
   - RAM 6 Kb.
2. Interfaces:
   - Contact (ISO 7816) up to 200 Kbps;
   - Contactless (ISO 14443) up to 848 Kbps;

  – Support of Mifare protocol for transport applications.
3. Cryptographic protection:
  – Cryptographic coprocessors supporting block cipher: DES, 3DES, AES;
  – Cryptographic coprocessor using residue number system for detection and verification of EDS according to algorithms ECDSA, RSA.
4. Means of engineering protection:
  – Cryptographic coprocessors;
  – Tamper protection hardware;
  – EEPROM integrity hardware check;
  – OS software protection;
  – Transaction support at OS level.

Production capacity and technological capabilities of Mikron allow fabrication of ~50 million microcontrollers of electronic ID type annually. This value doesn't include contactless tickets for public transport and other kinds of chips.

Technologies for manufacturing of military-class products on this equipment include:

- SOI (memory, microcontroller, and DAC/ADC microcircuits) with design rules of 0.25 μm and 0.18 μm (since 2012);
- BiCMOS SiGe (for radio frequency applications SHF and UHF up to 10 GHz—navigation systems GPS, GLONASS, radar system with APA, synthesizers, satellite communication) with design rules of 0.25 μm (since 2011) and 0.18 μm (since 2012).

Based on 0.18 μm CMOS process HCMOS8D Mikron (foundry) has produced prototypes and now master series production of 1892BM12T microprocessor microcircuits (designed at fabless OJSC Research and Development Center ELVEES) with SpaceWire channel and GigaSpace Wire gigabyte channel. It is planned to use 1892BM12T microcircuit as a general-purpose microprocessor that is designed for various onboard applications, including application as a network element of distributed control and data processing systems in modern networks with packet data transmission, including spacecraft onboard equipment, and as a controller of gigabyte solid-state drives.

The main technical characteristics of the 1892BM12T microcircuit are as follows:

- Architecture: MIPS32-compatible processor with 32/64-bit floating-point accelerator;
- 32/64-bit external memory port: SRAM, SDRAM, FLASH, ROM;
- Two SpaceWire ports (ECSS-E-50-12C) with data transmission rate from 2 to 300 Mbps within temperature range;
- Two serial ports (based on GigaSpaceWire, project of SpaceWire standard) with bandwidth of at least 1.25 Gbps within temperature range, with additional discrete choices of 5–125 Mbps transmission rate;
- Two UART ports;

- Embedded RAM: 1Mb;
- Two multichannel buffered serial ports (MFBSP) (12S (serial) serial peripheral interface (SPI) Super Harvard Architecture Single Chip Computer (SHARC) local port (LPORT)/general purpose input/output (GPIO) with direct memory access (DMA);
- Embedded input frequency multiplier/divider;
- Interval timer, real-time timer, watchdog timer;
- Joint test action group (JTAG) port, embedded program debugger (OnCD);
- Operating frequency: 120 MHz (under normal conditions) and 100 MHz (within temperature range);
- Temperature range: from −60 to +85°C;
- Package: CQFP-240, planar metal-ceramic package according to GOST 17467-88; package size including lead frame: 75.6 × 75.6 × 3.4 mm; package size after forming process: 40.3 × 40.2 mm; size of ceramic header: 34.05 × 34.05 mm.

Another example of Mikron microprocessors based on the previously mentioned 0.18 μm CMOS-process HCMOS8D that was earlier manufactured at a foreign company, TSMC, is К5512БП1Ф microcircuit (developed by design center KM211, Zelenograd; ordered by OJSC Progress Microelectronic Research Institute, Moscow). К5512БП1Ф is a semicustom VLSIC (with masked ROM) of system-on-a-chip (SOC) type with a QUARK embedded microprocessor core. It is used for designing VLSICs for equipment of national application (communication equipment, radar location, identification; control systems for power generation and power consuming objects; and systems for control of operation and motion of ground and air vehicles).

Composition and main technical characteristics of К5512БП1Ф microcircuits:

- 32-bit processor core QUARK (see Table 4.7) with processor speed of up to 150 MHz;
- 1024-bit coprocessor using residue number system with processor speed of 50 MHz;

Table 4.7 Comparative Characteristics of QUARK Core of Various Implementations

Implementation	Process/Device	Area and Number of Gates	Remark	Performance
PLD	Xilinx Virtex 4	2000 slices	No cache memory and memory management unit (MMU)	60 MHz
	Altera Cyclone IV	7150 LEs		
Silicon	TSMC, Taiwan CLN90G, 90 nm	0.2 mm^2 (72 Kgates)		> 400 MHz
		2.2 mm^2 (780 Kgates)	With cache memory and MMU 48 Kb (32 Kb + 16 Kb)	
	JSC Mikron, HCMOS8D, 180 nm	Cell area: 0.41 mm^2 (33.7 Kgates); Layout: 0.51 mm^2 (40 Kgates)	No cache memory and memory management unit (MMU)	150 MHz

- Four uncommitted logic array (ULA) blocks (300K gates) (for hardware implementation of special functions by ULA personalization);
- Masked ROM 128K × 32;
- Statistical RAM 64K × 32;
- SPI, I²C, GPIO interfaces;
- Metal-ceramic package of CPGA325 type;
- Number of pins: 226;
- 1.8/3.3V power supply voltage;
- Range of operating temperatures: –40 to +125°C;
- Average power consumption: less than 400 mW

Further modification of К5512БП1Ф for JSC NIIMA Progress is radiation-tolerant VLSIC Almaz-9 manufactured using CMOS SOI technology with rules of 0.24 *μ*m:

- Triple-redundant 32-bit processor core of MIPS type with direct memory access (DMA) support and (memory management unit MMU);
- Triple-redundant coprocessor using residue number system;
- Masked ROM 64K × 32;
- Statistical RAM 32K × 32;
- PLL for frequency of up to 150 MHz;
- Number of pins: 325;
- 3.3V power supply voltage;
- Average power consumption: less than 500 mW.

Mikron also cooperates with ICC Milandr and other fabless companies. Design centers of various levels actively concentrate around new Mikron fabrication. These are companies that develop microprocessors and microcontrollers:

- OJSC Research and Development Center ELVEES, Zelenograd;
- Scientific Production Enterprise Digital Solutions LLC, Moscow;
- CJSC ICC Milandr, Zelenograd;
- OJSC Scientific-Research Institute of Electronic Technology (OAO NIIET), Voronezh;
- OJSC NIIMA Progress, Moscow;
- CJSC Research Center Module, Moscow;
- Design Center KM211, Ltd., Zelenograd;
- Design Center Soyuz, Ltd., Zelenograd;
- Technology Center of National Research University of Electronic Technology (MIET), Zelenograd;
- SiBiS Ltd., Novosibirsk;
- OJSC Scientific Research Institute Submikron, Zelenograd;
- Private Enterprise NTLab-Systems, Minsk.

It is planned to create a fabrication facility on the basis of OJSC NIIME and Mikron that will have design rules of 65–45 nm on Ø 300 mm wafers at Alabushevo (1.5 km away from the main Mikron facility in Zelenograd). A piece of land has been reserved for the project, rent is paid, and a government order has been issued. Funding is still under question and the project is planned to be actively developed after Mikron has achieved project capacity with design rules of 90 nm.

4.3.3 Scientific Research Institute of System Analysis of Russian Academy of Sciences (NIISI RAN)

Scientific Research Institute of System Analysis of Russian Academy of Sciences has developed computers for important applications for more than 20 years and has mastered the whole cycle of complex system construction, including development of IC, computer modules, and system software. Results of its work are applied for development of microprocessor architectures (see Table 4.8), VLSICs (fabless), and microelectronic fabrication (foundry).

Development of ICs started in 1993–1994 when a license for R3000 microprocessor and R3010 floating point unit (MIPS32 core with *possibility of further architecture development*) was acquired form MIPS Technology. Licensed projects were redesigned and adapted to 1.2 *μ*m process of OJSC Angstrem. Adaptation included layout scaling and circuit characterization. At the moment these microcircuits

Table 4.8 Main Microprocessors Developed by NIISI RAN and Their Parameters

IC Type	Main Parameters
1890BM1T *Komdiv 32*	Single-chip microprocessor with MIPS-1 architecture, 32-bit, 30–50 MHz, 3.3V, −60 to +85°C, CMOS, 0.5 *μ*m, 3 layer metal, 1.5 million transistors, 2003.
1890BM2T *Komdiv 32*	Single-chip microprocessor with MIPS-1 architecture for processing 32-bit numbers with redundancy on block level, self-healing after fault, 80–100 MHz, 3.3V, −60–+90°C, CMOS, 0.35 *μ*m, 4 layer metal, 1.7 million transistors, 2005
1890BM3T *Komdiv 64-SMP*	Single-chip microprocessor with MIPS-64 architecture, 64-bit, 80–100 MHz, 3.3V, −60–+85°C, CMOS, 0.18 *μ*m, 2005
1890BM5Ф *Komdiv 64-SMP*	64-bit microprocessor with MIPS-architecture, 280 MHz, 3.3V, −60–+85°C, CMOS, 0.18 *μ*m, 2007
1890BM7Я *Komdiv 64-RIO*	64-bit microprocessor with RapidIO channels and 128-bit coprocessor with SIMD architecture, 3.3V, −60–+85°C, CMOS, 0.18 *μ*m, 2010
1890BM6Я *Komdiv 64-RIO*	64-bit microprocessor MIPS-architecture and RapidIO interfaces, 3.3V, −60–+85°C, CMOS, 0.18 *μ*m, 2011
5890BE1T *Komdiv 32-S*	32-bit SoC based on RISC microprocessor with advanced tolerance, 33 MHz, 3.3V, −60–+85°C, CMOS SOI, 0.5 *μ*m, 240-pin package of QFP type, 2007–2009
5890BE1T *Komdiv 32-R*	32-bit radiation-tolerant RISC-microprocessor for construction of backed up fault tolerant computers, 33 MHz, 3.3V, −60–+125°C, CMOS SOI, 0.5 *μ*m, 108-pin package of QFP type, 2010 (R&D work, code Kvartal-OVS)
1900BM2T *Reserv-32*	32-bit SoC for construction of backed up fault tolerant computers, 66 MHz, 3.3V, −60–+125°C, CMOS SOI, 0.35 *μ*m, 108-pin package of QFP type, 2010–2012
1907BM1T	32-bit SoC, SpaceWire interface, 100 MHz, 3.3V, CMOS SOI, 0.25 *μ*m, 240-pin package of QFP type, 2012–2014

(Л1876ВМ1 and Л1876М2 for R3000 and R3010, respectively) are manufactured at OJSC Angstrem.

NIISI RAN continued development of microprocessors of MIPS architecture (KOMDIV series) to improve technology (0.5 μm, CMOS SOI, three metals in 2003; 0.35 μm, four metals in 2005) on its line 1X1 and utilized 0.18 μm process at foreign facilities and Mikron. The institute developed a 64-bit version of architecture, added synergistic coprocessors and additional instructions, and now produces its own license-clear MIPS-type microprocessors. The initial R3000/R3010 MIPS project was redesigned starting from microarchitecture level while preserving full software compatibility. Since logic circuits were designed as Verilog RTL synthesizable model with necessary documentation and verification modules, they can be manufactured at various facilities (at the institute's 1X1 or at foreign facilities) with the help of various production processes, which ensures technology independence and possibility of multiple project use.

As far as line 1X1 is concerned, it is not designed for mass fabrication of VLSICs. Technology level of 0.5 and 0.35 μm was reached. In 2015 the line for 0.25 μm process was upgraded by means of supplementing/replacing with necessary equipment. Production output is several wafers a day (i.e., the line is in fact super low-volume and designed for production of small development batches). The institute fabricates only military-class products and doesn't take commercial orders.

This line had been the only country's line with the level of 0.5–0.35 μm until new technology was introduced at Mikron. Fabrication of its own microprocessor batches (NIISI RAN designs) and that of partners (ELVEES, Milandr, NIIET, Research Center Module) was planned half a year in advance due to low-volume production. Now the situation has changed: partners have reoriented to Mikron (foreign facilities are still employed).

4.3.4 Federal State Unitary Enterprise Federal Research and Production Center Measuring Systems Research Institute Named after Y. Y. Sedakov (Nizhny Novgorod)

Measuring Systems Research Institute named after Yu. Ye. Sedakov is a federal research and production center specializing in electronic engineering that is part of Rosatom State Atomic Energy Corporation. It has its own design center and a small 1X2 line for manufacturing radiation-tolerant LSICs and VLSICs in CMOS, CMOS SOS, and CMOS SOI process for national needs. Initially it was planned to create the line 1X2 with design rules of 0.35 μm, but it hasn't been achieved yet.

Designed projects of microprocessor and microcontroller class are manufactured at national and/or foreign facilities. Thus, navigation channel controller 1339ВП1Т was produced at 1X1 line at NIISI RAN, and microprocessor set of 1825 series was manufactured at the Institute's own line 1X2 (see Table 4.9).

4.3.5 Svetlana-Semiconductors CJSC (Saint Petersburg)

The company produces low-power high frequency bipolar transistors, integrated logic circuits, integrated circuits of analog switches and multiplexers, and radiation-tolerant integrated circuits of analog switches and multiplexers. It conducts the whole

Table 4.9   Composition of Microprocessor Set of 1825 Series and Controller 1339ВП1Т

Composition/Parameters	Technology	Design Rules, μm	Power supply Voltage, V	Implementation	Tolerance to Special Exposure Factors
1825ВРЗН2НИ 16-bit Arithmetic multiplier	CMOS SOS	3.0	From 4.5 to 7.5	Unpackaged, on flexible carrier with ribbon leads	Increased
1825ВС3Н2НИ 16-bit microprocessor slice		3.0			
1825ИР1Н2НИ Multifunctional register		3.2			
1825ВБ1Н2НИ Configurable synchronizer		3.6			
1825ВА3Н2НИ Bus transceiver	CMOS SOS	3.1	From 4.5 to 7.5	Unpackaged, on flexible carrier with ribbon leads	Increased
1825ВС3Н2НИ Match gate		3.0			
1339ВП1Т Navigation channel controller (digital correlator)	CMOS SOI	0.35	3.3 ± 10%	4245.240-6.01 ТАСФ. 30117 6.004 TU	Increased

cycle of research and production activities: research, development, fabrication, and sales of products—it is a combined company (fabless plus foundry).

The following products in the sphere of MP and MC are currently developed:

- Complex functional blocks for construction of systems-on-a-chip;
- Microcircuit set for digital data networks;
- Digital input/output channel adapter 1875ВВ1Т (there's no analog, current consumption $I_{occ} \leq 50$ mA, package type 4229.132-3);
- Controller that is functional analog of microcircuit 80C186EC;
- 32-bit microcontroller 1875ВД2Т (functional analog of 80386EXTB25, current consumption $I_{occ} \leq 90$ mA, package type 4229.132-3).

At the moment there isn't more detailed information and technical characteristics of these products.

Design flow is based on modern CAD systems of Cadence company. The design process includes stages from RTL-description to transferring information to factory. The level of the latest developed microcircuits: CMOS process with design rules of 0.35 μm at the factory of NIISI RAN (not at own factory). LIB_NIISI_035 library developed by NIISI RAN is used for designing. At the moment designing based on library with minimum element size of 0.18 μm is under development. One of tasks for the near future is to expand activities in order to develop microcircuits that contain analog units.

4.3.6 OJSC Scientific-Research Institute of Electronic Technology (OJSC NIIET) (Voronezh)

Scientific-Research Institute of Electronic Technology (fabless company) develops complex microelectronic products of the following types:

- 8-bit microcontrollers of 1830, 1882, and 1887 series with embedded nonvolatile memory and well-developed peripherals of MCS-1 and AVR RISC architectures;
- 16-bit microcontrollers of 1874 and 1887 series for embedded control, communication, and data processing systems of MCS-96 (including upgraded one) and C166 architectures;
- 16- and 32-bit fixed and floating point digital signal processor (DSP) for digital computer systems and control system of various applications (analogs of a range of ICs of TMS320Cxx family produced by Texas Instruments);
- 32-bit microcontrollers based on ARM Cortex-M4F core;
- Powerful HF and SHF field and bipolar transistors (more than 50 part types, perspective products on SiC and GaN are under development) and complex products (modules) on their basis for transmitting equipment of various communication means, avionics, navigation and radar systems, and instrument landing system, for APA of various RSs and others.

Tables 4.10–4.13 show main parameters of DSP, microprocessors, and microcontrollers that are produced and developed by OJSC NIIET at the moment. There's a number of developed products that haven't been fabricated due to various reasons (not contained in the tables).

OJSC NIIET and OJSC INTEGRAL are developing and manufacturing 8-bit microcontrollers of MCS-51 architecture. Comparative characteristics of these MCs are given in Table 4.13.

Table 4.13 shows that there isn't an absolutely best product in this segment of microcontrollers in terms of all parameters. Each MC can occupy a certain niche due to certain peripherals, interfaces, electric, and time parameters. Debugging aids and means designed for system debugging and programming that are developed on the base of microcontrollers are of special importance.

Apart from microcontrollers of MCS-51 architecture (1830, 1882 series), JSC NIIET produces widely applied microcontrollers and circuits for digital signal processing of the following architectures: MCS-96, AVR ATMega, C166, ARM Cortex-M4F (1867, 1874, 1887 series).

4.3.7 CJCS ICC Milandr (Zelenograd)

One of priority fields of ICC Milandr (fabless company that doesn't have its own semiconductor fabrication) is development of microcontroller large scale integrated circuits (LSICs) and processor LSICs for digital signal processors (DSP). A wide range of microcontrollers of various digit capacity has been designed over a short period of time (design works started in 2003) (Table 4.14).

Technical characteristics (including brief ones) aren't provided here due to their large volume (they can be found on the website and in catalogues of ICC Milandr).

Table 4.10 Microcontrollers of OJSC NIIET

Product Type	Main Parameters
1890BM1T H1830BE31 H1830BE51	Single-chip microprocessor with MIPS-I architecture, 32-bit, 8-bit microcontroller, clock frequency 12 MHz, RAM 128 × 8, ROM 4K × 8 (in H1830BE51 only), two 16-bit timers, UART, power consumption 132mW, power supply voltage 5V ± 10%
1882BE53У	8-bit microcontroller, clock frequency 24 MHz, RAM 256 × 8, ROM 12K × 8, EEPROM data memory 2K × 8, three 16-bit timers, UART ports, SPI, watchdog timer (WDT), power supply voltage 5V ± 10%
1887BE4У 0.35 μm	8-bit RISC microcontroller, 8 MHz, ROM 8K × 8 Flash, EEPROM 1K × 8, RAM 512 × 8, UART, SPI, ADC (8 channels, 8/10 bit), WDT, analog comparator, calibrated RC generator, power supply voltage 5V ± 10%
Л1874BE36 1874BE6	16-bit microcontroller, clock frequency 20 MHz, RAM 232 × 8, ROM 8K × 8, two 16-bit timers, ADC, PWM, power consumption 300 mW, power supply voltage 5V ± 10%
1874BE06T 1874BE76T	16-bit, 20 MHz, RAM 488 × 8, OTP ROM 16K × 8 (in 1874BE76T only), ADC, 3 PWM blocks, PTS, HSIO,UART,WDT
1874BE16T 1874BE86T	16-bit motor control type, clock frequency 16 MHz, RAM 488 × 8, OTP ROM 16K × 8 (in 1874BE86T only), two 16-bit timers, ADC, PWM, EPA, PTS, 3 phase waveform generator (WFG)
1874BE66T	16-bit microcontroller, 8–16 MHz, OTP ROM 16K × 8, ADC (14 channels, 8/10 bit, PTS, EPA, up to 64 input/output lines), 3 phase waveform generator (WFG), programmable frequency generator, provides motor control functions
1887BE3T	16-bit, RISC, 40 MHz, ROM 256 Kb Flash, RAM 15 Kb, ADC (16 channels, 8/10 bit), two CAPCOM modules, PWM, USART (2), SPI (2), I²C, Twin CAN, WDT, OCDS, JTAG interface; design based on licensed core C166SV1.2 Infineon Technologies.

Table 4.11 Radiation-Tolerant Microcontrollers of JSC NIIET

Product Type	Main Parameters
1830BE32У CMOS SOI 0.5 μm	8-bit microcontroller without ROM, 12 MHz, RAM 256 × 8, timers 3 × 16 bit, UART, 7 interrupt sources, programmable counter array (5 channels), instruction cycle 1000 ns, U_{CC} = 5V ± 5%.
1830BE32AУ CMOS SOI 0.5 μm	8-bit microcontroller without ROM, 16MHz, RAM 256 × 8, timers 3 × 16 bit, UART, 7 interrupt sources, programmable counter array (5 channels), instruction cycle 750 ns, U_{CC} = 3.3V ± 0.3%.
1874BE05T CMOS SOI 0.5 μm	16-bit microcontroller without ROM and ADC, 20 MHz, RAM 488 × 8, UART, WDT, HSIO, 3 PWM channels, PTS.

Table 4.12 DSPs of JSC NIIET

Product Type	Main Parameters
Л1867ВМ2 1867ВМ2Т	16-bit with fixed point, 40 MHz, 10 MIPS, RAM5 44 × 16, ROM 4K × 16, 32 input/output ports, timer, serial port. Analog: TMS320C25.
1867ВЦ2АТ	16-bit with fixed point, 57 MHz, 28.5 MIPS, RAM 9K × 16, ROM 2K × 16, serial port TDM, timer, JTAG-interface. Analog: TMS320C50.
1867ВЦ5Т	16-bit, 20 MHz, 20 MIPS, program Flash ROM 16K × 16, RAM 544 × 16, dual 10-bit 16-channel ADC, PWM (12 channels), SPI and SCI ports, DMA controller, watchdog timer, JTAG-interface. Provides electric motor control functions.
1867ВЦ6Ф 1867ВЦ6АФ	32-bit with floating point, 40MHz, 40 MFLOPS, ROM 4K × 32, RAM 2K × 32, RAM cache 64 × 32, two serial ports, two 32-bit timers, DMA, multiprocessor interface, U_{cc} = 5V ± 10% (ВЦ6Ф), U_{cc} = 5V ± 5% (ВЦ6АФ). Analog: TMS320C30
1867ВЦ4Т	16-bit DSP with fixed point, speed 40 MIPS, clock frequency 40 MHz, RAM 10K × 16, program ROM 2K × 16, HPI, BSP and TDM communication ports, JTAG-interface, power supply voltage 5V ± 10%. Analog: TMS320C542

Table 4.13 Current Developments of OJSC NIIET

Product Type, R&D Work Code	Main Parameters
NIIET-MC01 (ARM Cortex-M4F)	32-bit microcontroller based on ARM Cortex-M4F core with peripherals. Manufacturing factory: TSMC (Taiwan). Technology: CE018G (CMOS, 0.18 μm, embedded flash). Prototypes have been manufactured. Co-contractor: Scientific Production Enterprise Digital Solutions (Moscow).
1882ВМ1Т (Slozhnost-2)	Multiinterface 8-bit coprocessor with low power consumption based on MCS-51 architecture with embedded hardware implementation of data encoding/decoding algorithms. R&D work was delivered in November 2013.
1887ВЕ7Т, 1887ВЕ8Т (Slozhnost-5)	Series of 8-bit AVR RISC microcontrollers with speed of 16 MIPS with increased program and data memory capacity.
Obrabotka-3	Developing and mastering fabrication of radiation-tolerant digital signal processor with fixed point of TMS320VC54 type.
1874ВЕ7Т (Obrabotka-4)	Developing and mastering fabrication of specially hardened microcontroller of 1874ВЕ05Т type with interface GOST R 52070-2003, ADC and PWM. R&D work was delivered in November 2013.
1867ВМ7Т (Obrabotka-5)	Developing and mastering fabrication of radiation-tolerant digital signal processor with fixed point of 1867ВМ2 (TMS320C25) type. R&D work was delivered in November 2013.
Obrabotka-6	Developing and mastering fabrication of radiation-tolerant 32-bit digital signal processor with floating point.
1887ВЕ6Т (Obrabotka-7)	Developing and mastering fabrication of 16-bit high performance RISC microcontroller with increased resistance to special exposure effects based on C166 processor core.

Table 4.14   Microcontrollers and DSPs of ICC Milandr Design

Digit Capacity	Series	IC Part Types	Remarks
8-bit microcontrollers	1886	1886ВЕ1У, 1886ВЕ2У, 1886ВЕ3У, 1886ВЕ4У, 1886ВЕ5У, 1886ВЕ6У, 1886ВЕ6У1, 1886ВЕ7У	R&D works were delivered in 2006–2011, supplies are arranged
16-bit DSPs	1967	1967ВЦ1Т. Analog—TMS320C546A	R&D work was delivered on November 30, 2011
32-bit microcontrollers	1986	1986BE9x (ARM Cortex-M3 core), 1986BEx air, 1986BE2x (ARM Cortex-M0 core)	R&D works were delivered in 2010–2013, supplies are arranged
dual-core microcontrollers (RISC+DSP)	1901	1901ВЦ1Т	R&D work was delivered on August 19, 2011, supplies are arranged
64-bit DSP	1967	1967ВЦ2Т, 1967ВЦ3Ф	Development stage, term of delivery—2016

Designed products are fabricated at foreign factories: XFAB Semiconductor Foundries AG (Germany), TSMC (Taiwan), CSMC (China), H-NEC (China), and at domestic factory Mikron.

First circuits of 1886 series with PIC core (1886BE1 with masked ROM, analog of PIC17C756-331/L and 1886BE2 with EEPROM Flash, analog of PIC17C756A) were developed by describing LSIC operation according to microchip datasheets in VHDL and then by register transfer level (RTL) modeling and creating a new layout: it was decided to stop copying layout. But when a design is described in VHDL or Verilog it is impossible to reproduce a functional duplicate in principle (which is important for customers, especially regarding strategies of import substitution). Specifications of the 1886ВЕ1У (2У) had to contain detailed descriptions of dissimilarity from PIC17C756A. Microcontrollers of the 1886 series are PIC-compatible (though not fully compatible) and have a PIC-17 core with different peripherals.

Functional analog can be fully reproduced if:

1. Reference design is fully copied by cloning its layout and binding it to technology often without understanding the range of solutions (for what and why) made by developers; or
2. There is license for the microcontroller's reproducible core.

The second approach was chosen by ICC Milandr for further developments: In 2008 a license for the ARM Cortex-M3 processor core was acquired from Advanced RISC Machines (ARM) of Great Britain). At the moment 32-bit microcontrollers with ARM2 processor core are widespread around the world (just like MCS-51 for 8-bit microcontrollers in its time). Such microcontrollers are produced by many renowned companies: Atmel, Samsung, Intel, Motorola, Cirrus Logic, Oki, and others.

The company ARM provides customers with microcircuit designers or manufacturers with licenses for cores, which include:

- Behavioral description in synthesized language subset VHDL or Verilog (such cores are known as synthesized cores);
- Circuit diagram of processor core;
- Layout macrocell (description of core layout for implementation as part of integrated microcircuit).

Initial core description at RTL level ensures complete flexibility of its application. In particular, there's no strict binding to a certain technology and, consequently, manufacturing factory. By using a certain representation manufacturers create various modifications of microcontrollers by supplementing a processor core with a set of necessary peripherals. To produce its own microcontroller models, ICC Milandr utilizes modern microelectronic technology that ensures minimum size of 0.25, 0.18, and 0.13 μm that is not bound to a specific factory.

In 2010 the company Milandr acquired the ARM Cortex-M0 core, which is a light version of the Cortex-M3 core. Microcontrollers of 1986BE2x series are now developed based on this core. They are designed for systems of data collection and processing of resistive and capacitive sensors and have an embedded 24-bit delta-sigma ADC. Acquisition of a more powerful ARM Cortex-A9 core for advanced microcontrollers (of aerospace application) has been negotiated. Close cooperation between ICC Milandr and ARM allows acquiring new licenses on special terms.

4.3.8 Research and Development Center OJSC ELVEES (Zelenograd)

OJSC ELVEES originated from NPO ELAS, which was a renowned research and production association of the USSR Ministry of Electronic Industry and its only system enterprise. NPO ELAS developed not only equipment and systems, but also a corresponding component base. Products of NPO ELAS (space electronic systems mainly) were based on cutting-edge microelectronic technologies. Up until the end of the 1980s the microelectronic fabrication of ELAS (plant Komponent) had an equal technology level with Angstrem and Mikron. It was at NPO ELAS that the country's first CMOS microprocessor set was created in 1974. Today Open Joint Stock Company Research and Development Center Electronic and Computer Information Systems (OJSC R&D center ELVEES) is one of the leading fabless centers of VLSIC design in Russia.

OJSC R&D center ELVEES aims at producing conceptually new import-substituting and export quality systems-on-a-chip (SOC) based on the company's own design platform multicore within the framework of VLSIC projects. A microcircuit database allows creating new advanced systems of dual-purpose radio-electronic equipment with fundamentally new properties and a 10–15-year life cycle for both telecommunication and space applications.

Processor microcircuits of the multicore series are single-chip programmable multiprocessor SOCs based on IP-core platform multicore (IP stands for intellectual property). Processors of this series combine best characteristics of two types of devices: microcontrollers and DSPs, which allows solving two tasks simultaneously with the help of a small-size device: data (including signals and images) control and high precision processing (i.e., the feature of the processors is that they are multicore).

One microcircuit can contain only one CPU RISC core of MIPS32 architecture that serves as a central processing unit and one or more cores of digital signal processing coprocessor-accelerator with floating/fixed point ELcore-xxx (Elcore = Elvees's core) based on a Harvard architecture. The CPU core is the controlling core in microcircuit configuration and manages the main program. The CPU core has access to resources of the DSP core, which is a slave in relation to CPU core. The CPU microcircuits support Linux 2.6.19 or hard real-time operating system QNX 6.3 (Neutrino).

The microcircuits are designed by specialists of R&D center OJSC ELVEES together with design center of JSC Angstrem (layout design and PLL block design). Information on microcircuits of the multicore series is provided in Table 4.15.

The microcircuits have internal memory of 2–8 Mb depending on model, peripheral SHARC-compatible links and serial ports 12C, 12S, USB, Ethernet, PCI, UART, and JTAG, as well as hyperlinks of SpaceWire and Serial RapidIO types.

SHARC is the architecture of program and hardware compatible 32-bit DPSs of the company Analog Devices.

Macrolibraries and core libraries are developed to be suitable for the best foreign electronic fabrication factories. This allows creating the first IC pilot prototypes abroad. (R&D center ELVEES cooperates with factory TSMC, Taiwan) and then manufacturing them at Mikron in Russia. All projects remain proprietary to the foreign manufacturer.

Apart from developed and mastered microcircuit of this series, several other ICs are now under development:

1. 1892BM12T (MCT-03P): radiation-tolerant microprocessor microcircuit with SpaceWire channel and GigaSpaceWire gigabyte channel manufactured

Table 4.15   Microcircuits of the Multicore Series Developed by R&D Center ELVEES OJSC

Microcircuit	1892BM3T (MC-12)	1892BM2Я (MC-24)	1892BM5Я (MC-0226)	1892BM10Я (NVCom-02T)	1892BM7Я (MC-0428)
Year of design	2004	2004	2006	2012	2012
Design rules, μm	0.25	0.25	0.25	0.13	0.13
Chip size, mm × mm	10 × 10	10 × 10	12.3 × 12.6	8.8 × 9.5	11.7 × 11.9
Integration, mil transistors	~ 18	~ 18	~ 26	~ 50.2	~ 81
Package	PQFP240	HSBGA292	HSBGA416	HSBGA400	HSBGA765
Multiprocessor MIMD architecture	2 processors: RISCore32 + Elcore-14	2 processors: RISCore32 + Elcore-24	3 processors: RISCore32 + 2 × Elcore-26	3 processors: RISCore32 (with FPU) + 2 × Elcore-30	5 processors: RISCore32 (with FPU) + 4 × Elcore-28
Operating frequency, MHz	80	80	100	250	210
Peak performance, MFLOPs, 32 bit	240	480	1200	4000	6720

in 0.18 μm CMOS process HCMOS8D at Micron; its parameters and characteristics were examined earlier;

2. 1892BM14Я (MCom-02): microcircuit of multicore signal processor (SOC):
   - Manufactured in CMOS 40LP process at factory TSMC (Taiwan);
   - Chip size: 8.0 × 8.0 mm;
   - Core power supply voltage 1.0÷1.2V, customizable power supply voltage of peripherals 1.8/2.5/3.3V;
   - Maximum operating frequency: 1200 (CPU)/800 (DSP)/500 (VELCore) MHz under normal conditions;
   - Multicore heterogeneous MIMD-architecture (up to 8 processor and accelerator cores) based on standard processor and synergistic processor cores:
     - Central processor unit: dual cortex-A9 (CPU 0-1) with FPU accelerator and NEON SIMD accelerator (ARM);
     - Cluster based on DSP cores of new generation (8–64 bit) with controlled resources of microcircuit and external memory compatible with DELCore series with floating (single and double) and fixed point;
     - Embedded core of hardware and software graphic accelerator (MSLI-300, ARM);
     - Multichannel GLONASS/GPS correlator core;
   - Well-developed system of general-purpose ports and serial interfaces;
   - Package 784L HFCBGA, 19 × 19 mm, 0.65 pin pitch.

3. 1892BM15Φ (MC-30SF6): radiation tolerant microcircuit of signal processor with 6 unified serial ports for SpaceFiber/GigaSpaceWire:
   - Manufactured in 0.18 μm CMOS process HCMOS8D at Mikron;
   - Design technology: on the basis of radiation-tolerant libraries MK180RT of OJSC R&D center ELVEES design and IP-libraries of multicore platform;
   - CPU: MIPS32-compatible processor with 32/64-bit floating point unit (FPU);
   - Two DSP cores (120 MHz) supplemented with hardware and software aids for accelerating Fourier transform and filter processing in floating point format (160 MHz);
   - Hardware FFT accelerator with 64 GFLOPs performance;
   - 64-bit external memory port: SRAM, SDRAM, FLASH, ROM;
   - Two 32-bit DDR memory ports, 200 MHz;
   - Two UART ports of 16550A type;
   - Embedded RAM with 128 Kb capacity;
   - Embedded input frequency multipliers;
   - Temperature range: from −60 to +85°C;
   - Expected radiation resistance parameters: up to 200 Krad in terms of cumulative dose and heavy ions exposure; up to 60 meV cm^2/mg in terms of threshold value of linear energy transfer (LET) effect at maximum temperature of 65°C;
   - Package: metal-ceramic CPGA-720.

Multicore isn't just about microcircuits. Multicore is also a modern module system of software (SW) tools (MultiCoreStudio, MCStudio) for ICs designed on the basis of the platform including hardware and software aids for program design, debugging, and verification: C-compilers, assembler programming means, program simulators for cores and IC platform, JTAG-debuggers, debugging modules with FPGA prototypes of designed SOC cores, and others.

4.3.9   CJSC Research Center Module (Moscow)

The main three fields of activity of the Research Center Module(RC Module) are:

- Designing semicustom digital and analog-digital integrated circuits (fabless microelectronic design service);
- Development and fabrication of embedded and onboard computers for important applications (control machines for spacecraft, onboard equipment and avionics);
- Development and fabrication of hardware and software systems for image recognition, and radar signal processing.

As far as the first field is concerned, RC Module is famous for its high-performance processor cores NeuroMatrix with DSP/RISC architecture (1879BMxx series) and VLSICs of system-on-a-chip (SOC) type, including circuits with multichannel high precision and performance ADC and DAC in very large-scale integration (up to 15 million gates). The products are implemented based on semiconductor technologies of the world's leading companies—they are manufactured at foreign factories (Samsung and Fujitsu Microelectronics Europe GmbH provide RC Module with their fabrication facilities for medium-volume production of processors).

DSPs of the Neuro-Matrix family and 1979BMxx series shown in Table 4.16 are representative of vector-pipeline DSPs. They feature high-performance processing of big data flows while hardware costs and power consumption are low. Due to hardware support of matrix and vector operations and the possibility to increase performance during processing of data of lower width, processors have a wide application in the field of video processing, pattern recognition, signal processing, radio location, telecommunication, navigation, and other. Embedded means for construction multiprocessor systems allows using them as basic blocks for constructing parallel computing systems.

RC Module has developed microcircuits К1879ХК1Я, К1879ХБ1Я, and 1879ВЯ1Я to enter such segments of mass market as satellite navigation, digital television equipment, and multimedia devices for receiving and decoding digital signals (see Table 4.17).

While closely cooperating with Fujitsu Microelectronics Europe GmbH (FME), RC Module utilizes its own bit slicing technology for designing semicustom VLSICs. Modern methods of complex project management and high technology of FME have allowed developing high-performance VLSICs of 1879 series. Chip layout is developed at RC Module together with FME engineers, who deliver final netlist and time model for final verification. Only after verification has been completed at RC Module is an approval signed to start fabrication of VLSICs at FME's factory. RC Module also offers post-manufacturing support to integrate VLSIC at the customer's system level.

Table 4.16 Processors of NeuroMatrix family Produced by RC Module

Product Type, Year of Design, Manufacturer	Main Parameters
1879BM1 (NM6403) NMC core 1998 CMOS 0.5 μm Samsung	High-performance application-specific microprocessor that combines features of two modern architectures: very long instruction word (VLIW) and single instruction multiple data (SIMD). Clock frequency: 40 MHz, power supply voltage: 3.0–3.6V; Power consumption: 1.3W Chip size: 9 × 9 mm. Integration: 100,000 gates. Contains 32-bit CPU RISC-core with 5-stage pipeline and 64-bit vector DSP coprocessor. Package type: BGA256.
1879BM2 (NM6404) NMC2 core 2006 CMOS 0.25 μm Fujitsu	Modification of 1879BM1 (second generation DSP), fully compatible with 1879BM1 in terms of instruction set, both modifications have identical structural arrangement and program modules of separate units. Clock frequency: 80 MHz, processor power supply voltage: 2.3–2.7V and peripherals: 3.0–3.6V; power consumption: ≤2W. Contains embedded RAM with 2 Mb capacity (unlike 1879BM1). Chip size is also 9 × 9 mm. Package type: BGA576. A JTAG-port was added to significantly facilitate hardware testing and debugging of application SW.
1879BM3 2007 CMOS 0.25 μm Fujitsu	High-performance programmable controller with embedded analog-to-digital (ADC) and digital-to-analog (DAC) converters without DSP core NeuroMatrix. Option of interface with processors 1879BM1 and 1879BM2. Instruction bit length is 128 bit. Package type: BGA576. Power supply voltage: 3.3 and 2.5V; power consumption: 4.2W. Contains four 8-bit DACs (300 MHz sampling frequency), two 6-bit ADCs (600-MHz sampling frequency), clock frequency of digital interface is 150 MHz. Integration: 2,230,000 equivalent gates. Designed for preprocessing broadband analog signals, forming data flow for reprocessing by digital signal processor (DSP), signal recovery after reprocessing.
1879BM4 (NM6405) NMC2 core 2009 CMOS 0.25 μm Fujitsu	Further development of 1879BM1 and 1879BM2 (third generation DSP). Manufactured with the use of the same technology as 1879BM2 (same electric and climatic parameters) but clock frequency is up to 150 MHz due to a deeper pipeline. Being program compatible with its predecessors and having their architecture features, processor 1879BM4 has a processor core with advanced architecture, a range of differences allows increasing performance at the same clock frequency. The 1879BM4 is practically identical to 1879BM1 and 1879BM2 in terms of structural arrangement and program models. Two programmable DMA channels of memory-to-memory type and two communication ports of CP0/CP1 type have been added, which has allowed increasing bit rate to up to 75–150 Mbps for each port (unlike the 20 Mbps of the 1892BM2). Package: BGA576.
1879BM5Я (NM6406) NMC3 core 2012 CMOS 90 nm Fujitsu	Further development of processors of NeuroMatrix family (fourth generation). It has the same processor core as 1879BM4 but operates at frequencies of up to 300 MHz. Moreover, internal memory capacity has been doubled (4 Mb versus 2 Mb of 1879BM2 and 1879BM4). Power supply voltage: 1.2V (core) and 3.3V (input-output buffer), power consumption: no more than 1.2W. Package: PBGA416. Operating temperature range for microcircuit family: (−55 to +85)°C

Table 4.17 RC Module Processors for Satellite Navigation and Digital Television

Product Type, Year of Design, Manufacturer	Main Parameters
К1879ХК1Я 2011 CMOS 90 nm Fujitsu	Digital unified program receiver of SoC class. Receives analog signals, converts them in digital code, and performs program digital processing. VLSIC design to be used as a basis for digital receiver input channel. *Ordered by and manufactured together with Design Bureau Navis.* Contains two NMC3 cores that are responsible for processing data that are received from a signal preprocessing block, which is in turn connected with ADC (12 bit, 85 MHz), ARM1176 is central processing unit. Embedded RAM: 16 Mb. System clock signal frequency is 81.92 MHz. Power supply voltage of internal digital circuit: 1.2 ± 0.1V. Power supply voltage of external buffers: 3.3 ± 0.3V, 2.5 ± 0.2V. Power supply voltage of analog units: 1.2V and 3.3V. Power consumption (depending on operation mode): 0.5–2.0W. Chip area: 72 mm². Package BGA484. Operating temperature range: −40–+70°C.
К1879ХБ1Я 2012 CMOS 90 nm Fujitsu	Digital television signal decoder of SoC class. VLSIC for digital set-top box of standard and high definition is designed for receiving and decoding television signals of satellite and ground broadcasting, cablecasting as well as of IP-television with the use of cutting-edge technology of audio and video compression. Contains the following cores: NMC3 (operating frequency 324 MHz), which decodes audio signals, ARM1176JZF-S (operating frequency 324 MHz) and specialized devices. Embedded RAM 8 Mb. System bus frequency: 162 MHz. Power consumption: ≤ 2W. Power supply voltage (microcircuit core): 1.2V. Power supply voltage (DDR2 interfaces): 1.8V. Power supply voltage (peripherals): 3.3 V. Package PBGA544. Chip size 8 × 8 mm. Operating temperature range −40–+85°C.
1879ВЯ1Я CMOS 90 nm Fujitsu	Digital unified program receiver of SoC class. Identical to К1879ХК1Я in terms of composition and parameters. There's no information on differences between 1879ВЯ1Я and К1879ХК1Я yet.

4.3.10 CJSC MCST (Moscow)

CJSC Moscow Center of SPARC Technologies (CJSC MCST) was founded in 1992 on the basis of Lebedev Institute of Precision Mechanics and Computer Engineering (IPMCE), the leader of national electronic engineering. Abbreviation SPARC in the name of the company is explained by the fact that at the time CJSC was considering renowned American company Sun Microsystems as its general partner. Sun Microsystems promotes computing aids with scalable processor architecture (SPARC) in the world market and gives considerable benefits to collaborating companies that operate under its brand.

At the moment CJSC MCST is the leading Russian designer of high-performance general-purpose microprocessors of the Elbrus family (ELBRUS stands for explicit basic resources utilization scheduling) that works in two design domains:

1. SPARC-compatible microprocessors;
2. Microprocessors with unique VLIW-EPIC architecture (VLIW stands for very long instruction word and EPIC stands for explicitly parallel instruction computing).

The company's main activity is centered around the following fields of modern computer technologies:

- Research and development of microprocessor architectures;
- Microprocessor design with deep submicron rules;
- Design of microprocessor sets, memory systems, and controllers, including development of manufacturing documentation for factories;
- Design of computers (fault-tolerant multiprocessor computer systems) based on proprietary microprocessor designs of various classes (server, work station, personal computer) and hardware version (stationary, redeployable, embedded, laptop);
- Design of computer modules and logic devices.

In recent years specialists of CJSC MCST have developed unique general-purpose high performance microprocessors (see Table 4.18) with architecture platforms

Table 4.18 Technical Characteristics of the First Generation of CJSC MCST Microprocessors

MP name	MCST-R100*	1891BM1 (MCST-R150)	1891BM2 (MCST-R500)	1891BM3 (MCST-R500S)	1891BM4Я (Elbrus E3M)
Technology	0.5 μm	0.35 μm	0.13 μm	0.13 μm	0.13 μm
Chip size, mm × mm	13 × 13	10 × 10	5 × 5	9 × 9	12.6 × 15
Number of metal layers	No data	4	8	8	8
Integration, million transistors	~2.1	~2.8	~4.9	~45	~75.8
Number of cores	1	1	1	2	1
Frequency, MHz	80	150	500	500	300
Performance MIPS/ MFLOPs	62/22	140/63	520/200	1100/400	23100/4800
Power supply voltage, V	5	3.3	1.0/2.5	1.0/2.5/3.3	1.05/3.3
Power consumption, W	3	5	1	5	6
Package	304-pin PQFP	480-pin BGA	376-pin BGA	900-pin FCBGA	900-pin FCBGA
Production year	1998	2001	2004	2007	2008
Manufacturing factory	Atmel ES2 France	Tower Semi Israel	TSMC Taiwan	TSMC Taiwan	TSMC Taiwan

*Note: MCST-R100 received state approval; however, it wasn't put into serial production. MCST-R100 was redesigned with 0.35 μm technology to create MCST-R150 (i.e., MCST-R100 was prototype of production processor MCST-R150).

MCST-R and Elbrus, on the basis of which computer systems have been developed (Elbrus-90 micro, Elbrus-3M1, computer system of AD S-300 and S-400).

Microprocessors MCST-R500 and MCST-R500S implement a 32-bit version of the SPARC (v8) architecture, MCST-R1000 implements a 64-bit version of SPARC v9 and MPs of Elbrus architecture operate with 32/64-bit architecture.

Since 2006 Bruk Institute of Electronic Controlling Machines (JSC INEUM) has actively taken part in projects of JSC MCST. JSC MCST and JSC R&D Center ELVEES have concluded a strategic partnership in the sphere of scientific research and development of high-performance multiprocessor programmable microcircuits and systems. (The partnership resulted in development of the 1891ВМ7Я.)

Homogenous four-core microprocessor Elbrus 2S (development of 1891ВМ5Я, Elbrus S) based on 65 nm technology that will operate at frequency of 1 GHz is now under active development. As further development of CJSC MCST designs it is planned to implement microprocessors based on 45, 32, and 22 nm technology and more advanced technologies as they emerge (see Table 4.19). At the same time it is planned to optimize logical and physical design and achieve clock frequency of 4 GHz.

4.3.11 OJSC NIIMA Progress (Moscow)

Open Joint-Stock Company Progress Microelectronic Research Institute is a leading Russian design center (fabless) that develops specialized microelectronic components. It is a cross-industry center for designing VLSI type of SOC, and developer and manufacturer of GLONASS/GPS navigation receivers. OJSC NIIMA Progress is the directing microelectronic agency of state corporation Rostec.

Table 4.19 Technical Characteristics of Second Generation Microprocessors Produced by JSC MCST

MP Name	1891ВМ5Я (Elbrus S)	1891ВМ6Я (MCST-R1000)	1891ВМ7Я (Elbrus 2C+)
Technology (design rules)	90 nm	90 nm	90 nm
Chip area, mm²	142 mm²	128 mm²	259 mm²
Number of metal layers	9	10	9
Integration, million transistors	~218	~180	~368
Number of cores	1	4 (Sparc)	6 (2 CPU + 4 DSP)*
Frequency, MHz	500	1000	500
Performance, GFLOPs	64 bit: 4.0 32 bit: 8.0	64 bit: 2.0 32 bit: 4.0	64 bit: 8.0 32 bit: 16.0
Power supply voltage, V	1.1/1.8/2.5	1.1/1.8/2.5	1.1/1.8/2.5
Power Consumption, W	13W: typical 20W: maximum	20W	~25W
Package	1156-pin HFC BGA	1156-pin HFC BGA	1296-pin HFC BGA
Production year	2010	2011	2012

*Note: 1891ВМ7Я has two CPU cores with VLIW architecture (development of Elbrus-2000, JSC MCST) and four digital signal processor (DSP) cores of R&D center ELVEES.

Being cross-industry center for VLSI type SOC design, NIIMA Progress arranges cooperation of system centers for equipment design, centers for chip development, and (national and foreign) VLSI manufacturers. Thus, NIIMA Progress has organized cooperation to create semicustom VLSIC К551БП1Ф with embedded microprocessor core (core developed by design center KM211, manufactured by JSC Mikron), and radiation tolerant VLSIC Almaz-9; now the institute works with R&D center ELVEES to create a digital television chipset.

NIIMA Progress develops separate IP blocks, SOC, and IP-block prototypes based on PLDs and gate arrays (ULA) in cooperation with OJSC Mikron, OJSC NIIET, andOJSC Angstrem. Fabrication of VLSIC SOC requires the use of ready-made IP blocks that have been developed according to standard rules, verified, and certified. Designing of IP blocks for reuse (multiple use) is 2–3 times more expensive than designing them for single use. At the same time reuse of IP blocks for VSLI SOC is 10–100 times less expensive than designing from scratch. NIIMA Progress supports manufacturing facilities in terms of developing libraries of standard elements, memory, and also adaptation of computer-aided design (CAD) for development of digital, analog, and mixed signal microcircuits.

Though not directly developing proprietary microprocessor or microcontroller LSICs, NIIMA Progress takes part in their development by cooperating and interacting with other developers and manufacturers.

Sections 4.3.1–4.3.11 have reviewed the main Russian developers (fabless) and manufacturers (foundry) of microprocessors and microcontrollers. Apart from them, Russia has a wide network of design centers (more than 30) that perform professional fabless activities. We shall omit the details of the role of microelectronic design centers in the national electronic industry. Instead we shall briefly review the main design centers related to microprocessors and microcontrollers.

4.3.12 Design Center KM211, Ltd. (Zelenograd)

The design center is part of international group KM Core. The center's main field of activity is development of digital parts of microcircuits (IP blocks) actively using its proprietary cores. Microprocessor cores and proprietary IP blocks together with strong software support of microprocessor solutions combined with low administrative costs give considerable advantages over foreign and Russian counterparts. KM211 develops three processor fields:

- RISC-core KVARK, 32 bit, Linux;
- Microcontroller platform KROLIK (8–16–32);
- Multimedia platform HYDRA (32-bit DSP-optimized cores).

The 32-bit processor core KVARK is implemented in microcircuit Л5512БП1Ф and manufactured at Mikron based on 0.18 CMOS process HCMOS8D. Characteristics of microcircuit Л5512БП1Ф are given in Section 4.2, Table 4.3.

Architecture of the KROLIK family for microcontroller application is represented by 8-bit KMX8 and 32-bit KMX32 cores and designed for production of low power (minimum size core), low performance, and protected microcircuits with high reliability requirements. The KROLIK microprocessor family is oriented toward high efficiency of C-compiled code and high requirements for microcircuit tolerance and data security. Cores are designed in accordance with requirements for stable operation in case of unstable power supply, clock signal skew, and other factors. Architecture and instruction set are well-adapted to features of C-language, which provides high code density while core size and power consumption remain small.

Table 4.20 shows results of synthesis of core's maximum configuration address space with register file.

Main application domains of the KROLIK microprocessors:

- General-purpose microcontrollers;
- Embedded control;
- Smart cards;
- Bank cards;
- SIM cards;

Table 4.20   Characteristics of KROLIK Family Cores, KMX Version

		KMX8	KMX32	Unit of Measure
General characteristics	Performance efficiency, Dhrystone		2.1	DMIPS/MHz
	Performance efficiency, Coremark		2.1	Coremark/ MHz
	Operands size	8	32	bits
	Code size (generated from C sources as a percentage of ARM Thumb-2)		97	%
90 nm, area optimized	Area (utilization 70%)	0.026	0.09	mm²
	Gatecount	10	33	K gates
	Dynamic consumption	12	28	μW/MHz
	Static leakage	1.25	2.8	μA
90 nm, performance optimized	Clock frequency, not less than	100	100	MHz
	Area (utilization 70%)	0.035	0.12	mm²
	Gatecount	13	36	K gates
	Dynamic consumption		40	μW/MHz
	Static leakage		5	μA
180 nm, area optimized	Area (utilization 70%)	0.178	0.53	mm²
	Dynamic consumption		192	μW/MHz
	Static leakage		15	μA
180 nm, performance optimized	Clock frequency, not less than	50	50	MHz
	Area (utilization 70%)	0.266	0.94	mm²
	Dynamic consumption		296	μW/MHz
	Static leakage		33	μA
Altera Cyclone EP4CE115	Gatecount of FPGA Altera Logic Elements	3000	3500	LE

Note: TSMC 90 nm LP, JSC MIKRON 180 nm CMOSF8 (for KMX8), and OJSC MIKRON HCMOS8D (for KMX32) processes were used for this table.

- ID, passport, and visa documents;
- Ultra-low power microcircuits.

The HYDRA platform is a proprietary video and audio processing architecture. Multiprocessor platform includes:

- KM211_KVARC MPU or another vendor's CPU as head processor with operating system;
- HYDRA engine, represented by a number of:
  - Bit-stream processors for efficient real-time data parsing or compressing (BS_CPU);
  - DSP processors, FPU and fixed point, 4xSuperscalar, 4xSIMD, 4xVector (DC_CPU) put on a high-performance HYDRA bus.

IP blocks of KM211 design are given in Table 4.21.

Table 4.21   Characteristics of Proprietary IP Blocks of KM211

Brief Characteristics of IP Block	Maturity
Math coprocessors and accelerators	
16×16 multiplier, accumulator	In silicon
32 bit CRC with any generator polynominal up to 32 bit	Ready
16 bit CRC with generator polynominal $x^{16} + x^{12} + x^5 + 1$	In silicon
Smart cards and RFID	
ISO14443 interface type A	In silicon
I/O block ISO7816	In silicon
ISO14443 type A/B interface	In design
Controllers	
Ethernet controller, MAC sublayer	In silicon
1-Wire controller	Ready
AC97 Host controller	In silicon
SATA2 controller	Ready
TFT LCD controller	In silicon
Intellectual DMA controller	In silicon
Resistive touch screen controller	In silicon
USB Device 2.0 low speed/full speed controller	In design
USB Host 2.0 low speed/full speed controller	In design
OFNI compliant flash controller (asynchronous mode)	In design
Peripheral blocks	
Watchdog timer	In silicon
Timers	In silicon
Real time clock	In silicon
JTAG controller	In silicon
PWM controller	In silicon

4.3.13 Scientific Production Enterprise Digital Solutions (Moscow)

The enterprise was founded by the graduates and on the base of Moscow Higher Technical School named after N. Baumann. It is a design center that develops digital VLSICs and fabricates them at national and foreign silicon factories. The enterprise develops 3–4 VLSICs per year, including system-on-a-chip (SOC).

One of the projects implemented by SPE Digital Solutions is development of VLSI 32-bit microcontroller based on ARM Cortex-M4F core specialized for induction motor and electric drive control. The microcontroller was ordered by OJSC NIIET (Table 4.12). The microcontroller contains a large set of both analog and digital peripherals, and VLSIC was manufactured at TSMC factory (Taiwan). Technology CE018G (CMOS 0.18 μm, embedded flash) was used. IP blocks of OJSC NIIET and SPE Digital Solutions as well as of foreign companies were used. As a result, cooperation was established with the following manufacturers of analog and digital IP blocks: ARM, TSMC, S3 Group, Avnet ASIC Israel, and Arasan.

SPE Digital Solutions developed the layout design of the first national multicellular processor MultiClet P1 MCp0411100101. It was ordered by OJSC Multiclet. Initial data for layout design was a Verilog netlist transferred by the customer. Layout design of the multicellular processor was developed for silicon factory SilTerra (Malaysia) process: CL180G (CMOS 0.18 μm).

It is efficient to acquire ready-made IP blocks for VLSIC design to reduce design stage as well risk of faulty microcircuits. SPE Digital Solutions, applies IP blocks of both foreign suppliers and proprietary designs.

4.3.14 OJSC Multiclet (Yekaterinburg)

OJSC Multiclet is a Russian fabless company that develops high-performance and fault-tolerant processor cores and low-power processors of multicellular architecture as well as devices based on them. Since 2011 the company has been a resident of Skolkovo Foundation Space Cluster.

Multicellular architecture is a developing field of computing that provides a new solution to mathematical problem of parallelization. Multiclet is based on the so-called post-von Neumann architecture with instructions being executed simultaneously on several computing devices (cells) and the main question is how to distribute instructions among cells.

At the moment there are two models of four-cell processor core:

1. MCp0411100101: oriented toward achieving maximum performance;
2. MCc0402100000: oriented toward low power consumption.

The first core (MCp0411100101) has already been manufactured at silicon factory SilTerra (Malaysia) in CL180G (CMOS 0.18 μm) process and assembled in 240-pin metal-ceramic packages CQFP-240 (4245.240-6) at JSC Voronezh Semiconductor Plant–Assembly (VZPP-S).

Main technical characteristics of the multicellular processor ICs:

- Core: MCp0411100101, made in Russia;
- Number of cores: 1 (four cells);

- Manufacturing process: 0.18 μm, CMOS;
- Capacity: 32/64 bit;
- Performance: 2.4 Gflops;
- Data memory: 128 Kb ($4 \times 4K \times 64$);
- Program memory: 128 Kb ($4 \times 4K \times 64$);
- Power supply voltage: 1.8V (core) and 3.3V (peripherals);
- Power consumption: 1.08W (maximum);
- Floating point: yes;
- Package: metal-ceramic CQFP-240 (4245.240-6);
- Temperature range: −60 to +90°C (acceptance 5).
- End of mastering: 2014

The main customers of MCp0411100101 are companies that develop and manufacture navigation equipment. These are Granit Concern, NII KP (Scientific-Research Institute of Space Device Engineering), RPC RPA Mars, PBMC of Avionica, and Dalpribor. At the same time this product can have general-purpose industrial grade application.

Short-term and long-term goals of Multiclet include:

- Manufacturing a 16-cell processor on a FPGA with layout rules of 45 nm, performance of 18 Gflops, and ultra-low power consumption for audio and video signal processing;
- Development of 64-cell cores with 22 nm technology that will have performance of 384 GFLOPs and be used to create a desktop supercomputer;
- Expanding specifications of processors adapted primarily to mass tasks and development of SOC with the use of multicellular cores.

4.3.15 Scientific-Research Institute of Multiprocessing Computing Systems Named after Kalyaev of South Federal University, Taganrog (NII MVS YuFU, Taganrog)

Scientific-Research Institute of Multiprocessing Computing Systems named after Kalyaev of South Federal University, Taganrog (NII MVS YuFU), was founded almost 40 years ago as a base institute of the USSR Ministry of Higher and Vocational Education under the Taganrog Radio Engineering Institute in the sphere of design and development of computer aids. The institute conducts fundamental and applied research in various fields and development of high performance multiprocessor computers with programmable reconfigurable architecture is its flagman field.

It should be noted that it was in cooperation with NII MVS that in the 1980s the microprocessor 1815ВФ3 for fast Fourier transform (R&D work, code Devis-3) was developed as a part of a microprocessor set of the 1815 series (designed and manufactured by INTEGRAL). In 1989 the problem-oriented computing suite (PVK-460) Trassa was developed and several items were manufactured by order of central design bureau Almaz. Trassa was a parallel computer system with large-scale parallelism and programmable architecture. PVK-460 Trassa contained 512 microprocessors 1815ВФ3 with programmable structure and hardware implementation of macrooperations.

In 1991 two items of problem-oriented computing suite (PVK-1600) Module-8 with large-scale parallelism and programmable architecture were developed and manufactured by order of NPO Antej. PVK-1600 Module-8 contains 2,048 parallel microprocessors 1815ВФ3 with programmable structure, 1,486 microswitches 1029КП2, and 1,024 microcircuits of register memory 1517ИР1 and 1517ИР2 developed at NII MVS and serially manufactured at enterprises of Soviet Ministry of Electronics Industry (as mentioned earlier, 1815ВФ3 was manufactured at Integral).

There were some disappointing results. Single-chip floating point unit (R&D work, code Devis-2) developed together with NII MVS wasn't put into serial production.

Modern basic elements (high-performance PLD with frequencies of up to hundreds of MHz) allow implementing parallel general-purpose high-performance computer systems with programmable architecture and structural-procedural organization of computing. The use of application-specific integration circuits (ASIC) designs is becoming important for NII MVS.

4.3.16 Scientific-Industrial Complex Technology Center and JSC Plant PROTON-MIET

Scientific-Industrial Complex Technology Center is a classic example of a university research center that has a processing line allowing research and development of micro- and nanoelectronics, system equipment and electronic equipment.

Technology capabilities of the center allow producing wafers using the following technologies: CMOS LSIC (1.2 μm) process, bipolar LSIC (1.5 μm), piezoresistive microsensors (3 μm) for serial production and R&D works. as well as CMOS, BIP, BiCMOS (1.0 μm) for research. Production capacity is up to 2000 wafers of 100 mm diameter a month.

Today the technical capabilities of MIET do not allow producing microcircuits of microprocessor and microcontroller class.

Several Belarusian private companies that work in this field should also be mentioned.

4.3.17 Private Enterprise NT Lab Systems, Minsk

Private Enterprise (PE) NTLab-systems is part of the NTLab group of companies founded in 1989 and a resident of Hi-Tech Park (HTP). The company specializes mainly in microelectronic designs: embedded software for microcircuits, electronic blocks and systems, software models of IP blocks of microelectronic units, and computer-aided design software for designing VLSICs: memory compilers, compilers, and debuggers of microprocessors and microcontrollers.

High research and technology potential of the company is proved by its participation in projects of component development for Russian navigation systems GPS/GLONASS/Galileo/Beiduo, in joint Russian and Belarusian program *Trajektoriya* (microcircuits for Russian RSs) as well as by the fact that NTLab is developer of microcircuits for digital television receivers of DVB-T/H/SH standards, Russian biotmetric passports, social and medical plastic cards, and so on. Russian companies (including Angstrem OJSC and Mikron OJSC) actively cooperate with PE NTLab systems to enter markets for high-tech products.

NTLab works together with many large silicon factories around the world and designs products with 0.35, 0.18, 0.13, and 0.09 μm technology.

4.3.18 Scientific and Technical Center DELS (Minsk)

Scientific and Technical Center DELS was founded in 1992 on the basis of the Minsk Higher Engineering Anti-Aircraft Missile Defense School of Air Defense (MVIZRU PVO) and works in the field of micro and radioelectronics. The main field of center's development is designing VLSICs that are comparatively simple in terms of functionality and are accepted by customers for special-purpose radio-electronic equipment ordered by enterprises and institutions of defense-industrial sector. These are various kinds of signal conditioners (timers with programmable parameters, eight-channel buffer gates with digital filtering of impulse noise, eight-channel buffer gates with tri-state outputs, tri-state bus drivers), and other circuits. The designed VLSICs are manufactured mainly at OJSC Integral (Minsk).

Microcircuits are designed according to end-to-end principle based on hardware description language (HDL). This method has allowed creating a hierarchical library of synthesizable HDL-descriptions (IP blocks) of digital devices that can be used as library macrocells for designing new microcircuits including SOCs.

4.4   MP and MC Development and Debugging Tools

By definition a microprocessor is a software programmable device designed for digital data processing.

The first programs for early microprocessors and microcontrollers were written in machine code in assembler language. Disadvantage of assembler is that it is hard to write programs, and it takes time to master programming skills in such languages. The instruction set for one core can be very different from another. For example, the instruction set of the MCS-51 family is very different from the AVR instruction set, although they perform similar functions.

Today source code is written mainly in high-level languages like Pascal, C/C++, and others. Special compiling programs transform descriptions from high-level languages to machine code of a certain MP or MC. Hardware developers must provide the possibility for fast design of custom control systems for the use of MPs and MCs. Therefore, advanced development and debugging tools, primarily programming systems, are needed.

The main debugging tools include:

- In-circuit emulators;
- Software simulators;
- Development boards;
- Debug monitors;
- ROM emulators.

This list doesn't contain all existing debugging tools. Apart from the indicated ones, there are combined software and hardware that compensate for disadvantages of the main tools applied separately.

Any of the listed tools contains several interacting (software or hardware) functional modules. Each of them provides a certain range of services during development and debugging. Some modules are specific to a certain type of development tools; others are used for all kinds of program development systems of microprocessors and microcontrollers.

Any development system contains a minimum set of functional blocks that includes:

- Debugger;
- Microcontroller emulation unit;
- Emulation memory;
- Breakpoint subsystem.

More advanced models may additionally include:

- Breakpoint processor;
- Tracing program;
- Profiler (analyzes efficiency of program code);
- Real-time timer;
- Hardware and software that allow reading and modifying resources of emulated processor on the fly (i.e., during real-time program execution);
- Hardware and software that allow synchronous control required for emulation in multiprocessor systems;
- Integrated development environment.

An in-circuit emulator is a hardware and software means that replaces an emulated (modeled) processor in a real circuit. The in-circuit emulator is connected to the system to be debugged with a cable that has a special emulator connector that plugs into the system in place of the MC. In case the MC cannot be taken out of the system, the emulator can be used only provided that the microcontroller has a debugging mode such that all its pins are in a unresponsive (third) state. In such case a special adapter is connected to the pins of the emulated MC to connect the emulator.

A simulator is software that imitates operation of a microcontroller and its memory. As a rule, a simulator contains a debugger, a model of the processor core, and its memory. More advanced simulators also contain models of embedded peripherals such as timers, ports, ADCs, and interrupt systems.

A development board is typically a printed circuit board containing a MC and its standard peripherals. It also has logic to couple with an external computer. There is also area where custom application circuits can be mounted. Sometimes there are ready-made interconnections for installing additional devices recommended by a company. These are, for example, ROM, RAM. LCD-display, keyboard, ADC, and others.

A ROM emulator is a hardware and software means that allows replacing ROM on the board to be debugged with RAM, where a program can be downloaded from the computer via one of the standard communication links. This helps users to avoid a number of cycles of ROM reprogramming. In terms of complexity and cost this device is comparable with development boards and its universal application is a big advantage. ROM emulators are suitable for any kind of MCs.

All major developers (fabless companies) and manufacturers (foundries) of microprocessor and microcontroller ICs offer not only MP and MC microcircuits, but also hardware and software for their debugging. Today the market needs not only a set of microcircuits, but a complete solution that includes recommendations on application and tools (reference boards at the very least).

It doesn't mean that semiconductor companies deal with these issues themselves. There are a number of big and small companies specializing in development, manufacturing, and supplying with tools for microprocessor and microcontroller families. For example, in Russia one of the most prominent companies is Phyton. However, in order to cooperate with such companies, a customer company should have application-specific departments with highly qualified staff: programmers, systems engineers, designers, and mechanical engineers (who should not be confused with true circuit designers and programmers of IC layouts).

If your microprocessor or microcontroller is a complete functional analog of a foreign product in terms of all parameters and has a widely known architecture with well-developed software and a large number of applications, there's no need for proprietary debugging tools. In such cases customers are advised to acquire existing tools, including foreign ones. Customers usually do not make inquiries since they are familiar with, for example, development environment Keil uVision, in-circuit emulator PICE-52, and hardware and software package CodeMaster-52.

In case your microprocessor or microcontroller is even slightly different from its analog, then it doesn't have ready-made hardware and software, and it needs to be either developed by your company or ordered at other companies. MC operating modes in each particular case (for each customer) are specified by microcircuit developers and according to such specifications MC operation instructions are extended and improved. At the same time, a developer of MP or MC microcircuits cannot produce full documentation with recommendations on operation at system level for complex products of such class without experience in development of modules and end devices.

In Soviet times in the 1980s each enterprise of the Ministry of Electronic Industry that manufactured MPs and MCs had a consulting maintenance center (KTTs) that were created to facilitate large-scale application of microprocessors and microcontrollers into the national economy. Head KTTs were created at the Central Scientific Research Institute Tsyklon, which was the main Institute of Ministry of Electronic Industry specializing in application of electronic engineering products.

Today each company addresses such issues in its own way.

CJSC ICC Milandr has a special department for development of electronic modules that is in charge of technical support. The department employs highly qualified specialists in development of electronic units, devices, and modules for various applications. The department's fields of activity include:

- Technical support for microcircuits produced by CJSC Milandr;
- Development of software and hardware needed for development and debugging projects which utilize such microcircuits;

- Providing hardware developers with recommendations and reference documentation;
- Assistance in addressing problems that occur during microcircuit operation;
- Holding consultations and seminars;
- Development of product demonstration prototypes;
- Assistance in implementation of projects, joint projects, and cooperation with companies acquiring products.

Nowadays the customer's choice of a microcontroller for equipment largely depends not on controller itself or its characteristics, but on presence of a debugging board, development environment, and suitable software. Therefore, in order to win the market, the MP developer usually supplies customers with debugging hardware and software (often free of charge) and also cooperates with universities to include laboratory courses on microcontrollers of Cortex-M architecture in curriculum (and to provide universities with debugging systems), so that future specialists can manage their professional tasks after graduation.

4.5   Development Trends of Modern Microcontrollers

4.5.1   Customer Transition to 32-Bit Microcontrollers

Today most applied MCs are based on 8-bit processor architecture. There are also 16-bit MCs but their share of the market is still small compared to 8-bit MCs. Recent years have seen active transition of developers of embedded systems to 32-bit microcontroller architectures. The main reason for transition to 32-bit MCs is sophistication of embedded systems due to market requirements. Since embedded products are becoming more and more functionally complex, 8- and 16-bit MCs cannot provide necessary *performance*. Even though 8- and 16-bit MCs meet the requirements of today's projects, they provide little possibility for future upgrading and reusing program code for other designs. Users switch to 32-bit MCs instead of 16-bit ones, since there's small difference in cost whereas choice of peripherals, *development tools*, and chip suppliers is wider.

Another reason is that transition to 32-bit MCs ensures not only more than tenfold increase in *performance*, but also reduces *power consumption* and *program size*, accelerates development of *software* and provides possibility of its multiple usage.

MCS-51 is the most widely applied architecture among 8-bit architectures (PIC, ARV, MCS-51). This architecture is popular since it's offered by a number of suppliers, it is easy to use, and there are many software development tools.

However, MCS-51 architecture has already reached its limits. Not only is work with multidigit data (16- and 32-bit) slower due to byte-at-a-time processing, but also a number of physical constraints emerge that complicate usage of the architecture. For example, program size is limited to 64 KB. It is possible to increase program memory by memory paging or storing it off-chip (1880BE1У, Dvina 51AC operates with external memory only), but memory paging has a lot of disadvantages:

- Unproductive expenses regarding size of program code and instruction execution time increase;
- Memory is wasted since memory banks aren't used to their fullest;
- Special add-in code for data access by other programs is needed and it occupies stack memory, which is rather limited;
- There's no standardized memory bank switching, which can cause problems with compilers and debuggers and also complicates transition to other products;
- In case memory paging is performed externally, the number of available input/output ports reduces and the cost of system increases while its reliability decreases due to additional connections on the printed board.

MCS-51 architecture is designed for operation with 16-bit addresses, 16-bit program counters, and 16-bit data pointers are used, and the instruction set is designed to support 64 Kb address space. If memory of more than 64 Kb is needed, hardware and instruction overheads are required to generate additional address bits.

Another important constraint is internal memory space. The stack can be located only in internal memory, which is limited to 256 Kb. The first 32 bits are used for the workspace register (4 register banks from R0 to R7) and some internal memory can be used for variable data (e.g., when memory with bit addressing of Boolean data type is used it can be located in RAM). As a result, maximum capacity of stack memory is limited. The number of special function registers (SFR) and data pointers is limited, too. MCs contain data pointers (DPTR), but there's no standardized programming model for DPTR, which complicates application code porting to different MCS-51 chips.

Operation of the MCS-51 processor is largely based on the use of accumulator (ACC) and data pointer (DPTR) for data movement and processing. Therefore, instructions for data movement to/from ACC and DPTR are needed, which leads to increased code size and more instruction cycles.

Memory interface also limits performance of 8- and 16-bit processors. For example, many instructions of the MCS-51 processor have a length of several bytes. Since the program memory interface is 8-bit, instruction fetching requires several read cycles.

As far as the world market of microcontrollers is concerned, it is characterized by the following trends:

- 8-bit MC segment historically had the largest sales volume but since 2010 the 32-bit segment has prevailed. In 2010 the difference between sales of 4-/8-bit and 32-bit MCs constituted $235 million USD but the gap has considerably increased since that time. In 2013 the share of 32-bit MCs reached the value of $6.9 billion USD, which outsells the 4-/8-bit segment by 57%;
- In terms of volume of supply 16-bit MCs became the largest segment in 2011 outselling 8-bit devices for the first time. 16-bit MC volume increased by 11% in 2011 after a 23% increase in 2011, partially due to automobile applications;
- In 2013 volume of 16-bit MCs grew by 9% up to 7.9 billion pcs, supplies of 4-/8-bit MCs increased by 6% up to 6.7 billion pcs, and volume of 32-bit MCs increased by 20% up to 4.5 billion pcs.

4.5.2   Licensed Architectures MIPS32 and Cortex-M

A wide variety of possible applications is a feature of microcontroller market; therefore, dozens of architectures are in demand. Renesas (SuperH), NEC (V850x), and Freescale (68000) were the leading architectures on the market of 32-bit MCs up to a certain moment. Today the leading architectures are MIPS32 (developed by the company MIPS Technologies) and especially Cortex-M (developed by the company ARM), which covers 75% of the market. MIPS and ARM sell licenses for their architectures and now are represented on the market by various versions from different manufacturers. There are about 3 dozen manufacturers that offer more than two hundred microcontrollers.

Such companies as AMD, Broadcom, Infineon, Realtek, Sharp, Sony, Toshiba, and Microchip use MIPS architectures. Atmel, Cirrus Logic, Fujitsu, Intel, Freescale (Motorola), National Semiconductor, NXP (Philips), ST Microelectronics, and Texas Instruments acquire Cortex-M architectures from ARM. They are also used by Russian companies.

Unified architecture of processor cores allows customers to if not reduce, then at least slow down the increase of expenses for development tools, standard libraries and drivers, and software development. According to different estimates, software development for systems that use microprocessors and microcontrollers accounts for about 60–80% of all expenditures.

In the 32-bit MC segment developers rarely work with assembler and use high level languages instead. In case there's need to change manufacturer (due to disruption of supplies, absence of necessary libraries, new functional requirements, price increase, and so on), core program code won't require alterations. Thus, if program code of peripherals is written independently from the code of core, fast transition from one manufacturer to another can be ensured.

MIPS32 and Cortex-M architectures have become de facto world standards. It is more profitable for a company to acquire a license for fabrication of MPs and MCs with a renowned architecture than MPs and MCs with new architecture and new instruction set (it hardly stands a chance on the market) or reproduce a duplicate by copying its layout. (Complex design rules, metal layers, multicore architecture, and know-hows of modern microcircuits cannot be fully reproduced). As far as new software is concerned, modern translations of programming languages produce millions of lines of code, which requiring a lot of highly skilled labor.

We shall briefly compare ARM and Cortex-M architectures without going into much technical detail: both MIPS and ARM Cortex-M cores have a 32-bit wide data bus and are a reduced instruction set computer (RISC). But this is where similarities end. MIPS is oriented toward high-performance applications whereas Cortex-M is oriented toward entry-level cell phones with a focus on control.

4.5.3   Features of Designing and Organizing Production of Microcircuits of Space Application

The so-called *space microelectronics* domain provides a highly reliable electronic component base (ECB): microcircuits, discrete semiconductor devices, sensors, gyroscopes, SOC, and systems in package (SiP)) for radio-electronic equipment of rocket and space technology.

Unlike the so-called commercial or industrial (or dual-purpose) ECB, ECB for space application has a range of conceptual characteristics [5–14] that shall be examined more thoroughly. First, these characteristics include:

1. Strict reliability requirements (faultless operation, long operation life, high working capacity, and so on). In particular, fractions of faulty operation shall constitute 10^{-9}–10^{-10} and 10^{-11}–10^{-12} for active and passive components of ECB, respectively.

2. Exceptionally low level of demand and consequently low level of production, which can be regarded as low-volume or even single piece (unique, one-off) rather than serial production. Low-volume production and all its consequences are typical of space electronics. A specific space project usually requires no more than 10–30 microcircuits or discrete semiconductor devices. The maximum number of space microelectronic parts for a large national space project generally doesn't exceed 10,000–50,000 for the whole life cycle of designed spacecraft, from 5 to a maximum of 15 years.

3. Exceptionally wide and varying specifications of ECB, which results from the variety of tasks performed by radio-electronic equipment of space application. Full counts of microcircuit specifications only range from 1,000 (for specific simple space projects like the space shuttle) up to 10,000 (for large scientific space programs like lunar or Martian expeditions *Skylab* and *Messenger*) types, according to open sources (periodicals, reports of NASA, and other national space agencies including the European Space Components Information Exchange System).

4. High tolerance to radiation exposure of various types, both outer space radiation, which affects manned and unmanned spacecraft, and residual radiation from portable nuclear reactors of propulsion systems that are sometimes placed onboard spacecraft in violation of international conventions. According to official data, such products shall have at least 50 Krad radiation tolerance, whereas special SVs (military communication satellites, navigation and reconnaissance satellites, and so on) shall remain operational for a specified period of time (from 20 minutes to 20 hours) even in case of a nuclear explosion, which can be theoretically utilized by an adversary (or terrorists) to disable spacecraft, as military doctrines of many NATO members state. It should be emphasized that space microelectronic products are required to be tolerant not only to radiation exposure that is typical of ground equipment as well, but also to other specific, exotic types of destabilizing factors that are present in outer space and hardly occur on the Earth.

5. Wide range of operation and storage (sleeping mode) temperatures. ECB of space application is usually required to operate from low (from -70 to $-60°C$) to high (from $+125$ to $150°C$) temperatures. Further, microelectronic products for certain SC and applications are to operate in outer space vacuum.

6. Guaranteed long-term (minimum of 15 years) faultless operation under conditions of space exposure factors including ionizing and electromagnetic

effects, ambient temperature drop, as well as extreme periodic or one-off acceleration, impacts, effects of zero-gravity and moisture, space mold, and many other extreme factors that aren't typical of functionally equivalent ECB specifications for military applications.

The main characteristics of ECB of space application allow concluding that features of space microelectronics development not only considerably diverge from the roadmap of civil (industrial) microelectronics development, but also have stricter requirements than military microelectronic products do.

The trend of constant reduction of LSIC design rules in consumer and industry microelectronics is primarily explained by the possibility to obtain more and more yield from one semiconductor wafer, the cost of which now reaches $500–$2000 USD. It is connected with the so-called consequences of Moore's law, according to which a twofold reduction of microcircuit chip area ensures at least a fourfold increase in yield from a wafer of the same diameter.

4.6   Peculiarities of Using a Fabless Model for Designing Microelectronic Products for Space Application

In this section we shall attempt to describe principles of organization and implementation of this truly modern and rapidly developing microelectronic business as well as its technical, methodological, economic, and other aspects. Understanding the essence of these aspects is necessary not only for managers and technical specialists of microelectronic enterprises that design and/or manufacture microelectronic components of space application, but also for customers of microelectronic components that either have already started or are starting to apply this developing business model to cooperate with national and foreign microelectronic design centers.

4.6.1   Preconditions of the Fabless Business Model Emergence and Development

The fabless model as a field of microelectronic business for manufacturing products of industrial application (IA) started actively developing due to explosive cost growth of semiconductor fabrication plants. Today $2–4 billion USD, not including maintenance costs of at least $2–4 million USD monthly are needed to create even not the most advanced semiconductor factory. Thus, to ensure minimum profitability of the construction project, the investor's annual profit must constitute about $200–400 million USD, which can be achieved provided that sales volume is above $2 billion USD annually. Therefore, questions rise: what is to be manufactured, and to whom and where it can be sold to get those $2 billion USD?

Figure 4.11 shows cost dynamics of standard semiconductor fabrication plant construction over a quarter of a century (from 1983 to 2011) based on diameters of manufactured wafers and minimum values of design rules [1].

As shown in Figure 4.11, only $200 million USD were needed in 1983 to arrange production of microcircuits with design rules of 1.2 μm and wafers with 100 mm whereas 2001 saw the beginning of hard times for microelectronic business: at least

$3 billion USD were to be invested in order to found a factory for 200 mm wafers and design rules of 0.25 μm.

This is the reason why many microelectronic companies that used to perform the whole cycle of microcircuit designing and manufacturing (known as integrated device manufacturers according to common classification) stopped upgrading their manufacturing process in 2008–2010 due to a dramatic increase in costs of new technology acquisition and plant construction, and started actively employing the so-called silicon foundries instead (TMCS, UMC, and other). At the same time a number of fabless companies emerged that designed microcircuits only and then place orders for their manufacturing at specialized factories—fabs.

Division of labor in world microelectronics developed even further: apart from silicon foundries and fabless design centers, new assembly facilities and test foundries emerged. Such division of functions during microelectronic production allows substantial savings and reduction of time needed to enter the market with competitive products.

4.6.2 Structure and Features of Fabless Business

Two fields can be identified in the structure of the fabless business [15] (Figure 4.12): a device-oriented field that focuses on production of end microelectronic products and a virtual component field oriented toward creation of basic libraries and IP-blocks. Both fields are naturally interconnected in fabless business, but they shall be examined separately to understand features of business organization.

The article "Development of Industry of Semiconductor Virtual Components" (*Komponenty I tekhnologii* [*Components & Technologies*], No. 5, 2012, p.172) by Timur Paltashev and Mikhail Alekseev gives a detailed and qualified review of the

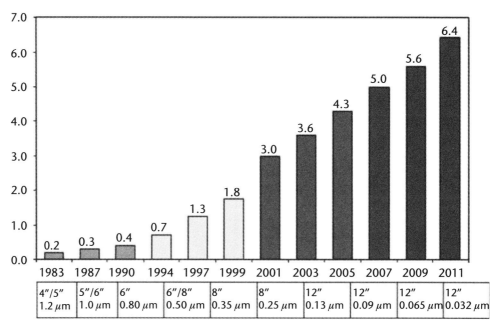

Figure 4.11   Cost increase of semiconductor production due to transition to tighter minimum design rules and growth of wafer diameter.

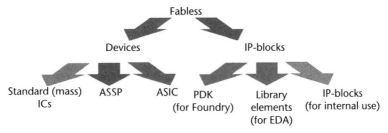

Figure 4.12   Fields of fabless business.

main features of a company that represents the virtual component field of fabless business; therefore, we shall pay more attention to device-oriented field.

Three independent subfields of device-oriented fabless model can be identified in terms of type of microelectronic product:

1. Fabless design of custom products (ASIC);
2. Fabless design of application specific standard products (ASSP);
3. Fabless design of standard products.

In any case final (market) price of a product is the decisive factor for the choice of fabless model field. In [15] we offer some simple formulas to estimate value of product price limit (P) for all three main types of business model:

$$\text{For foundry: } P_{FB} = W_F(x) + \frac{D + \Delta_D}{Q} \tag{4.1}$$

$$\text{For IDM: } P_{IDM} = W_F(x) + \frac{D(x)}{Q} \tag{4.2}$$

$$\text{For fabless: } P_{FL} = W_F + \Delta_F + \frac{D(x)}{Q} + L(Q). \tag{4.3}$$

where

W_f is cost of product fabrication;

D is cost of product design;

Δ_F is profit of outsourced factory (foundry);

Δ_D is profit of outsourced design center;

Q is sales volume;

L is logistical costs,

(x) is controllability (ability to control expenses).

Analysis of these simplified equations allows concluding that all other conditions being equal, it is almost impossible to gain profit from production of standard (mass) products on fabless model since cost of fabrication at foundries equals or exceeds average market prices. The only obvious niche for fabless business is custom/

specialized designing (when there's no duplicate). The end client (customer) realizes and accepts the fact that the price includes not only cost of product manufacturing, but also additional payment to developers for product uniqueness (included in D (design cost)).

We shall review the main preconditions and components of this kind of business and provide its brief characteristics.

4.6.3 Peculiarities of the Choice of Customer for Fabless Project

A key aspect of fabless business is to find a customer (consumer) and arrange tight cooperation with them. Experience shows that at the first stage of business development it is necessary for a fabless design center to cooperate with the customer for a prolonged period of time (several years). This time is required for a company to integrate in scientific and production policy of the customer (to master end-to-end design) and for a client to become familiar with approaches and technical solutions as well as for establishing a certain level of trust with understanding between the customer and design center. It is especially important when products of space application are concerned due to their specific requirements. It is understood that as a rule a whole range of hard preliminary R&D works that does not guarantee commercial success (usually at the expenses of the fabless center) needs to be performed to produce truly breakthrough technical solutions.

Figure 4.13 shows general model of interaction between a design center and the customer in case of designing custom (ASIC) (Figure 4.13(a)) and specialized (ASSP) (Figure 4.13(b)) microcircuits.

The major difference is that there are usually several customers of ASSPs so the fabless company itself usually initiates production, whereas in case of ASIC, the customer sets the project in motion by identifying and specifying their needs in a technically competent way.

In any case the ordering company is interested in reduction of expenses and/or achieving competitive advantage over its rivals in terms of consumer-oriented characteristics.

Considering the pay-off period of R&D work, a customer should be a rock-solid company, since about 5–10% of annual sales volume is intended for the customer. However, a customer (or customers) is a necessary (compulsory), but not sufficient (as mathematicians put it) condition of fabless project success.

We shall examine *technical aspects* of fabless business organization based on experience of our company (OJSC INTEGRAL) and international practices.

4.6.4 Technical Prerequisites and Conditions of Implementation of Standard Fabless Projects, Software, and Design Libraries

To utilize the fabless model a company shall have licensed software that complies with requirements (regulations) of contract microcircuit manufacturers (foundries) and supports PDK libraries. Such software is produced by Mentor, Cadence, Synopsys, and other companies. License acquisition and support is estimated at the value of $500,000–1.5 million USD (depending on SW contents and functionality). As far as design libraries are concerned, there are two approaches: some part of PDK

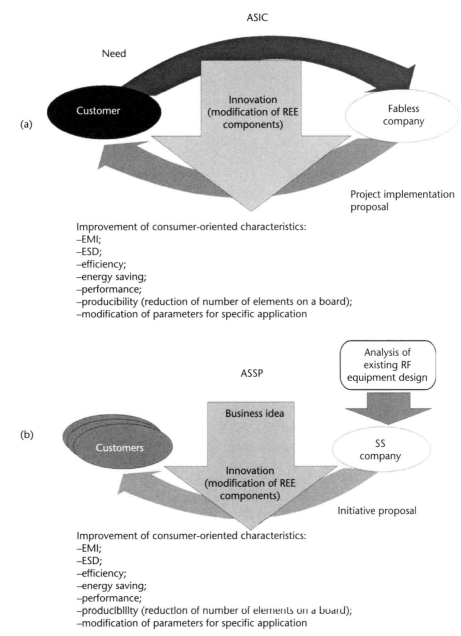

Figure 4.13   Models of interaction between design center and customer [15].

is distributed free of charge for manufacturing processes that aren't know-how, and the other part is sold based on volume of future foundry orders for special processes, especially for manufacturing products of military and aerospace class. Unfortunately, free PDKs are oriented toward standard (mass) products and are ill-suited for fabless business. The second group of (potential) PDKs is viewed as IP by foundry companies. It can be stated from experience that they are hardly willing to share it with design centers associated with other semiconductor enterprises (also under payment, as a rule).

4.6.5 Choice of Manufacturing Process

At the first stage of project implementation, it is necessary to make sure that the design can be implemented by your contract manufacturer using his standard process. However, it should be understood that such design won't bring much profit: either a digital or rather slow analog-digital microcircuit can be produced in such process; it should be taken into account that *if your design really has some value, it will be immediately replicated by your rivals.*

If you opt for complex manufacturing process, for example, to manufacture radio frequency circuits (RFCMOS or RFBiCMOS), competition will be less but you should be prepared that standard cells and elements of manufacturer might be not accurate enough to ensure necessary product quality at the first attempt. Therefore, a couple of additional iterations for each new design might be needed, which are highly expensive.

4.6.6 Need for Own Analytical Base

Experience shows that one cannot fully rely on manufacturer's models. *One should use one's own analytical and measurement techniques for optimization of design for the chosen manufacturing process.* In particular, specialists in the field of extraction of structure and process-dependent parameters (including radio frequency ones) should be able to work with modern software and hardware like Agilent's IC-CAP. In any case. specialists of the analytical department need to undergo additional training.

The so-called optimization for manufacturer factor should be taken into account. For example, when a design has been successfully finished, a GDSII-file for radio frequency and analog-digital circuits (and complex digital ones) has been generated and sent to a silicon factory (foundry), the manufacturer usually replies with a large list of design rule violations that are to be corrected. Such corrections take a while, though not requiring additional expenses since they are usually made by place and route specialists who were initially involved in the project. To avoid manufacturer's sanctions against failure to meet time constraints specified during planning, it is advisable to send GDSII-file to fab long before the planned date of wafer carrier shuttle launch.

4.6.7 Manufacturing of Prototypes

To manufacture prototypes within a foundry order, the so-called shuttle (MPW) is usually used, which allows dividing cost of production among several clients of the foundry that utilize the same manufacturing process.

A small number of produced prototypes allows you to make sure that design and circuit solutions have been chosen appropriately, but such number of chips isn't enough to perform all necessary tests in accordance with approved quality standards (to prove that design is suitable for mass production). One shuttle costs about $20,000–50,000 USD (90 nm or 130 nm process). In case your prototypes aren't operational, it'll take several months until there are clients for the next shuttle. Potential photomask problems should be also taken into account.

In case the microcircuits are functional, the customer has tested them and provided a report (which is rare in practice), a set of photomasks (metal-mask pattern) is ordered. Cost of such set for one microcircuit varies from $55,000 USD (0.25 μm process) and $300,000 USD (0.13 μm process) up to $1 million USD (65 nm process). In case a design needs improvement (which is usually the case), you have to pay again. Typical financing schemes look like this: the customer pays the fabless company and the latter settles accounts with the foundry. It should be noted that the trust factor plays an important role, since money is paid in advance. If there's no prepayment, the fabless company has to settle accounts itself. (In the best-case scenario expenses are covered by payment for previous stages.)

When orders are placed, the relationship between design center and foundry matters. In case of a large order that employs up to 30% of factory capacity, the customer will obtain VIP status and best prices (Δ_F in formula (4.3) will have minimum value). And vice versa if that's not the case. If you order about two hundred wafers monthly, you aren't a profitable client for a factory; therefore, your microcircuits will be manufactured residually at artificially high prices. In case a VIP client wants to order additional 500–1000 wafers a month at Taiwan Semiconductor Manufacturing Company, smaller orders will be simply pushed back and clients will have to put up with that. Provided that the design is simple and processing-independent, an alternative factory can be found (though this may result in failure to meet time constraints). However, it won't work in case of a complex analog-digital high-frequency circuits device since its design includes the PDK of a specific manufacturer so is optimized for a specific manufacturing process.

Special attention should be paid to testing of designed product. Suppose wafers with your product have been manufactured and now need to be tested. Since companies that manufacture wafers usually do not perform testing, it can be done either at your company or at test foundry. In both cases the developer needs to write a test program, but if measurements take place at test foundry, a highly qualified specialist is required who is familiar with the foundry's testing equipment (which is different at each facility). Then wafers need to be cut, and chips are to be assembled into packages. This costs from $0.20 to $5 USD, depending on the number of pins. In most cases assembly foundries can perform testing, too, but testing of functionally complex RF and digital-analog circuits is a complicated process and the number of operational circuits may reduce by 30% merely due to faults of testing personnel (e.g., failure to set up test probes properly). Outsourced testers often don't understand the nature of problems because they aren't quite as familiar with product as developers.

To avoid such situations, efficient fabless companies usually assign a group of specialists that tightly cooperate with testers. (Knowledge of at least one foreign language is an essential requirement for such specialists.)

4.6.8   Financial Paradigm of Fabless Projects

When a viable design has been developed, a company wants to produce batches of at least 200 wafers a month, each wafer containing from 1,000 to 20,000 chips depending on their area (the larger the area is, the more assembling costs). You need to pay for photomasks, wafer manufacturing, assembling, chip testing, and logistics while sales haven't even started. Furthermore, staff shall be well paid in accordance with their qualification. Ideally for a fabless company, these expenses are covered by the customer, but such large amount of money can be given only to a proven company that has successfully implemented a range of projects, and the customer is assured of company's reputation, technical competence, and financial stability. Most importantly, the customer should afford financing the fabless project without detriment to their main business. It is possible only provided that customer is a large company with big turnover (or it is a state order contractor). In practice, all financial gaps are covered by fabless companies. Such a financial paradigm answers the question of why a large number of companies don't engage in modern fabless businesses although they have highly qualified developers, advanced software, hardware and analytical equipment, and so on.

Other problems may occur even after a project's successful implementation. For example, it's been two years after first yield was obtained. Products sell well, but 5–10% of your microcircuits are discarded during incoming control (such situations aren't that rare: Consider recent well-known and not yet settled cases of AMD and Global Foundries regarding accelerated processing units with design rules of 32 and 28 nm^3 or the case of Ramtron and Intel [16]).

As a result of hard work in a cutting-edge and well-equipped reverse engineering laboratory that has electron microscopes and equipment for chemical and plasma etching, your company's domestic experts in failure analysis find out what causes the faults of a fab's manufacturing process. It is necessary to inform the factory that there's a problem in their manufacturing process that has a negative effect on the company's profit and the clients' nervous system. But since you're just a cog in the machine of a large factory, your information will be regarded with sympathy, but you will be told that unfortunately the microcircuit errors are caused by faults of your developers or assemblers, since other customers have no problems with this manufacturing process. To our regret, these aren't some abstract assumptions, but outcomes derived from experience. Foundries have neither the desire nor the possibility to deal even with justified remarks on technical aspects from small clients.

4.6.9 General Recommendations on Work Organization in Fabless Mode

Several general conclusions can be made based on the previously mentioned facts. First, the fabless business is a rather complicated business in the field of design.

The main niche of fabless business is designing and supplying custom (ASIC) and specialized (ASSP) products, the price of which includes payment for uniqueness.

The customer is a major player in fabless business. They are essential for the start of a fabless project. In the end, it is the customer who pays for product design (uniqueness) and manufacturing expenses.

This is possible only provided that the designed product fits the customer's needs (gives technical, financial, or other advantages), (i.e., technical characteristics shall be negotiated with customer at the first stage of design).

Working with silicon foundries far abroad poses rigid constraints in terms of CAD (license and support). There's distinct correlation: the more standardized manufacturing process a foundry provides, the less unique the designs of a fabless company are and the smaller its profit is. At the same time, specific manufacturing processes of silicon foundries, as a rule, utilize raw PDKs, which may result in additional expenses.

The multiproject wafer (MPW) approach allows cutting expenditures on prototype manufacturing; however, its disadvantages are dependence on schedule of MPW launches of a certain foundry and a small number of launches.

Cooperation with a foundry can be rather uneasy, but it becomes more efficient provided that volume and rate of orders grows. Outsourcing to test foundries provides a possibility to cut expenses on testing but bears certain risks.

A design center is required to have significant current assets, either its own or raised (customer/bank loan) to implement a fabless project.

4.7   Special Considerations Relating to the Selection of Foreign-Made ECB for the Design of Domestic SC

4.7.1   Basic Premises for the Necessity of Using Foreign-Made ECB

After the collapse of the Soviet Union, use of foreign-made electronic components in the design of domestic radio-electronic devices has become a constantly growing trend [17–23]. The proportion of foreign-made electronic components in the total volume of components used is steadily increasing. These trends are also characteristic for the process of creating electronics for strategically important sectors, including rocket and space technology, which require higher quality of electronic components used.

Due to substantial support from the state, in Soviet times, the electronics industry, in spite of falling behind the development in some other countries, provided the domestic component base meeting most of the electronic industry's needs, including the needs of the Soviet military-industrial complex (Figure 4.14).

Thus, all electronic components of launch vehicles and spacecraft were designed only using the home-produced ECB. The upper stages and boosters of the Russian *Fregat, Briz-M, Briz-KM (D, DM), Vesuvij*, and others already use 10 to 20% imported ECB. Since 2000–2005, for the design of new space objects and

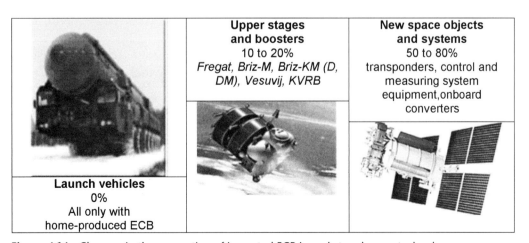

Figure 4.14   Changes in the proportion of imported ECB in rocket and space technology.

systems (except launch vehicles) the proportion of imported ECB has increased up to 50–80%. As of this writing, in Russia there are no ECB specifications required for the design of transponders, onboard converters, control and measuring system equipment, navigation, and so on.

It should be noted that, unfortunately, this trend will prevail in the foreseeable period of development, due to the following factors. First, the ECB product range to be used in future RST and weaponry and military equipment exceeds tens of thousands of part types. If the relevant services of the Ministry of Defense, the MIC, and Roscosmos organize and implement the labor-intensive process of harmonization and minimization of this list, it can be reduced in the short term to the quantity of 5,000–7,000. But even this quantity cannot be produced by our electronic industry in a short time due to a number of objective reasons.

Second, with the development of the Internet, designers of electronic components and systems have received access to an extensive database on the ECB for various purposes, including space applications. The level of representation and quality of the database make it possible not only to easily choose the right type of a microcircuit, but also to promptly design the electronic device that the designer needs, using specific examples of circuit design features of the detailed guidance on the application of this ECB.

Unfortunately, the similar domestic ECB databases are significantly inferior in this sense, in terms of quality of representation and scope of technical documentation, sometimes containing inaccuracies and errors, which hinder solving the developer's main tasks—to quickly and accurately design a required electronic device (unit, system).

Therefore, it's not considered possible now or in the near future to completely eliminate the use of imported ECB in the design and manufacture of not only the electronics for general civil applications, but also for space and nuclear industries, except for some particularly sensitive areas of military and technological security of the state.

In any case, it is necessary to create a modern domestic infrastructure that allows covering the minimum demands of these sensitive sectors for the advanced ECB—this applies both to the microelectronic technology (much has been done already in this field) and design technologies. To this end, this book presents materials on the most recent technology and circuit design methods, submicron ECB test methods that can be used by domestic experts to resolve this issue, which is of key importance for the sustainable development of the rocket and space industry.

In the process of REE creation, several stages in relation to the foreign-made ECB can be distinguished that significantly affect the quality and efficiency of electronic components [18]: selection of electronic components range, their procurement and organization of certification testing procedures, as well as their implementation as part of a specific electronic device. The most detailed considerations for these phases of ECB selection are described in [23], the key points of which are discussed later.

4.7.2 Considerations for the Foreign-Made ECB Selection Phase

The phase of ECB selection by the designers of electronics starts with drawing-up of the statement of work for systems engineering and circuit design, for both the equipment as a whole and ECB included in the equipment, and ends with the customer's decision concerning the acceptability to use the selected range of foreign-made ECB in such REE. The same decision should approve a model of external exposure factors, for which conformity certification tests of foreign-made components will be conducted. The success of the selection process of foreign-made electronic components is determined by the following factors [18]:

- Correct formation by the contractors of required quality (performance) of electronic components, which is necessary for the customer and meets the tactical and technical requirements for the sample specified in the SOW for R&D work on REE development, as well as correct modeling of exposure factors under which these electronic components will have to operate;
- Contractors' thorough knowledge of the features of the world market of required electronic components;
- Contractors' ability to deal with information on foreign components. Here, it is important that the most recent and reliable (not advertising) information is received from authentic sources.

It should be noted that the lack of clear definition of electronic component grade often leads to significant errors in initial determination of the quality of the foreign-made ECB. The most common error in determining the foreign-made ECB grade is a substitution of the concept of conditions of use for some derived parameter, such as operating temperature range. Thus, the name given by the foreign ECB manufacturer to the operating temperature range is often used to designate a specific grade of electronic components (i.e., their overall quality, without regard to other indicators of the exposure factors model, for which the product has been designed and manufactured).

Grade is understood as the generally recognized type of application conditions, for which a specific ECB part type is designed and manufactured (including the scope of control and testing during the manufacturing process).

The grades of foreign-made components are as follows:

- *Space* products (products for space applications) comply with the specifications and standards of government bodies involved in the assessment of levels of products quality and manufacturers. As of today, there are only four such internationally recognized government bodies, three of them operating in various space agencies and one being part of the U.S. Department of Defense. The qualification levels of manufacturers (manufacturing processes, ECB quality) can be learned from the documents of these bodies.
- *Military* products (products for military applications) comply with the specifications and standards of government bodies involved in assessment of levels of products quality and manufacturers. As of today, the military grade is qualified by only one government body, which operates under the U.S. Department of Defense.
- *Industrial* products comply with the standards of international professional associations of a certain industry. The quality of products is confirmed by

certificates of conformity issued to the manufacturer by these professional industrial associations.
- *Commercial* products comply with the manufacturer's specifications drawn up on the basis of the ECB market promotion policy; in this case the only guarantor of quality of these products is the company's own quality management system.

The process of selecting the electronic components is a multiobjective optimization task. The range of factors affecting the choice and their weight in the decision depend heavily on the specific conditions of a project under development. However, the general groups of factors can be identified that almost always accompany the selection of electronic components [15, 18]. These include, first of all, the parameters of a specific ECB market segment, the possibility of mass production of designed REE; technical and economic parameters required for R&D work. Let us take a closer look at each of these groups.

Parameters of Market Segment of Elements with Required Functionality

Here, two groups of factors should be considered: functionality of the designed REE and information about the recommended ECB manufacturer.

1. Product functionality includes the following components:
   - Specifications relating to the functionality of the following electronic components;
   - Model of exposure factors, for which this type of ECB has been designed and is manufactured;
   - Current limitations on the applications, in which it can be used.
     According to the product specifications and models of exposure factors for which it is designed, it is expedient to select the any products, which, as marketers say, have public recognition (state qualification centers, industry associations, and so on).
2. The manufacturer information should include the following data:
   - Quality system used in the production;
   - Technologies used, both currently used and planned for use in the medium term;
   - Documentation provided by the company during the components selection and delivery phases;
   - Supplier's warranty.
     With regard to the management of the company's quality and implemented technology ensuring the manufacture of the desired product, it is expedient to select any manufacturer that has any public recognition. Not unimportant for the effective use of electronic components is any additional documentation provided by the manufacturer, including reports on carried out and documented inspections and tests during production of required products.
3. Product prices (they are necessary for the calculation of the total project cost and, in particular, of REE prototype.

REE Series Production Prospects

Here, an important parameter is the period of production of a component that you have selected, which is declared by the manufacturer, as well as manufacturer's specific recommendations concerning the possible replacement or upgrading of components.

No less important for the designers is the information about any governmental limitations on the component distribution.

Technical and Economic Indicators of the R&D Work (Prepared by the REE Designer)

The largest contribution to this parameter group is made by the following [18]:

- Scheduling of jobs in order to optimize the list of foreign-made ECB as part of the R&D activities (the entire project);
- Ensuring the possibility for price/quality change (flexibility) with respect to the range of selected foreign-made ECB during the subsequent stages of R&D.

When selecting the electronic components, the practical skills of the project executors for working with primary sources of information on foreign-made ECB are of great importance. It is important to understand that there are four main groups of primary sources of information related to foreign-made electronic components:

- National (or international) centers for qualification of manufacturers, technologies, and products;
- Professional associations of industries manufacturing (using) electronic components;
- ECB manufacturers;
- Various professional associations of electronic components distributors.

At this stage, when electronic components are selected for the project, the first three groups are the most informative (information from the distributors' associations is more often used at the subsequent stages of work with foreign-made electronic components). Each of these groups can provide additional information about certain component-specific and manufacturer specific aspects, which can be used at the subsequent stages—during negotiations on deliveries, guarantees, and so on.

4.7.3   Special Aspects of Certification Tests of Foreign-Made ECB

In terms of significance, the first priority for the customer is information on qualified electronic components provided by the national (or international) qualification centers, especially when it comes to high-tech electronic components base, primarily for military and space applications. Currently, the following authoritative organizations play the most important role in the process of ECB quality evaluation:

- U.S. Defense Supply Center Columbus (DSCC);
- NASA's Goddard Space Flight Center (GSFC);

- European Space Components Coordination (ESCC);
- Japan Aerospace Exploration Agency (JAXA).

The Defense Supply Center Columbus is part of Defense Logistics Agency (DLA); with its current functions it was established in 1996 and maintains a database on technical regulations and deals with quality only in respect of military supplies. Now the center supports the development, coordination, and revision of standards, regulations, and specifications at the following levels [18–23]:

- Handbooks on items for military and space applications;
- Federal/military standards;
- All types of federal/military/detail/performance specifications;
- Standard microcircuit drawings and vendor item drawings.

The ECB quality assessment issues are dealt with as part of the following international programs:

- Quality assessment of manufacturing technology with the issuing of a qualified manufacturers list (QML). The work is done with respect to manufacturers (manufacturing technologies) of semiconductor devices (FSC-5961) and microcircuits (FSC-5962) of all types (integrated and hybrid);
- Quality assessment of manufactured products with the issuing of a qualified parts list (QPL). Within only the Electrical and Electronic Equipment Components supply group, the DSCC maintains the QPLs for 17 supply classes, including resistors (FSC-5905), capacitors (FSC-5910), relays, solenoids (FSC-5945), coils, transformers (FSC-5950), and so on. The database includes more than 200 lists of qualified products;
- Quality assessment of services provided by distributors, with the issuing of a qualified suppliers list of distributors (QSLD). The work is done mainly with respect to distributors—secondary suppliers of semiconductor devices (FSC-5961) and microcircuits (FSC-5962). It should be noted that certain attention is also paid to other distributors of other classes of products, which is reflected in the issued qualified parts lists.

The leading body in the electronic components and packages program is GSFC, which is a part of the National Aeronautics and Space Administration.

GSFC maintains a database of standards, regulations, and specifications and deals with the quality of electronic components for space applications. The center coordinates its activities with ESCC, Canadian Space Agency, JAXA, U.S. DoD, and other organizations supervising the design of electronic components. Now GSFC supports the development, coordination, and revision of standards, regulations, and specifications at the following levels: handbooks on items for space applications, NASA standards, and specifications for certain types of products intended for space applications.

The center also issues a catalog of QPLs. This catalog includes only products manufactured to the center's specifications.

The center supports the NASA website with lists of selected components.

ESCC acts as part of European Space Agency. ESCC ensures coordination in the field of electronic components between organizations and companies involved in space research, including the national space agencies of the UK, Germany, Italy, Ireland, and France. ESCC activities are regulated by the standardization documents that are developed by ESA's specialized body—the European Standards Organization for Space Activities.

ESCC also maintains a database of standards, regulations, and specifications and deals with quality of electronic components for space applications. ESCC supports the development, coordination, and revision of standards, regulations, and specifications at the following levels:

- Handbooks on items designed for space applications;
- ESCC standards;
- ESCC specifications for certain types of products intended for space applications.

ESCC also issues the following documents: ESCC REP001—a list of published ESCC documents and specifications (the ninth edition of this list, dated December 2011, is currently effective):

- ESCC REP002: a list of discontinued ESCC documents and specifications (the seventh edition of this list, dated December 2011, is currently effective);
- ESCC REP005: an ESCC QPL. The edition of this list dated 15 December 2011 is currently effective. In accordance with this edition, the qualified parts for space applications are manufactured within 11 classes;
- ESCC REP006: ESCC qualified manufacturers list;
- EPPL: electronic preferred parts list.

JAXA supports the development, coordination, and revision of standards, regulations, and specifications at the following levels:

- Specifications for qualification tests with respect to certain types of electronic components designed for space applications;
- Application data sheets for certain types of electronic components designed for space applications.

The agency also maintains the following electronic components quality assessment databases:

- Quality assessment of manufacturing technology with creation of a qualified manufacturers database;
- Quality assessment of manufactured products with creation of a qualified parts database;
- Lists of specifications for qualification tests of certain types of electronic components designed for space applications.

In the documents processed by JAXA, we can clearly see the consistency with normative and technical documentation approved in the United States by the Goddard Space Flight Center and the Defense Supply Center Columbus.

All of the previously mentioned state organizations (DSCC, GSFC, ESCC, and JAXA) have practically similar methodology and organizational ways of dealing with quality of electronic components. This also applies to the ECB quality control management. There are differences almost exclusively in the European Space Agency with regard to certain aspects related to the assigned ECB quality levels, as well as presence of ESCC resident inspectors in all companies certified under QML and QPL programs.

The second group of authentic sources related to the electronic components of industrial grade is professional associations of industry producing (using) electronic components. The most important and most frequently referred to are three organizations listed next.

The International Electrotechnical Commission is a nonprofit organization for standardization in the field of electrical, electronic, and related technologies. Some of its standards are developed jointly with the International Organization for Standardization (ISO). IEC promotes the development and dissemination of standards and maintains a database of standards related to electronic components testing (IEC 60068 series of standards), as well as general and group specifications for all classes of electronic components (detailed specifications for components are developed by the manufacturers in accordance with requirements of IEC standards).

The Joint Electron Device Engineering Council (JEDEC) is an engineering standardization council of the Electronic Industries Alliance—an industrial association representing all branches of electronic industry.

JEDEC was founded in 1958 to develop standards for semiconductor devices. JEDEC is known for its coordination activities in the development of standards for computer memory and has a good reputation and serious influence on industrial production. The world's leading manufacturers and developers of electronic components participate in the JEDEC activities. The best known are standard for stress test driven qualification of integrated circuits (JESD47) and standards with test methods (JESD22 series).

The Automotive Electronics Committee is a developer of standards for the electronic components used in the automotive industry. Its best known standards are for microcircuits (AEC-Q100), semiconductor devices (AEC-Q101), and passive elements (AEC-Q200). Groups of authentic sources of information, which include manufacturing companies, provide a wider scope and more comprehensive information.

4.7.4 Considerations on How to Perform the Analysis of Separate Lists of Foreign-Made ECB

All of the matters considered and how deeply they are studied during selection of elements are reflected in the part types lists allowed for use. In Russia, the parent organization—the RNII electrostandard test center—periodically performs rapid analyses of foreign-made ECB part types lists. To perform the analyses, they use foreign-made ECB part types lists allowed for use and received for certification testing. The main areas of foreign-made ECB lists analyses are as follows [20]:

• Presence of repeating part types in the foreign-made ECB list provided by one customer;

- Production period (phase) of selected electronic components;
- Quality level of the selected electronic components;
- The possibility to improve the quality of elements within the product of the selected manufacturer;
- The possibility of standardization of electronic components with partial replacement of the selected manufacturers.

Figure 4.15 shows the results of the duplication analysis of part types of electronic components for the five supply classes of components in the lists. As we can see, the duplicability of elements is 6 to 26%. The presence of repeating part types in the ECB list suggests that the orders placed by ECB customers have not been processed by the main contractor. It is important to understand that the repeating ECB part types would lead to delivery of nonuniform lots of components (reduced reliability) and increased certification tests.

Periodic (phased) production of selected electronic components is determined according to manufacturers' information posted on the websites. While performing the analysis of information from ECB manufacturers, attention should be paid to the following aspects:

1. The selected ECB part types are currently available and manufactured. During the analysis, its results can be reduced to one of four outcomes, enabling the customer to make a decision in the future:
   - This ECB part type is currently manufactured;
   - The manufactured ECB is of the same series as the selected part type, but has a different modification;
   - The ECB part type is discontinued; there is no information about this ECB part type on the manufacturer's website;
   - There is no information about this ECB part type on the manufacturer's website. The most probable reasons for the lack of product information on the manufacturer's website are as follows:
     - The product is custom-made (one-off) and is not produced in series;

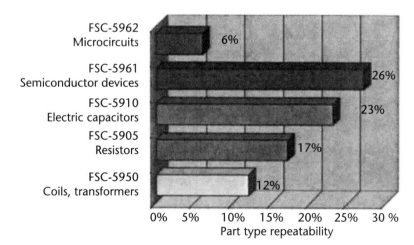

Figure 4.15   Foreign-made ECB by part duplicability.

- The product has long been discontinued, and the company does not support its online information anymore;
- The product is counterfeited.

2. Specific manufacturer's limitations and recommendations exist concerning the use of preselected ECB part types. The limitations are mainly two types of the company's warnings:
   - This ECB part type is not recommended for new design—more often it means that these components will be discontinued soon;
   - The final sales of this ECB part type are under way—in this case, for items to be discontinued, the end date of acceptance of ECB manufacturing orders and date of the planned shipment are often specified.

The manufacturers' recommendations include restrictions on the use of specific elements for critical applications (under the conditions of use associated with exposure factors, life safety, and so on). Manufacturers' recommendations were also considered regarding the substitution of the ECB part types in the event of discontinuation of production or limitations on the use in a new design.

The results of the analysis of manufacturers' information on microcircuits and semiconductor devices are shown in Figure 4.15. As we can see, the proportion of components in active production constitutes 60–73% depending on the class of components. A large proportion of part types of electronic components that are discontinued, not recommended for new design, or the information on which is unavailable on the manufacturer's website are a source of future problems with the procurement of components and financial problems (losses) due to counterfeit goods.

The results of quality analysis of selected elements such as microcircuits (FSC-5962) and semiconductor devices (FSC-5961) are shown in Figure 4.16. As we can see, here the most representative group is that of commercial products—more than 80%. On the basis of work performed it can be predicted that in this group problems may arise may arise in the future with obtaining accurate specifications and technical

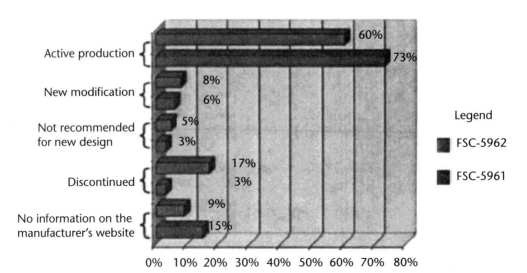

Figure 4.16 Foreign-made ECB list by component lifecycles.

documentation on these elements, as well as possible problems with their life cycle (especially in the production of series) and problems with counterfeit products.

The percentage of *industrial* electronic components is very small—only 5–8% of the devices produced according to the technical documentation complying with the standards of professional industry associations. Qualified element base for military and space applications accounts for 10–11%, with the proportion of the radiation tolerant components not exceeding 0.5% of the total number of components.

The analysis of possibility of increasing the quality of components from *commercial* and *industrial* to *military* within one manufacturer is shown in Figure 4.17. The figure shows that this is possible not only for 8–9% of the active components, but also for 32–35% of passive components. The main reasons for such a low probability of quality improvements are as follows:

- Choice of manufacturer, who is not involved in the technology/product qualification programs of one of the state (interstate) qualification centers;
- Choice of part types series that do not have a higher quality level version.

The possibility of electronic components standardization when changing of some of the selected manufacturers can be estimated using the example of FSC-5905 resistors selection that were on the list in one of the projects. As part of the list optimization activities, a list of resistors was generated on the basis of products made by one manufacturer. Thus, the products of five manufacturers (EVER OHMS Technology Co., Uniroyal Electronics Industry Co., Vishay Intertechnology, YAGEO Corporation, and Phycomp) have been reduced to production of a single manufacturer Vishay Intertechnology working under the QPL program (Figure 4.18).

The main results of this work have been as follows:

- The total number of part types of resistors was reduced by 27% because no resistors of the same size and rating were ordered from different manufacturers

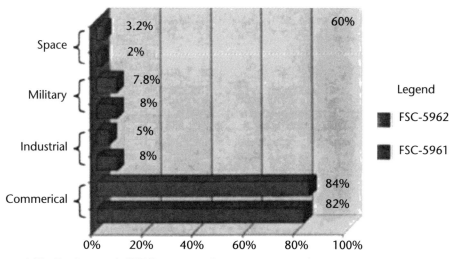

Figure 4.17 Foreign-made ECB list contents by component grade.

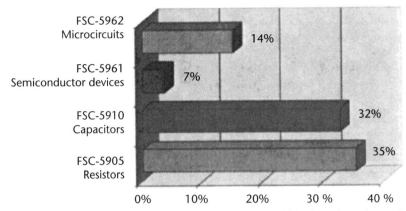

Figure 4.18 Possibility of components' quality improvement within the selected manufacturers.

(this reduction in the number of part types leads also to an increase in batch size of one part type of resistors).

- A uniform group of one part type of resistors was sorted in terms of quality, and selection was done without loss of product quality. Tolerances and temperature resistance coefficients were used as the quality indicators; when making a group, these values were selected by the maximum value in the sorted group.
- The ability to improve the quality level of the whole range of resistors to military and space versions. Thus, all part types of resistors from the list of one manufacturer can be replaced by the components of higher quality level.

Thus, we can make the following conclusions:

1. The selection phase is a key to the quality level of the foreign-made electronic components. The efficiency of work at the stage of selecting the foreign-made ECB depends on how well the matter was studied by both individual REE designers and experts of the chief designer's team;
2. Analysis of some lists of foreign-made ECB part types allowed for use has shown the following:
   - Foreign-made ECB intended for use in the equipment of strategically important facilities contains an unacceptably large proportion of *commercial* level components;
   - The lists contain electronic components with the manufacturer's restrictions on use in new designs, as well as in critical applications.

4.8   Features of Homemade Element Base for Spacecraft Power Supply Systems

The level of development of military and space equipment largely depends on the efficiency of power supply systems used. Modern power engineering in general and power supply systems (PSS) in particular are complex and interconnected electric power supply systems by enterprises and organizations of various industries, the army, and the navy. PSS include a wide range of devices for power generation, conversion, transmission, distribution, and switching.

In the past 15 years, the world market of secondary power sources (SPC)—the main component of the PSS—has been characterized by constantly high growth rates (12–15% per year). The market volume of AC/DC, DC/DC, and DC/AC converters in 2010 in the United States alone amounted to more than $15 billion USD, of which DC/DC-converters account for $5 billion USD.

As the SPC market grows, the competition becomes greater, which, in turn, leads to product improvements and is possible only when using advanced methods of PSS design, highly efficient technical solutions, new electronic components, improved manufacturing processes, and production automation.

The current situation with the global development of space systems, weapons, and military equipment (WME) is characterized by a significant increase in their power loading. Thus, the problem of WME power supply has become one of the most important elements in the new conceptual approaches to the development of warfare means.

The main features of modern SPC are:

- Uniformity of functional tasks performed by all types of electronics (electrical energy conversion, output voltage stabilization, equipment overload protection, interference suppression, galvanic isolation of input and output circuits, and so on);
- Extensive use of standard designs, as well as of a unified component base while creating the required range of such products;
- The possibility of modular SPC design, standardization of envelope and installation dimensions, thus making easier the estimation of their interchangeability during operation.

In the Soviet Union's industry, the SPC were designed mainly by the enterprises for themselves to meet their own power supply needs. As a rule, these designs were not standardized; they were expensive and suitable only for special-purpose applications in specific areas of military radio electronic means. At that time, the Soviet industry did not have specialized manufacture of military SPCs, in contrast to the leading foreign countries.

Until the 2000s, the primary PSS of the spacecraft and weapons systems continued to be designed based on traditional principles corresponding to the performance standards of the 1980s. In addition, at the time when foreign manufacturers made huge progress in their electronics, the appearance and performance of the USSR's homemade power supply systems remained practically unchanged for more than 15–20 years.

At the same time, a significant number of performance characteristics of special-purpose power supply equipment made in the USSR are lower than the best Western counterparts (e.g., the specific SPC power output is 2–3 times lower and number of switchgear actuations is more than 10 times lower).

Then-existing homemade SPCs used to have a much lower technical level than their foreign counterparts and did not meet the requirements of customers like

Roskosmos and the Ministry of Defense in terms of power quality, reliability, design, values of output electrical parameters, power consumption, and efficiency. However, most sensitive to the SPC quality and reliability are advanced computerized radios and electronics widely used in weapons, control, guidance, navigation, radar systems, and so on, requiring a highly stable and uninterrupted power supply.

Therefore, the main WME and space system designers received a special authorization to acquire and use imported parts and components.

Currently, the PSSs account for 30–40% of weight and volume of home-produced weapons and special equipment. In many respects, this explains the fact that up to 50% of faults and failures of space and military equipment occur because of the secondary power systems of radio-electronic equipment. Therefore, dealing with reduction of weight and size of the SPC and their reliability is of prime importance for the development of military and especially aerospace equipment.

One of the main areas for improving the quality of power supply systems and to design reliable high-performance power supply systems is their intertype unification and modular design, the main principles of which were defined back in the 1980s by leading experts of the Ministry of Defense (MoD), Ministry of Electronics Industry (MEI), Ministry of Radio Industry (MRI), and other industries of the USSR. Today, these principles are crucial for the manufacturing of standardized modules for special-purpose secondary power supply systems.

It should be noted that there was a large gap in this field between home and foreign SPC samples, and serious work in this sector was started only in the late 1990s, while the introduction of modular design principles of REE abroad resulted in the power sources becoming virtually a new class of components in 1970s.

A typical example of this gap was described in a report by United States military aviation experts published in 1997. The report contained an opinion on the technical level of the *MiG-25P Foxbat* jet interceptor flown by Senior Lieutenant Victor Belenko in 1976 to Japan. On one hand the experts were surprised by the highest level of engine/airframe integration, but on the other hand they described the difference between MiG-25P and F-4 Phantom II electronics as that between phonograph and transistor radio.

To eliminate this disadvantage of domestic space and special-purpose electronics power supply systems, the key problems that require immediate solution within the national long-term dual- and special-purpose ECB development programs are as follows:

1. Design, production, and stable provision of the national aerospace and defense complexes with a wide range of semiconductor devices and integrated microcircuits for the design of advanced SPCs able to operate both on land and in deep space under continuous and impulse radiation exposure;
2. Study of the reliability of the whole range of ECB for the PSS of space and military REE under real operating conditions;
3. High-voltage semiconductor transistors with an operating voltage of 600–1200V for AC/DC SPC, powerful high-speed semiconductor transistors with an operating voltage of 60–200V for DC/DC SPC, high-speed diodes and transistors for high-frequency rectifiers, operation controllers for powerful key elements of the power supply systems, 1.25; and 2.5V reference voltage

sources with an accuracy of ±1% are key components that have to be used in modern secondary power sources in order to ensure stable SPC operation.

For this purpose, in particular, it is necessary to design and put into mass production powerful new compact integrated circuits for compensation voltage regulators with low line drop on the control transistor and wide range of output voltages; high-power n-MOSFET, p-MOSFET and bipolar transistors (IGBT); power management PWM controller ICs; and high-speed Schottky diodes.

A successful solution to this problem may allow the designers to reduce the electrical losses in the SPC several-fold, to increase reliability and converter frequency, and, consequently, to reduce the SPC weight and dimensions to obtain an advantage in both cost and performance.

Among the main consumers of components for SPC manufacture, with the exception of distributors and leader-followers, we can mention Russian SPC manufacturers such as JSC MMP-Irbis, Aleksander Electrik, Kontinent, and others. These companies were the first to realize the advantages and opportunities of unified SPC modules market even in the time of collapsing industry of the 1990s, when most companies were dissolving their subdivisions dealing with SPCs and power sources in general. Another manufacturer of unified SPCs is the JSC SPE ElTom, which in the late 1990s selected the design and manufacture of module SPCs as the basis for their further development. Unlike other companies, ElTom has focused only on SPC for special-purpose equipment, and only after a long period was the ElTom's path followed by some other Russian companies, such as JSC GC Electroninvest, JSC SKTB RT, JSC Aleksander-Electric, and JSC MMP-Irbis.

It should be mentioned that JSC ElTom was the first Russian company that, using components produced by Minsk plant Transistor, designed the SPC and successfully passed the certification by the Russian MoD completely for the national component base, having ensured the quality of the MoD standards and becoming the lead designers and manufacturers of *military* SPCs.

Due to this fact, SPE ElTom, in cooperation with the plant Transistor, has altered the then-existing Russian MoD's opinion that the manufacture of advanced special-purpose SPC using home-produced component base was absolutely impossible and that only foreign-made semiconductor components would have to be used.

4.9   Resistance of Onboard Electronic Equipment to Ionizing Space Radiation

Special Aspects of Requirements for RST ECB

In the past two or three years, the assurance of quality and reliability of rocket and space equipment has been given much attention. This includes meetings attended by top public officials and the relevant decisionmakers of federal agencies and departments. Also, several publications on this subject were made in industry sources [24, 25].

As mentioned earlier, the main features of the ECB used in outer space are as follows [9]:

- Wide functional range (about 1,500 items, according to ESCIES);
- Extremely small quantities produced (from 10–15 pcs to 100,000 units throughout the product lifecycle);
- High demands for reliability (flawless operation, durability, storability): $\lambda = 10^{-9} - 10^{-10}$ for active elements and $\lambda = 10^{-11} - 10^{-12}$ for passive elements despite low-volume ECB manufacture;
- Resistance to ionizing space radiation (not less than 50 krad, according to escies.org) and to other specific destabilizing space factors (DSF);
- Extended temperature range (from −60 to 125°C);
- The need to ensure long periods of flawless operation (15 years and longer).

Even this short list of features of the RST ECB shows that the way space microelectronics develops is different from the general industrial electronics roadmap defined by ITRS [26]. It deals with the extremely low production volumes required.

With regard to testing, its cost is determined by sets of standards and customers' additional requirements. The majority of ECB radiation tolerance test methods under the existing standards correspond to the effects of a nuclear explosion. But outer space is quite different. It is characterized by a relatively low intensity electron and proton radiation coming from the Earth's natural radiation belts and acting on the spacecraft for long periods of time as well as by heavy charged particles of the outer space and high-energy protons.

The sufficiently complete international classification of effects caused by DSF, having not only radiation nature, has been currently adopted.

Figure 4.19 shows some statistics on the impact of various destabilizing factors on RST equipment. However, a sensitive issue is still the lack of provision on the *space* quality ECB, which takes into account the basic features of functioning in DSF conditions.

In other countries, much attention is paid to writing and updating regulatory documents. In particular, there is an extensive system of standards and specifications of the U.S. Department of Defense, which describes both the requirements and methods of design and manufacture of space ECB. Figure 4.20 shows the basics of them.

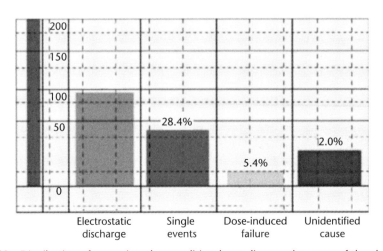

Figure 4.19 Distribution of operation abnormalities depending on the cause of the abnormality.

> **US Department of Defense main standards for space electronics:**
> MIL-STD-883. Test methods standard. Microcircuits
> MIL-STD-750. Test method standard. Semiconductor devices
> MIGSTD-15478. Electronic parts, materials, and processes for space and launch vehicles
> DOD-E-8983. Electronic Equipment, Aerospace, Extended Space Environment, General Specification
> MIL-PRF-19500 – Semiconductor Devices, General Specification For
> MIL-PRF-38534 – Hybrid Microcircuits, General Specification For
> MIL-PRF-38535 – Integrated Circuits (Microcircuits) Manufacturing, General Specification For
> QML-38535. Qualified Manufacturers List
> MIL-STD-1523. Age Controls of Age Sensitive Elastomeric Materials for Aerospace Applications
> MIL-STD-1580. Destructive Physical Analysis for Space Quality Parts
> MIL-STD-1540. Test Requirements for Launch & Space vehicles
> MIL-HDBK-217. Reliability Prediction of Electronic Equipment
> MIL-HDBK-263. Electrostatic Discharge Control Handbook for Protection of Electrical and Electronic Parts, Assemblies
> MIL-HDBK-343. Design, Construction, and Test Requirements for One-of-a-Kind Spacecraft

Figure 4.20   U.S. Department of Defense standards and specifications.

Most of them are freely available on the Internet from the official websites of the U.S. Department of Defense agencies and its contractors (such as snebulos.mit.edu/projects/reference, assist.daps.dla.mil/quicksearch, www.dscccols.com, and others). Relevant regulatory documentation is also developed by ESA via ESCIES (www.escies.org). There is also a special program within NASA (nepp.nasa.gov). Obviously, ISO 9001:2008 certification is mandatory for designers and manufacturers of space-quality ECB.

Figure 4.21 shows the certification scheme of a developer and manufacturer of space quality ECB with Aeroflex Colorado Springs (www.aeroflex.com) as an example. As we can see, test and certification procedures are carried out at all stages of the product lifecycle. Also note that the ECB for space applications, as a rule, has very long life cycle. For relatively serial components of service equipment (control and measuring systems, power management systems, telemetry, and so on), the device production period can exceed 20 years. Considering that the design takes a long time, which is determined primarily by the extensive testing of equipment, ground testing, and flight testing, we can say that the key components should be in production for a long time, which is contrary to the current state of microelectronics, where the turnover of microcontrollers and PLDs occurs every 2–3 years and even more often. There is a contradiction between the requirements for the equipment lifecycle and the actual turnover of ECB generations.

Therefore, despite the emerging tendency for the active use of industrial-grade ECB in RST equipment, there are several impeding factors. Although the storability, reliability, and radiation tolerance values guaranteed by the manufacturers are insufficient, the costs of certification and further testing of such ECB are very high and relatively similar to the cost of design and manufacture of special-purpose ECB.

Due to the extremely limited life cycle of the industrial grade ECB, very costly work is necessary to either reissue the design and technological documentation and

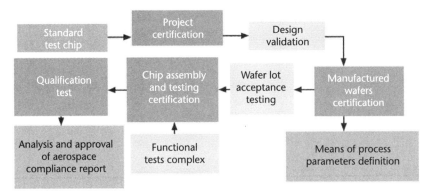

Figure 4.21 Diagram of certification of design and production of ECB for space applications (using the example of Aeroflex Colorado Springs).

conduct additional expensive equipment testing, or produce the similar products. Thus, the key problem for those who create RST ECB is to make the cost of small series production acceptable while unconditionally providing for a wide range, long life cycle, reliability, resistance to destabilizing factors, and fault tolerance (see Figure 4.22). Let us consider ways to address these problems in the current circumstances.

The modern approaches to improving the reliability of specialized DSF-resistant ECB can be divided into four large groups (Figure 4.23). The technological methods include the use of specialized processes for manufacture of VLSIC and special materials—technologies such as silicon-on-sapphire (SOS), silicon-on-insulator (SOI), specialized doping operations, and so on. All of these methods are extremely expensive, so are used only by a few manufacturers. In particular, the leading manufacturers of such structures are the U.S. companies Honeywell, Peregrine Semiconductors, and some others. The design methods for increasing resistance include the use of special packages, local protection, and so on.

The circuit design methods of radiation hardening, including the resistance to heavy charged particles (HCP), are use of cell libraries with majority voting at the level of gates, encoders, Hamming decoders enhanced cell libraries, selection of library cells, and a number of other techniques (Figure 4.23).

The majority of circuit design methods can be divided into two groups—control and error correction methods and redundancy methods. The first group includes redundant coding methods, such as parity checking or Hamming codes, majority voting, and others. All of them require permanent function of redundant equipment together with main equipment to detect and, if possible, correct the errors. The second group consists of methods, based on operability testing and automatic substitution of defective units with redundant ones.

The main advantage of circuit design methods for increasing radiation resistance is the possibility to implement them at the factories where standard, mass production technology is used. This approach has received an international name Rad-Hard By Design [8]. In this way the French company MNS provides the guaranteed tolerance of about 100 krad on bulk silicon. Aeroflex uses the same approach, applying conventional technology of the leading manufacturers.

With regard to Russia, such an approach allows using both existing production capacities of foreign factories and capacities of JSC Micron and JSC Angstrem coming

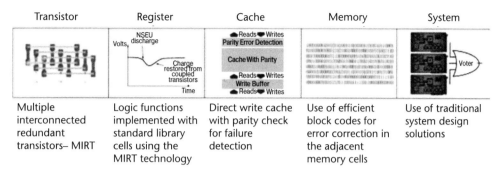

Transistor	Register	Cache	Memory	System
Multiple interconnected redundant transistors– MIRT	Logic functions implemented with standard library cells using the MIRT technology	Direct write cache with parity check for failure detection	Use of efficient block codes for error correction in the adjacent memory cells	Use of traditional system design solutions

Figure 4.22   Methods to ensure fault tolerance of digital system components.

into operation. The rad-hard by design method ensures increased resistance comparable to the use of special technology, but at a significantly (5–7 times) lower cost.

In particular, the rad-hard by design method was tested by JSC Russian Space Systems and NPP Digital Solutions while carrying out the R&D for the creation of reliable IP cores and VLSIC [18]. JSC Russian Space Systems designed the memory controller IP core, which uses the Reed-Solomon protection codes. This code has optimal code length (i.e., for a given CRC length the code fixes a maximum possible number of errors). Reed-Solomon codes are convenient when working with memory. Since the flash memory is accessed in page mode, it is advisable to encode an entire page; in this case, the checksum will be located at the end.

Another approach that allows enhancing the resistance to single events and controlling the VLSIC performance is designing VLSI on self-timing principles [27]. As a result of work carried out by NPP Digital Solutions, circuit and layout implementations of self-timed CPU IP cores and peripheral units were obtained. With the self-timing approach, every combinational block, once the transitional processes are complete, must generate the readiness signal, which is used for synchronization of the previous blocks, thereby ensuring the logical sequencing of events.

Self-timing has the following advantages:

- Maximum possible speed, determined only by technological parameters, operating conditions, and the type of information processed;
- The widest range of operating temperatures and supply voltages;
- The best power efficiency, since there are no power-consuming clock signal transmission circuits and there's automatic switching to power-saving mode of that part of a circuit where the input is not receiving data;
- No additional resources are required for distribution of high-speed clock signals across the circuit with demanding phase shift requirements, since the synchronization of individual blocks is local;
- Fluctuations of external conditions and process parameters only result in a change in speed and do not cause failures (provided that the switching capacity of the elements is preserved), which is why self-timed circuits are very stable.

An intermediate position between the technological and structural approaches is occupied by methods based on minor modifications to existing technology,

(Figure 4.23), which combine structural and technological solutions. These also include the increase in the resistance of master-slice LSIs.

As is well known [28], CMOS circuits are affected most of all by surface effects at the silicon-dielectric interface. Under exposure to ionizing radiation, electron-hole pairs are generated, while positively charged holes, due to their low mobility, are trapped in the dielectric traps. As the cumulative dose of proton radiation increases, the positive dielectric charge and the threshold voltage shift of the working and parasitic MOS transistors increase as well. To a greater extent, this applies to n-channel transistors controlled by positive potential. In this manner, the main task, while using conventional bulk silicon technology, is to obtain transistors with low threshold voltage shift and a high breakdown voltage. To do that, technological methods are used to improve gate oxide quality. As gate oxide thickness decreases, the effect of a cumulative dose reduces as well (this explains sufficiently the high performance of VLSI on bulk silicon with 0.18–0.25 μm design rules with respect to this parameter), and threshold voltage shift decreases, but breakdown voltage also drops. For example, the gate oxide annealing at approximately 900°C is introduced as part of the standard process flow, ensuring more uniform SiO_2 structure. However, as annealing temperature increases, the threshold voltage shift grows.

In addition, the breakdown voltage values are dependent on the distance between the guard ring and the drain region, as well as on the carrier concentration in the guard ring (which affects the threshold voltage level of parasitic transistors). By

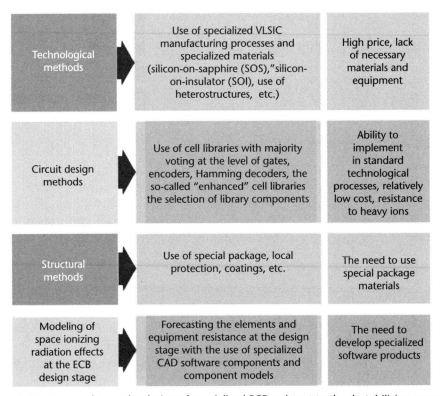

Figure 4.23 Approaches to the design of specialized ECB resistant to the destabilizing space factors.

varying these parameters within the standard process flow, the necessary resistance levels can be obtained.

Special attention should be given to modeling the effects of destabilizing space factors (DSF) at the VLSIC design stage. The existing standard design tools, in general [29], do not provide for the special procedures for assessment of potential DSF-resistance of IP blocks and VLSIC.

This issue still needs to be addressed. Obviously, domestic design tools and related models should be developed. Some steps in this direction have been taken by the Industry Center for VLSIC Design involving the domestic AVOCAD software developers (V. N. Perminov's team). This domestic AVOCAD software [30, 31] allows embedding custom models of semiconductor components for microelectronics, including models of elements that take into account the effects of space ionizing radiation. The Industry Center for VLSIC Design is already carrying out the integration of this system into a unified design flow of complex–equipment–components. The results obtained suggest that it is possible to establish a VLSIC design flow taking into account their behavior under DSF conditions. Along with the use of special circuit design techniques, this method allows significant reduction of both the number of iterations when developing the circuit design solutions and development costs.

Let us consider the production capacity and possible organizational measures in this respect. The main challenge the developers and manufacturers of ECB for space applications always face is how to combine ice and fire (i.e., to ensure a small-scale production with a wide product range). A wide product range makes it necessary to use a large number of processes, which are implemented using the expensive equipment and raw materials, require continuous maintenance, and so on.

Given that the production volumes of most space products are extremely low, it is necessary to provide for approaches ensuring the standardization of solutions. First, it is necessary to distinguish between spacecraft payload and support systems, as well as to have a clear idea of the estimated number of items required. The spacecraft support systems include telemetry, power supply, command equipment, and measurement equipment. Given that most spacecraft are built on standard platforms, it is safe to say that the support system equipment is standardized to a certain extent and performs the same functions regardless of the spacecraft's mission. So, the telemetry is not dependent on whether the spacecraft performs the remote sensing, communication, or navigation functions. Furthermore, the spacecraft platforms live long; therefore, the support systems and components for them should also have a long lifecycle. The production quantity of such systems is hundreds or thousands of devices. The requirements for support systems are not too stringent in terms of data processing speed, performance, and so on. So, in order to implement the components of support systems, no deep submicron technology is required.

The payload equipment is usually unique for each spacecraft. This is especially evident for the equipment of interplanetary spacecraft, where mass production is out of the question and the demand of each part type will hardly exceed tens of items. We can consider as mass-produced some of the satellite communication and navigation systems, consisting of a sufficiently large number of devices, such as GLONASS, GONETS, and others. But unlike the support systems equipment, the requirements for payload systems in terms of speed and performance are often quite stringent.

Let us look at the problem from the point of view of technological feasibility at our national production capacities. Their condition is currently well known and is often covered in the media. When it comes to design and production, currently available are 0.09, 0.18, 0.35, and 0.8 μm technologies at JSC Mikron, JSC Angstrem, GC NPK Technological Center MIET, and Integral Holding (Minsk). In addition, there are several silicon production facilities at the Ministry of Industry and Trade and Roscosmos enterprises.

Regarding the JSC Angstrem, in 2016 the processing line Angstrem-T was launched with an 0.13 μm technological process. Therefore (Figure 4.24), in the short term (until 2015) a substantial increase in the use of home-produced ECB in the SC support systems was possible. As for the payload, a considerable proportion of foreign-made ECB is stille used, often of industrial grade due to the lack of space grade ECB that would ensure the required characteristics (speed, logic capacity, memory size, and so on). Equipment designers have to take into account this situation, when making orders and lists the design of ECB.

Although Russia's first submicron production is emerging, it is necessary to be aware that a number of technologies will not appear in the short and medium terms. It suffices to analyze the text of the Russian federal target program Development of Electronic Component Base and Radio Electronics (e.g., www.fasi.gov.ru), in order to realize that we should not expect the radiation-tolerant high capacity memory, flash memory, antifuse PLD technology, and so on to appear in the next 5–10 years.

This means that, when designing the equipment, a reasonable balance between domestic and imported ECB should be maintained, and, when making part lists, the current capabilities and prospects of microelectronic industries development should be taken into account.

In view of that, when planning ECB design and production for high-reliability applications, it is possible to build a cooperation scheme according to which the

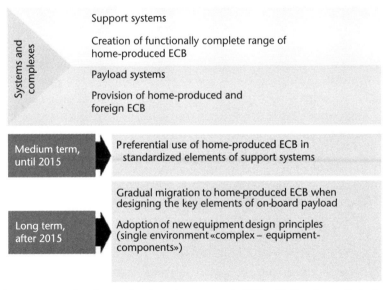

Figure 4.24 Prospects for the production of space grade ECB with the domestic technological base [3].

design and final assembly and testing operations will be carried out by the Russian producer, and the chips will be manufactured either at the Russian factories or abroad (Figure 4.25). Certain steps in this direction are taken by the Industry Center for VLSIC Design at JSC Russian Space Systems, which was specifically established to meet the demands of Roskosmos companies for special purpose VLSICs.

4.10   Power Semiconductor Devices for SC Electronic Systems

4.10.1   Basic Principles of Power Device Operation

Along with digital and analog IMCs, the SC electronic systems also widely use power semiconductor devices [32]. Let us consider the control units of SC onboard systems, so-called control-electronic units. The electronic control units convert the input signal received from a sensor into an electric signal. Then, to obtain the necessary control signals, this signal is processed by signal processing devices. And, finally, the signal processing unit sends the control signals to the controlled device or execution unit in order to obtain the necessary result.

The electronic control units carry out the computations and measurements; they always require sensors, microcontrollers, and power semiconductor devices. Figure 4.26 shows a general block diagram of an electronic control unit.

If we divide the electronic control unit into functional blocks, we can distinguish the power supply, data exchange interface, microcontroller with peripheral devices (memory and so on), post-processing devices (e.g., ADC), and execution units. Power semiconductor devices are used mainly in the execution units and power sources.

Electronic devices increasingly substitute for electromechanical apparatus, such as relays. Since they have to function as part of more complex systems, the reliability requirements for certain components (short circuit and overload protection) become more stringent. The diagnostic functions must facilitate fault detection in case of failure and increase the serviceability. That is why the control and stabilizer microcircuits within the electronic control units are often joined to distributed networks.

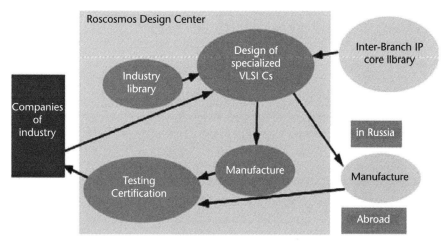

Figure 4.25   Possible cooperation in the design and manufacture of specialized ECB for space applications [9].

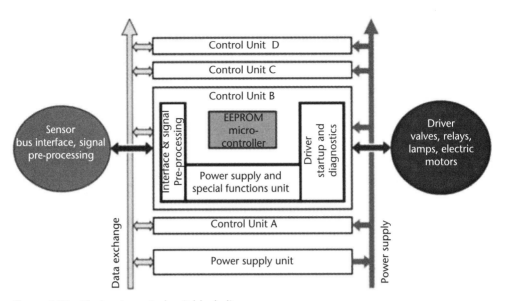

Figure 4.26   Electronic control unit block diagram.

They may have a very simple structure or, in some extreme cases, may function as quite complex controllers. In this view, the term *power semiconductor device* is to a large extent determined by the type of a device under consideration, because, depending on the application, the switched power may vary ten- to thousand-fold.

However, in any case, one of the functions of this kind of device is actuation of execution units, such as electric motors, lamps, heat resistors, and other electromagnetic drives. Another function is supply of power to a control device in general. In this case, after required functionality, the most important parameter is efficiency (efficiency factor). As the devices' functional integration increase, their power consumption is constantly reduced. If we consider it from a miniaturization point of view, we can find that the density of energy emitted by modern microprocessors (W/m^3) has recently reached values similar to density of energy emitted by a nuclear reactor core. This poses some new challenges for power electronics. The types of domestic power semiconductors, as well as their ranges and main design and technology solutions, are analyzed in [32].

As an example of a design and technology solution from a foreign manufacturer of power semiconductor microelectronics, Figure 4.27 shows the technology range of one of the world's largest semiconductor manufactures—Infinion Technologies.

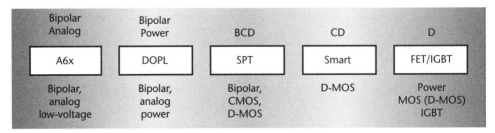

Figure 4.27   Manufacturing technology of semiconductor wafers for automotive electronics.

The company has the following subdivisions: automotive power, industrial power, microcontrollers, and advanced sensors, each of them dealing with its own market segment.

Unlike the domestic manufacturers, the customers of this company (e.g., for automotive microcircuits) receive an evaluation kit. It consists of a finished printed circuit made for standard application, test reports and reference guide, complete software package that may be needed to operate the device, and contacts of technical personnel responsible for solving technical issues. To categorize the power semiconductor devices by groups, let us consider their characteristics as a function of their level of integration (complexity) [32].

If we add an over-temperature protection circuit to MOSFET or IGBT, then in the simplest case we obtain a TEMPFET (overheat-protected MOSFET). Power semiconductor devices with additional intellectual controls are usually called SmartFET. Then we can mark out PROFET (protected FET), or high-side switches (located between the power source plus and the load, the other load terminal is connected to the ground), and HITFET (high-integration FET) or low-side switches.

If we combine some SmartFETs and add some extra functions, we obtain a power integrated circuit (PIC). This group includes multichannel switches, half-bridge and full-bridge microcircuits (a half-bridge microcircuit consists of two tandem switches located between the positive and negative rails of the power source), a power sources microcircuit, and a line drivers microcircuit.

Higher integration leads to the creation of power system microcircuits. Their manufacture and cost are application-specific (e.g., in automotive electronics, to control the airbags or anti-lock braking system). The highest integration is obtained by the use of embedded power devices. They are actually finished semiconductor systems on one chip (or on several chips in one package).

4.10.2 Key Processes for Power Devices

A great number of processes are used in power electronics. These include basic bipolar, MOS, complementary MOS (C-MOS), and power double diffused MOS (D-MOS), as well as their combinations [32]. Figure 4.28 shows how these processes relate to certain product groups.

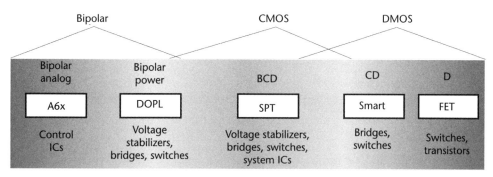

Figure 4.28 Overview of processes and related product groups.

CMOS Process

The CMOS process uses only p- and n-channel transistors, resistors, and capacitors. A transistor is made in the p- or n-type tubs with a polycrystalline silicon gate. As a resistor, the polysilicon layer can be used. The polysilicon layer and doped substrate serve as capacitor plates, and the oxide layer serves as a dielectric. The CMOS process is optimized for implementation of logic functions. In fact, using this technology it is possible to create devices with low supply voltages (5V; 3V; 1.8V), which allows using small-size, high-integration components. Analog functions may also be implemented on CMOS transistors. Infineon has a whole range of processes (C5, C6, ..., C11) for manufacture of logic circuits, some of which are used in the manufacture of EEPROM or flash memory microcircuits.

Bipolar Process

The bipolar process uses n-p-n and p-n-p bipolar transistors as active elements. Purely bipolar structures do not require polysilicon gates. Consequently, such processes involve fewer operations and therefore are very cost-efficient. The integration level depends on process class by breakdown voltage. The voltage class depends on the size of integral transistors. DOPL is a bipolar process developed by Infineon.

DMOS Process

DMOS transistors are transistors optimized for switching of high currents and are designed to operate at high voltages. This transistor has a long channel, ensuring a high breakdown voltage. It consists of parallel cells, which allow obtaining high currents (low resistance in open state) and high energy density. DMOS structures have a thicker gate oxide layer than logic structures, providing for higher reliability of devices. Infineon offers various DMOS processes optimized for certain devices, such as PFET and SFET.

If these basic processes are combined in a logical sequence, then we obtain the following interesting options, which by their characteristics are intended for specific applications.

BiCMOS Process

BiCMOS combines bipolar and CMOS components. Such a combination of components allows implementing various analog functions (e.g., high-precision sources of reference voltage). Infineon offers various BiCMOS processes optimized, for example, for use in HF devices.

CD Process

CD process is a combination of CMOP and DMOP components. This allows combining in one microcircuit logic functions, high power, and high currents. One of the examples of CD technology offered by Infineon is smart technology.

BCD Process

BCD is a process that combines bipolar, CMOS, and DMOS components. It can be used to manufacture components for various voltage classes. CMOS provides for high density of logic elements. Thus, it is possible, for example, to integrate a microcontroller. Bipolar and CMOS structures are used to create a precision reference voltage circuit. DMOS transistors allow switchig high currents and voltages (up to 20A and 80V).

In some cases, to achieve high degrees of integration in low voltage logic circuits, more than one gate oxide layer is used (submicron logic). It is also possible to create several resistive polysilicon layers. The advanced BCD process uses more than 25 photolithographies (photomasks). However, this makes it more expensive compared with more simple technologies, such as CMOS.

4.10.3 Power MOSFET

In most cases the Infineon power MOSFETs are used as switches. Only in some rare cases does their analog operating mode play any role. The following states of the transistor can be identified:

- *Transistor off.* The lowest possible current at the maximum applied voltage must flow through the transistor. Parameters associated with this operating mode of the transistor are breakdown voltage and leakage current.
- *Transistor on.* In this state, the transistor resistance (on-resistance) should be as small as possible at the highest possible current. Parameters associated with this operating mode of the transistor are drain are source on-resistance and maximum current.
- *Switching transistor ON or OFF.* Switching time should be as short as possible and the charge changes as small as possible. The parameters characterizing this operating mode of the transistor are switching time, transfer characteristic slope, and gate charge.

Modern technologies allow us to manufacture devices with maximum operating temperatures exceeding +200°C. High operating temperatures allow making silicon chips with smaller surface and reduced cooling costs. Additional parameters that affect the reliability and that should be mentioned are resistance to electromagnetic interference (EMI) and electrostatic discharge (ESD).

In 2013 Infineon developed, especially for the space industry, OPTIMOS technology, which allows manufacturing devices with the optimal settings for almost any application.

Figure 4.29 shows how the switching transistors were improved in terms of their key parameter—on-resistance R_{on}.

Figure 4.29 represents the latest four generations of switches with 18 mOhm or lower on-resistance, indicating the manufacturing technology and cost reduction. The price to performance ratio has improved significantly: cheaper and smaller nonmilitary microcircuits can be used for higher power switching. The appropriate

Changes of R_{on}, package type and cost–the case of high-side switch with R_{on}= 18 mOhm

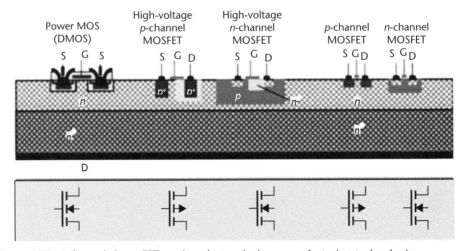

1990	1995	1999	2000
18 mOhm	18 mOhm	16 mOhm	11 mOhm
BTS 542	BTS 442	BTS 443	BTS 6143
TC218	TC220	DPAK15	DPAK15
SIP2	milliFET	S1FET	OptiMOS
Cost 100%	65%	45%	50%

Figure 4.29   Development of semiconductor switches—the case of the 18 mOhm high-side switch.

package for the devices can also be manufactured at a lower cost, since the power loss will be smaller.

4.10.4 Smart MOS Transistors (SmartFET)

If we take a device made by the power MOS process and some extra p- and n-type tubs, it will be possible to combine more functions in one device. MOSFETs are becoming smart. For example, it is possible to integrate the protection and device status control functions. Some of the components that may be obtained from this improvement are shown in Figure 4.30.

As we can see, bipolar structures are not shown in this figure. So, we are dealing with CD-process (CMOS and DMOS). The main parameters of devices manufactured by these processes are defined by the electronic switches specifications.

In the case of the CD process, of special interest is the number of additional components that may be integrated. This parameter is called layout rule: it determines the components size and distribution density; as a rule, it depends on gate dielectric thickness and additional manufacturing stages. In any case, the rule is

Figure 4.30   Infineon's SmartFET semiconductor devices manufacturing technologies.

that the process using smaller layout rules is more difficult to implement and, consequently, is more expensive.

As in the power MOSFET case, the current flows vertically through the transistor. Thus, all the components have a common substrate, which serves as the drain of the power structure. On-resistance will be smaller than in the structures with an insulated drain. Further, it can be reduced by grinding a wafer to several tens of microns. By doing so, it is possible to manufacture some power structures, where drains will be connected together. In such common-drain devices, the heat-sinking plate must be isolated from the external grounded radiator, because the drain potential often does not coincide with the system ground potential. Figure 4.31 shows a flow of electric current in the structure of this type.

Figure 4.32 shows applications of such structures in REE. In terms of application, the CD process can be used to create high-side and low-side switches. The

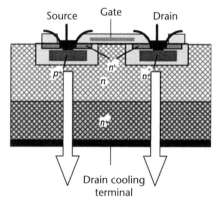

Figure 4.31   Current flow direction in common-drain structures.

Low-side switch:
–Load is connected to +V_s
–Drain is connected to ground

High-side switch:
–Load is connected to ground
–Drain is connected to +V_s

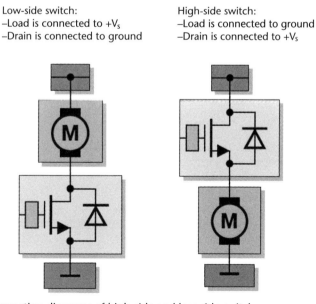

Figure 4.32   Connection diagrams of high-side and low-side switches.

terms *high-side switch* and *low-side switch* follows from their circuit diagram (Figure 4.32), which shows the switch and load connection topology. In general, a negative potential of the power source is connected to the system ground, as, for example, in onboard power sources.

If the load is connected to the system ground, it allows using one wire fewer, as the current will return to the power source through the device package. If the load is connected to the ground, the switch must be connected between the load and the positive supply; in the diagram shown in Figure 4.32, it will be on the top side. On the other hand, the low-side switch is always connected to the system ground. Figure 4.33 shows typical high-side and low-side switches manufactured by Infineon.

In addition to discrete electronic switches manufactured, bridge circuits are also manufactured, as the electronically controlled reversible drives are used increasingly. Figure 4.34 shows the bridge (H-bridge) circuit. This circuit can be used for connection of a load, typically a DC motor, with a power source of either polarity.

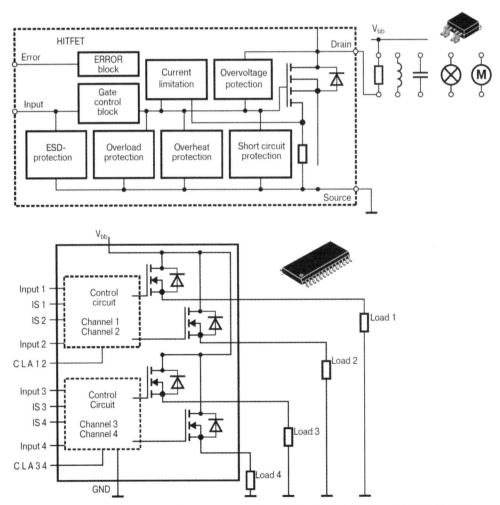

Figure 4.33 Block diagrams of Infineon BTS3150 low-side switch (above) and BTS5140 high-side switch (below).

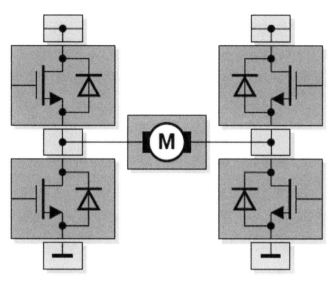

Figure 4.34 Bridge circuit for reversible electric drive.

The latest generation of Smart 5 industrial microcircuits has an extremely low on-resistance and significantly smaller layout rules (see Figure 4.35). This made it possible to manufacture high-power electronic switches with complex logic. Now, no more obstacles are left for the production of smart switches with complete diagnostics of their status.

In the near future, in the pursuit of the idea of silicon instead of relay, the devices will be created with stringent diagnostic requirements; in such applications, the only option is to use smart switches. A practical example is shown in Figure 4.36. ITo illustrate more clearly the effect of size reduction, this figure shows in scale microcircuits with the same set of functions. Above is the previous generation microcircuit, and below is its counterpart made using the latest Smart 5 technology.

Of course, in the case of CD technology, the appropriateness of the use of monolithic ICs must be verified. Infineon has taken this fact into account while developing their basic concepts: chip-on-chip (CoC) and chip-by-chip (CbC). In these cases, a standard MOS transistor is used as a master slice. Only a limited number of additional smart microcircuits require sophisticated (expensive) technology, when two

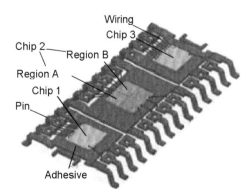

Figure 4.35 A standard microcircuit of Trilith IC family double diffused MOS.

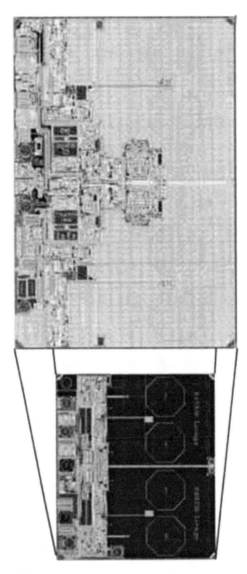

Figure 4.36   Reduction of IC chip size in Smart 5 generation.

chips are mounted one on another or one by another using a special technology. As a result, depending on the desired current or on-resistance, the high-current switches are made by CoC or CbC technology, determined by the cost of manufacturing.

Today, with on-resistance values of 20 to 50 mOhm, the CoC method is less expensive in production. At lower currents, it is possible to place several switches in a single package. This current range requires the use of technology with free-switching transistors manufactured by the smart power IC technology.

4.10.5 Smart Power ICs

CD technology is characterized by relatively low layout rules, so it cannot be used for making analog devices. This gap is filled with smart power technology (SPT), by which it is possible to develop ICs with powerful output stages, the PIC. Depending on the technology solution, the set of IC components can be much larger than the minimum set of components shown in Figure 4.37.

Using this technology, it is possible to obtain freely connectable power transistors. The current flow is shown in Figure 4.38. First, the current flows vertically down to the hidden integrated transistor layer. The hidden layer conducts the current toward the higher drain connection. The resistance of vertical current flowing toward the surface is low due to vertical n⁺ region, which allows quick making of the isolated power drain connections on the chip surface. However, this comes at the price of much higher on-resistance.

Consequently, the use of SPT devices is appropriate for low- to medium-level currents. To isolate the transistor drain from the rest of the circuit, the p-type substrate is used, which, while being connected to the ground, maintains the p-n

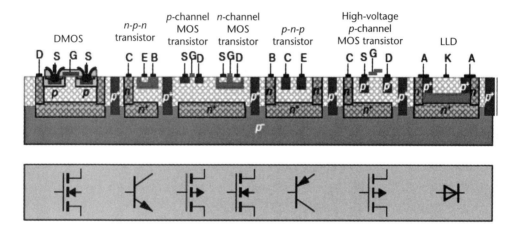

Figure 4.37   Family of the devices fabricated using SPT technology.

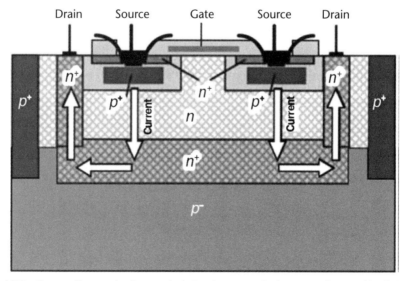

Figure 4.38   Current flow and substrate isolation in power devices manufactured by SPT technology.

junction between the drain and the substrate in the closed state. It is easy to notice that the p-n junction between the drain region and the components tubs will be closed only if the p-substrate will always have the lowest potential in the circuit.

For cooling purposes, the reverse side of the chip in the device may have a ground connection. In this case, no additional insulating screens are required. A wide variety of additional options is shown in Figure 4.39.

Apart from the high-side and low-side switches, this technology can be used for implementing a half-bridge circuit consisting of two series-connected high-speed MOS transistors.

Furthermore, there is a possibility of manufacturing virtually any required analog and digital circuit combinations. Typical examples are the TLE6263 CAN-transceiver microcircuit with multiple precision analog circuits and digital control interface (SPI), and the TLE6288 multichannel low-side switch with a lot of additional functions.

Infineon has been producing SPT devices for 20 years; today, they use the fifth-generation SPT5 technology. Figure 4.40 shows how the most important parameters—R_{on} and layout rules—were improved over time.

We can also see how the production technologies of the power MOS (above) and bipolar devices (below) were combined with CMOS technology, resulting in SPT technology.

The main advantages of SPT technology are presented on a generalized diagram (Figure 4.41). It shows the third, fourth, and fifth SPT generation ICs with a similar set of functions. In the figure, we can see that a new IC of the same size has considerably higher functionality.

The increasing complexity of electronic circuits generates certain requirements for the distribution of functions across the surface of a chip manufactured using

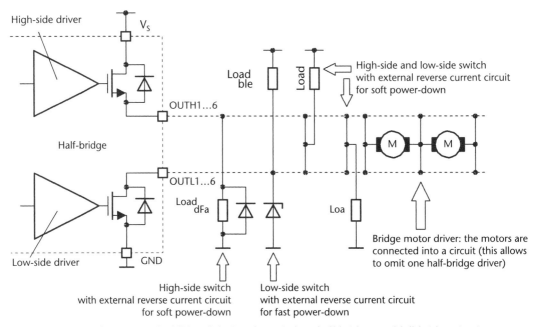

Figure 4.39 Configuration of additional devices for switches, half-bridge, and full-bridge circuits.

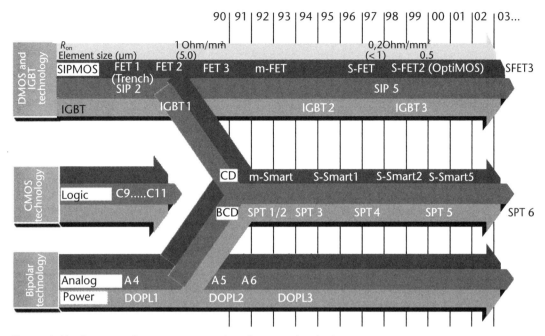

Figure 4.40   Progress of smart power semiconductor technologies.

modern technologies. Thus, in a standard SPT microcircuit, 30% of chip surface is occupied by power elements, about 40% by analog circuits, and the remaining 30% by digital circuits. If the complexity of each of these parts increases, the chip size will triple. The largest proportion of this increase will be due to the digital part, which will increase six times. To compensate for this increase, the layout rules in the next-generation technologies will have to be made smaller by rapid leaps. Accordingly, the chip surface will be distributed as follows: 46% will go to power components, 34% to analog circuits, and 20% to digital circuits. Thus, there will

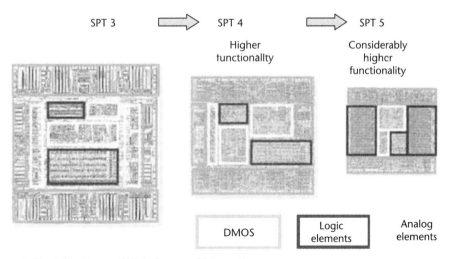

Figure 4.41   Advantages of high degree of integration.

be a further potential for integration as a more logical system functions in the smart power IC systems.

4.10.6 Smart ICs of Power Systems

Increasing integration of smart power supplies led to the emergence of smart ICs of power systems. To determine the optimal degree of integration of electronic control units, it is necessary to take into account the parameters of their functional blocks. Figure 4.42 shows the functional blocks of a typical electronic control unit. Here we can distinguish the logic (digital), analog, and power blocks.

Current, voltage, and temperature requirements determine what additional functions can be implemented and with what technology, as well as what types of packages are to be used. In particular, Figure 4.43 shows that the interface between the controller and the surrounding high-voltage power semiconductor devices is extremely complex and interesting.

In this case, to find the technical and economic optimum, a very close collaboration is required (e.g., between the REE manufacturer, its suppliers, and the electronics manufacturer).

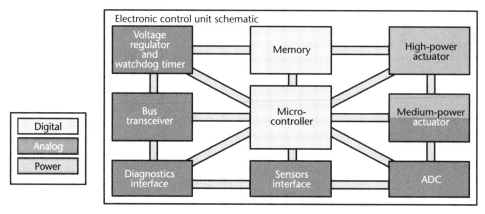

Figure 4.42 Functional blocks of electronic control unit.

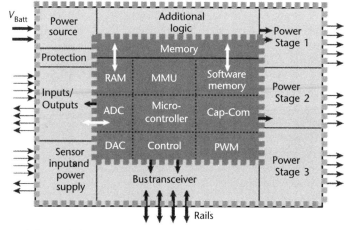

Figure 4.43 Breakdown of electronic control unit into smart functional blocks.

The requirements to be met by future technology are determined by characteristics that are improving, in some cases significantly, in conjunction with higher efficiency. This implies, on one hand, higher computing capability of controllers, and on the other hand, the demand for improved (with lower on-resistance) power semiconductor switches. Figure 4.44 shows the trends of some parameters of power devices and microcontrollers (temperature, electromagnetic compatibility, current, and voltage).It is obvious that the requirements are developing in different directions.

An analytical review of this typical situation is shown graphically in Figure 4.45. It is obvious that various technological options, in some cases equally promising, are possible based on a combination of switching and current characteristics.

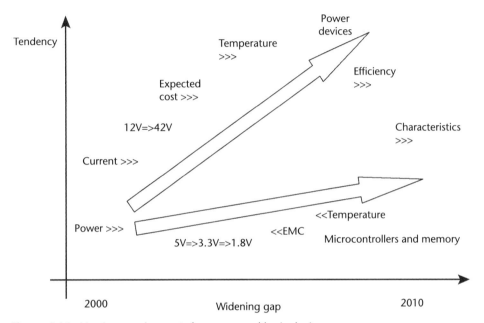

Figure 4.44 Varying requirements for power and logic devices.

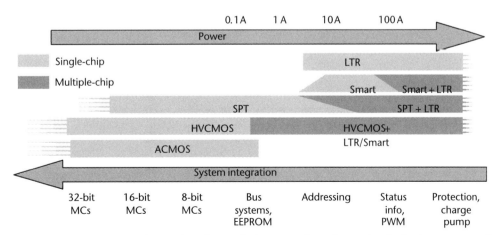

Figure 4.45 Comparing the semiconductor devices manufacturing technologies by current and number of logic functions.

The need to integrate comprehensive diagnostic functions into the complex systems becomes understandable if we consider the possibility of the whole system failure. Infineon's experts in automotive electronic components often produce the following argument while talking to customers: starting the engine of a sophisticated modern car requires more computing capability than was necessary for the flight to the Moon.

Another important factor is the creation of smart networks and related decentralization of control functions. This is another driving force behind mechatronization (i.e., complete integration of electronics into a mechanical system). An example is an electronic muscle, which develops mechanical force and can be used, for instance, as a robotic arm on a SC. This can be a solenoid or an electric motor. For example, a controller can be integrated into an electric motor, thus turning it into a smart actuator, as shown in Figure 4.46. In this case, the transmission of control instructions in one direction and diagnostic data in the reverse direction will be carried out via serial single-wire interface. The only things additionally required in this case are two power supply wires (plus and minus).

The next step may be transmission of control instructions via power line wiring, as is already used in industrial electronics. Without regard to the analog functions, one of the future technologies will be the combination of power switches with high-voltage CMOS devices. In particular, this method will be applied in systems of low complexity (e.g., in the mechatronic systems mentioned previously). However, high and average computing power will remain a prerogative only of CMOS devices on the short-term horizon.

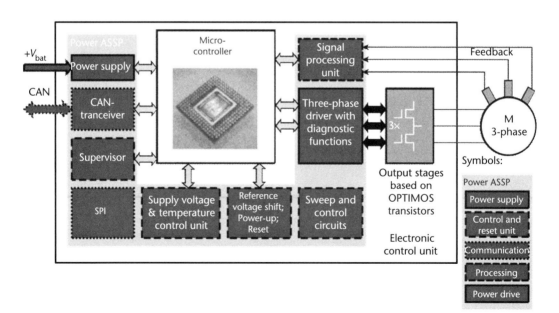

Figure 4.46   Control system of a brushless DC electric drive: electronic muscle.

References

[1] Belous, A. I., V. A. Emelyanov, and A. S. Turtsevich, *Fundamentals of Circuit Design of Microelectronic Devices*, M.: Technosfera, 2012.

[2] Belous, A. I., V. V. Zhurba, and O. V. Poddubny, *Digital Signal Processing LSIC Microprocessor Kit*, M.: Radio and Communications, 1992.

[3] www.ixbt.com/news/hard/index.shtml?10/03/68.

[4] www.actel.ru/catalog/SoC%20Catalog%202013.pdf.

[5] Bumagin, A., Y. Gulin., S. Zavodskov, V. Steshenko, et al., "Special-Purpose VLSIC for Space Applications: Design and Manufacture Issues," *Electronics*, NTB, No. 1, 2010, pp. 50–56.

[6] Koons, H. C., et al., "The Impact of the Space Environment on Space Systems," Technical Report, AD-A376872; TR99(1670)-1; SMC-TR-00-10, El Segundo Technical Operations.

[7] Kobzar, D., "Procedural Issues in the Use of Military Electronics: The Regulatory Framework and the Truth of Life," *Modern Automation Technology*, No. 3, 2007, p. 86–97.

[8] Lacol, R., "CMOS Scaling, Design Principles and Hardening-by-Design Methodologies," 2003 IEEE NSREC, Radiation Effects in Advanced Commercial Technologies, California, 2003, p. II-1-II-142.

[9] Steshenko, V. B., "Design and Manufacture of Special-Purpose ECB for Space Applications: Current State and Development Prospects, Part 1," *Components and Technologies*, No.11, 2010.

[10] Osipenko, P., "Single Failures: A Challenge for Moder n Microprocessors," *Electronic Components*, No. 7, 2009.

[11] Telets, B., S. Tsybin, A. Bystritskiy, and S. Podyapolskiy, "FPGAs for Space Applications. Architectural and Circuitry Features," *Electronics*, NTB, No. 6, 2005.

[12] www.russianelectronics.ru/developer-r/review/2189/doc43922.phtml.

[13] www.electronics.ru/files/article-8-329.pdf.

[14] cj.kubargo.ru/2012/02/pdf/39.pdf.

[15] Belous, A., and V. Solodukha, "Fabless Business Model in Microelectronics Company: Myths and Reality," *Components and Technologies*, No. 8, 2012, pp.14–18.

[16] "AMD Refuses Cooperation with Global Foundries," http://servernews.ru/news/595521.

[17] Khartov, V., "The Space Problems of Electronics: Shake Before Use," *Electronics*, NTB, No. 7, 2007.

[18] Danilin, N., and S. Belosludtsev, "Issues of Modern Foreign-Made Industrial Electronic Components Application in the Manufacture of Rocket-and-Space Equipment," *Modern Electronics*, No. 7, 2007.

[19] Berzichevsky, K. V., P. V. Kulik., and D. V. Nikitin, "Highly Reliable Electronic Components for Space Applications, *Petersburg Electronics*, No. 2, 2010, pp. 43 50.

[20] Zakharov, A. S., V. G. Malinin, and A. V. Sapieha, "Foreign-Made Microcircuits in Strategic REE," *Petersburg Electronics*, No. 3, 2012, pp. 37–44.

[21] "Spacecraft Consists of 75% Imported Components," http://vpk.name/news/Roscmos.

[22] Lysko, R., S. Glagolev, V. Makarov, S. Lukachev, and P. Pavlyuk, "Russia in the WTO: What Will Happen to the Electronics Industry," *Electronics*, NTB, No. 8, 2012. pp. 30–32.

[23] Golovachev, V., "And Then the Rocket Engineers Were Given 32 Tons of Gold," http://vpk.name/news.

[24] Basayev, A., and V. Grishin, "Space Instrument Engineering: The Main Thing Is Right Concept," *Electronics*, NTB, No. 8, 2009.

[25] Khartov, V., "The Space Problems of Electronics: Shake Before Use," *Electronics*, NTB, No. 7, 2007.

[26] Steshenko, V. B., et al., "Design of System-on-a-Chip VLSIC: Design Flow, Circuit Synthesis," *Electronic Components*, No. 1, 2009.

[27] Rutkevich, A., V. Steshenko, and G. Shishkin, "Self-Timed Electronics: Development Trends," *Electronics*, NTB, No. 8, 2009.

[28] Chumakov, A. I., *Effect of Space Radiation on Integrated Circuits*, M.: Radio I Svyaz, 2004.

[29] Rabai, J. M., A. Chandrakasan, and B. Nikolic, *Digital Integrated Circuits: Design Methodology*, 2nd Ed., M.: OOO Williams Publishers, 2007.

[30] Perminov, V. N., et al., "AVOCAD + CADENCE and SYNOPSYS, VLSIC CAD Software: Integration Based on Multilingual Translators Technology and Object Databases," *Electronics*, NTB, No. 4, 2005.

[31] Kokin, S. A., V. N. Perminov, and S. A. Makarov, "Latest Circuit Modeling Technologies: AVOSpice System by LLC 'UniqueICs,'" *Electronics*, NTB, No. 5, 2007.

[32] Belous, A. I., S. A. Efimenko, and A. S. Turtsevich, *Power Semiconductor Electronics*, M.: Technosfera, 2013.

CHAPTER 5

PDK Structure and Specific Uses in the Development of Products with Submicron Design Rules

5.1 PDK Development Process Flow, Standard PDK Structure

The standard PDK development currently includes the following main key phases (Figure 5.1) [1, 2]:

1. Process selection;
2. Obtaining of basic process data;
3. Basic elements identification for their incorporation to the PDK;
4. Review of the features and characteristics of the process for more accurate identification of the selected elements;

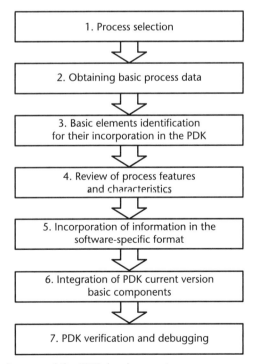

Figure 5.1 Simplified diagram of the PDK development process.

5. Obtaining and incorporation in PDK of information in the format determined by the software selected as a tool for technology/device/circuit/system design;
6. Integration of PDK current version basic components;
7. PDK verification and debugging.

Of course, if you create a PDK based only on these standard unified rules, the information obtained only by following the steps shown in Figure 5.1 will not be enough. Figure 5.2 shows a slightly expanded process flow diagram of PDK development, with account being taken of the accepted rules and standards. It should be noted that the flowchart areas marked by dashed lines are phases that allow us to standardize the PDK in process.

Thus, region A contains procedures and standard rules for description of individual elements to be integrated in the PDK. Using standard descriptions of their contents, values, and standardized translation mechanisms, the developer has the opportunity to create any other identical PDK components.

Region B contains definitions required for standardized description of PDK components in terms of their quality assurance.

In case of need to ensure full compliance of PDK components with the selected phases upon customer's request, the phases presented in the regions A and B must be carried out without fail.

Another well-known methodological approach to the PDK creating process, based on the design phases, is shown in Figure 5.3.

The PDK components subject to standardization may also be defined in a multidimensional space. Inside each dimension, the PDK components may be described using discrete categories.

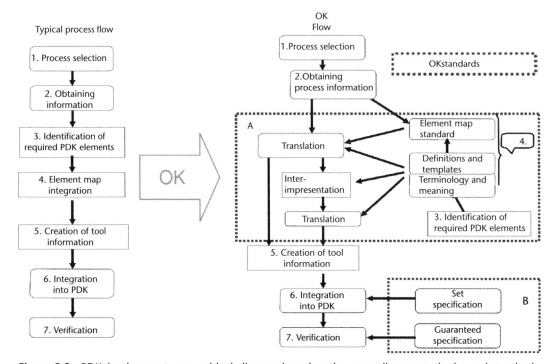

Figure 5.2 PDK development process block diagram based on the generally accepted rules and standards.

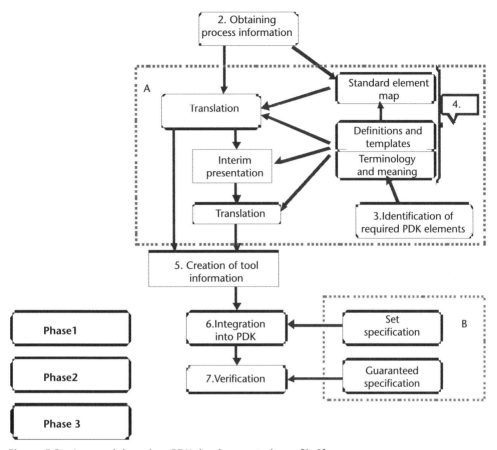

Figure 5.3   Approach based on PDK development phases [1, 2].

5.2   Terms and Definitions Used to Describe PDK Components

Next, we have shown in a convenient tabular form the standard terminology used to describe the PDK components, as well as PDK development process [2, 3]. Table 5.1 contains general terminology; Table. 5.2 contains terms used to describe the devices; Table. 5.3 contains terms relating directly to the modeling process; and Table. 5.4 contains only those terms that refer to the process of description of computer-aided microcircuit design phase tools.

It should be noted that these tables contain only basic terms most widely used by PDK developers.

5.3   PDK Standardization

Many IC design companies have faced the problem of unification and standardization of approaches to the creation of process design kit (PDK) development libraries, as well as the coordination and the use of these approaches for interaction with technologically different IC manufacturers.

Table 5.1   General Terminology

Term	Definition
Custom circuit design	Designing electrical characteristics, physical implementation, modeling, and physical verification of a circuit, where discrete elements are used as a base.
Device	Main element available for inclusion in the developed circuit and implemented for a specific process, such as transistors, resistors, and capacitors.
Structure	A unit formed as a result of some sequence of process steps (diffusion, implantation, metallization, and so on), which is used to create a device. The structure may also be regarded as a single device.
Supported device list	A set of supported devices, the characteristics of which have been described for the selected process.

Table 5.2   Terms Used to Describe the Devices

Term	Definition
Intentional device	Devices used for preparation of a device electric circuit, which are displayed as its contents on the PC screen.
Extracted device	Devices that are not displayed on the PC screen, but are included in the electrical circuit (netlist) in the process of modelling. These devices are typically included in the circuitry while analyzing the results of physical implementation of the designed device.
Main device element	Device element that performs the main function or the most significant behavioral role.
Parasitic device element	Device element resulting from physical implementation data analysis of a device designed on the basis of main components.
Principal device(s)	One or two devices, on which the architecture is built and the characteristics of which are fundamental in the optimization of process parameters, such as n-MOS and p-MOS transistors in digital CMOS circuits. It is important to note that the performance of the principal devices is affected by spread-specific requirements of the specification.
Primary device(s)	Elements included in the supported device list of the process, which are formed on the basis of principal devices and supplemented by certain additional structures. The inclusion of one or more specific well-controlled process step is usual. An example is the formation of polysilicon resistor on the basis of a polysilicon gate with a specially implanted and masked silicide area.
Secondary device	A device from the supported device list, which is formed from a structure created in the production of principal and/or primary devices. Often these devices are called free or intentionally parasitic (e.g., the use of interconnectors metal to form inductors or capacitors). Secondary devices tend to have greater tolerance to the parameter scattering described in the specification.
Device class (category)	Basic intended function of a device, such as MOS transistor, bipolar transistor, resistor, capacitor, and so on.
Device type (vertical)	Unique combination of vertical structures that make a device of a certain class or category (e.g., poly1 resistor, poly2 resistor).
Device style (lateral, horizontal)	Changes of horizontal dimensions and shape of a particular type of device, intended to control its characteristics, for example, changing the horizontal (lateral) position of the emitter, base, and collector layers of a bipolar transistor in order to obtain the optimum current-voltage configuration.
Device size	A special case of a device style when its functional characteristics are regarded as scalable, such as changing the gate width of a MOS transistor.

Table 5.3   Terms Relating to the Devices and Circuits Modeling Process

Term	Definition
Model	Presentation (description as a function) of functional parameters and operating characteristics of a device used in the modeling process; models are generated based on behavior characteristics and corresponding parameters.
Model general behavior	Part of model describing the basic functionality for the selected device class and type. It may be implemented on the basis of equations of compact models, behavioral models, and/or subcircuits.
Model process parameters	Part of model describing the functionality of a device based on the relationship of performance and process-specific parameters.
Instance parameters (allowed design variables)	Properties related to the instance of a device, which are transferred to the simulation software and to the model through the netlist and which directly describe the device parameters, such as MOS-transistor channel length and width.
DRC (Design Rule Check)	Verification of compliance of a developed layout with the rules laid down in the terms of reference.
LVS (Layout (netlist) Versus Schematic)	Validation and verification of compliance of the designed circuit layout against the results obtained during the schematic modeling.
(LPE) (Layout (netlist) Parasitic Extraction	The extraction of a netlist, as well as of capacitance and resistance nominal in order to check their compliance with the requirements laid down in the SOW.
RCX (RC extraction)	Extraction (search) of new nodes and parasitic elements based on the layout analysis of the designed IC.

Table 5.4   Terms Used to Describe EDA Tools

Term	Definition
Primitive	Minimum element used to build a circuit within the EDA software systems that is used to create a graphic image of the circuit.
Instance	Device placed in a position selected in accordance with the circuit design.
Property	Information associated with a specific primitive or instance and is intended to control their characteristics.
Instance property (direct, indirect)	User-managed (directly or indirectly) parameters associated with the instance of a device, set by the IC developer. Indirect properties can be calculated or derived from the direct ones.

When designing today's ICs, especially for special purposes and space applications, the need to take into account and make an in-depth analysis of the process used during the computer simulation is also very important.

With the emergence of designer-independent companies engaged in mass production of customer-ordered microelectronics (integrated circuits foundry), as well as the need to jointly use the design tools and IP-blocks (intellectual properties) from various companies, the design libraries have become a major link between the manufacturer and IC design teams [4].

Unfortunately, until now the unification of requirements and standards have hardly affected the sphere of PDK development, which needs standardization of such requirements as nomenclature, models used, interfaces (connection/integration rules), quality parameters (Q-factor) and, finally, the approaches to PDK presentation to end users. The solution to this problem will help to eliminate the confusion, as well as waste of time and money for the adaptation of existing PDKs, which unfortunately often occur in modern electronics industry. Manufacturers will be able to minimize time expenses for providing the developers with the design libraries adapted as much as possible to the software packages currently in use. This will reduce the cost and improve the quality of libraries, components, and IP blocks supplied by companies specializing in this area. For companies engaged in developing electronic design automation (EDA) software packages, the process of creating tools to describe components, as well as methods of their integration, will be simplified, thus eliminating the need to create the libraries' low-level adjustment tools.

The IC design engineers will get the major advantage due to simplification (unification) of the IC design process, the possibility of rapid and transparent transition to the new processes, as well as the re-use of IP-blocks [4].

The integrated microcircuits design process has long been ill-famed as inefficient because of great bottom-up dependence in data representation formats of different software packages. The design data traditionally depends on the tool used, since the software developers actively try to implement and make the de facto standard their own exclusive products and storage formats; this is especially characteristic of domestic design centers, with absolutely no account being taken of the independent developers' wishes and concerns.

The evolution of IC design technology has led to the fact that fabless companies (design centers), when buying EDA packages, have to purchase and install special data sets of available process-specific elements (separate devices or circuits). After some time, these data have been called process design kits, to be distinguished from the tools used to describe the process. Initially, such libraries were used by so-called integrated device manufacturers (IDMs) in order to allow mixed designing of ICs [4].

In recent years, due to well-defined division of the electronics industry companies (not including Intel) into foundries and design centers (see Section 3.6 for details), the PDKs are developed and debugged almost independently. The same is happening to the EDA and IP-blocks developers. These libraries are characterized by quite a wide range of names and acronyms (TDK, PDK, and others), but the traditional name is process design kit. However, due to lack of standards, all of them are different, although slightly, in composition and approaches to the content description.

Next we shall consider the established description of the standard PDK overall structure, the process of its creation, as well as proposals to unify the contents and structure for the design of integrated circuits with design rules under 130 nm. The choice here of specifically this level of the design rules is explained by the fact, that the integrated circuits with the design rules of 90, 60, and 45 nm, as a rule, are used in the space standards due to a lower level of the radiation resistance, primarily—to single event effects (SSE). The level of 180 nm does not always make it possible to solve the task of accommodating within the limits of one chip of the entire complex functional device.

The advantage of PDK standardization is the obvious fact that, once adopted by the major companies, a set of standardized requirements to the contents and the rules for description of PDK interfaces will reduce their development costs and increase their opportunities in terms of combined basic functionality, performance, and most importantly—predictability and success in the market of new products. IC manufacturers will be able to offer developers more flexible modifications in terms of opportunities and more filled PDKs in terms of contents that describe the latest technological processes. PDK suppliers will have to reduce prices and to simplify licensing conditions.

In turn, the companies involved in IC design will obtain three advantages by increasing the level of PDK standardization [4]. First, the comprehensiveness, consistency, and logic of process description allow the exclusion of minor errors occurring while the product is being materialized in silicon. Standardization implies a more comprehensive knowledge of the design process, as well as strict design version control and management.

Second, engineers of the units responsible for the support and use of IC computer aided design (CAD) will have fewer problems and more flexibility during adaptation of existing software to the requirements of and opportunities offered by the new PDKs, and the main thing is that the time required for the implementation of advanced technological solutions in the design process will be reduced substantially.

And third, heads of engineering departments will be able to take more financially efficient decisions while planning the work on a new product, because they will be much better protected against surprises arising from the use of new processes, due to better awareness of the advantages and disadvantages of implemented solutions.

Foundries such as X-FAB, TSMC, OJSC Integral, OJSC Micron, and IDMs like Intel will also benefit from greater PDK standardization.

This approach offers other obvious advantages. First, the transparency of technical cooperation, based on the use of standard terminology and data presentation, will improve cooperation among the developer teams from both sides and will reduce the time required for the implementation and transfer of new designs from designers to manufacturers.

Second, the use of standard representation and process description significantly reduces the cost of creating and maintaining by supplier companies and developers of a wide range of specialized technology/device/circuit/system design software, or electronics design automation (EDA). This is achieved by simplifying the representation process through the use of standardized data structures, parameter sets, and their estimated default values—the adoption of a single *baseline* level for data representation will facilitate access to the required data, which finally will be beneficial for both the developer and the customer.

The third advantage of standardizing the PDK representation methods is a simpler version control procedure due to standardization and documentation of the respective modules of CAD systems from the initial version of the process to its modified versions, which will have a positive impact on the life cycle of the designed solutions.

And finally, the fourth advantage, which major IC manufacturers and developers will get, is the possibility to directly exchange information on the opportunities and challenges associated with the implementation of new products based on advanced technological processes. This will reduce costs for both the manufacturer's developer support and the process completion when some faults and inaccuracies are detected.

With regard to the PDK developers, the standardization of requirements for the PDK will eventually minimize the differences in the component description for various manufacturing processes and manufacturers, as well as reduce the time for the creation of PDKs due to quick understanding of the characteristics and differences of the processes [5].

General requirements for PDKs will also significantly simplify the process of developing and testing the PDK components during their migration from one PDK to another.

On the other hand, the standardized PDK requirements provide at least three advantages for IC design software developers. First, standardization will accelerate the process of acceptance of new software products and their distribution while reducing costs of software adaptation to new technological processes. Besides, it offers excellent opportunities for significant changes and innovations in mixed and analog IC design tools.

Second, the need for in-depth and comprehensive testing of design tools during the process change will decrease, even when the information concerning such tools is not sufficient.

And third, the EDA developers will be able to increase the number of solutions offered, in terms of completeness of the supplied PDK and the contents of design tools, finally increasing the number of software users and creating conditions for developers of customer owned tooling (COT).

Thus, we can identify the following tasks to be solved as soon as possible in order to implement a standardized approach to be used for the creation of new and revision of existing PDKs by all interested market players:

1. Simplification of the PDK creating/generating/testing process;
2. Contribution to the PDK support/maintenance;
3. Inclusion of the modules and units from multiple sources;
4. Transferring the proven designs and libraries, including the basic cells layout, during transition from one design rule to another;
5. Simplification of replacement for software used in the design process by software of similar functionality;
6. Inclusion of new software tools with new features in the design process;
7. Comparison of the strengths, weaknesses, and opportunities of different processes during selection by the developer.

5.4   Mixed Analog/Digital Circuits Design Flow

The objective of PDK standardization is to achieve maximum compatibility between the CAD packages used by IC developers and data included in PDK, to enable the standardization of software modules coming from different vendors.

The mixed custom analog-digital circuit design flow shown in Figure 5.4 describes the basic components (shaded blocks) to be included in the PDK. It is important to remember that the PDKs are not a kind of separate software and are used only in connection with specific CAD software. The proposed design flow explains the procedure for interaction of the basic PDK components with applications that use them.

The design process usually starts with developer's general ideas of functionality and characteristics of the created IC. Until this moment there is no formal project description related to the employed PDK. Then, the IC designer creates a schematic representation of the electrical circuit, a corresponding list of interconnections of its elements (netlist) and its layout, from which the additional netlist is extracted. Furthermore, the designer develops a behavioral model (either manually or using specialized model generation software modules). A tool or a verification toolkit is used to verify compliance with the terms of reference and correct implementation of a netlist, layout, and, where appropriate, behavioral model, which provides description of the target characteristics of a product as a whole [1].

It should be noted that in practical applications the same modeling software is quite often used for verification of netlist and determining the microcircuit characteristics or parameters (characterization). The purpose of characterization is to create an effective model and data describing all temporal and electrical characteristics of a circuit at some abstract level. In addition, it is possible in some cases to translate (convert) the characterization results into data formats used at the subsequent design stages.

Typically, the basic input data for circuit modeling software tools are physico-mathematical and technological models of semiconductor devices, the content of which is the main subject of standardization and unification for various CAD systems in microelectronics.

In the flowchart shown in Figure 5.4, some data have already been described using the industry standards such as the language of circuit modeling and electronic devices description in modern EDA packages (SPICE), a high-level language intended for functional and logical designing of digital circuits (e.g., Verilog). However, just selection of identical data formats will not guarantee compatibility and mutual PDK usage across multiple design packages, which poses a big problem.

Figure 5.5 shows more complete information describing the custom analog-digital IC design flow at transistor level. In particular, the blocks hatched in Figure 5.4 are presented in more detail.

At this level of detail (Figure 5.5), functions that do not comply with the selected standards are becoming more apparent and require adequate response from the designer.

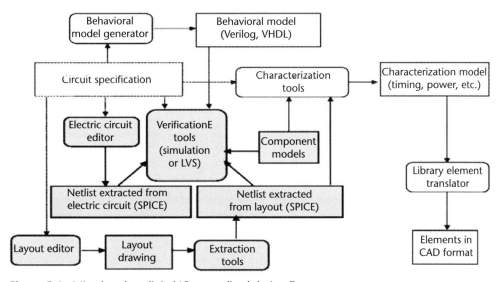

Figure 5.4   Mixed analog-digital IC generalized design flow.

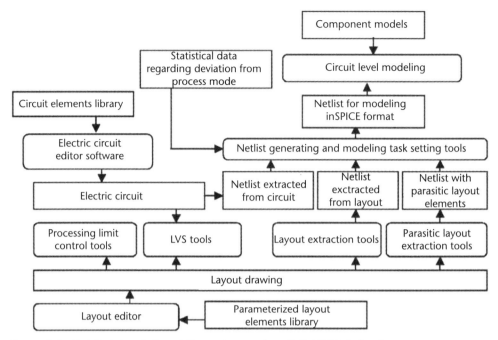

Figure 5.5 Detailed description of the custom analog-digital IMC design flow.

5.5 Generalized Information Model of a Mixed Analog-Digital IC Design

Figure 5.6 explains (in a language understandable to developers) a general approach to the device description and supported device list as part of a standard design process [2, 6].

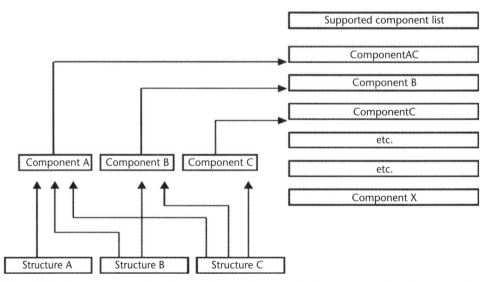

Figure 5.6 General approach to the description of designed IC and preparation of supported device list as part of design flow.

Figure 5.7 shows a general approach to the description of the designed IC.

Figure 5.8 shows the known relationship between the circuit, the layout, and the netlist, real and parasitic elements as part of the design process.

Figure 5.9 shows the ideology of the general approach to description of a specific instance of the designed device and its corresponding features and primitives needed for the IC layout design (primitive A and B, properties A–D).

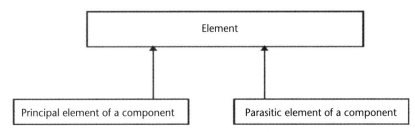

Figure 5.7   General approach to the device description problem.

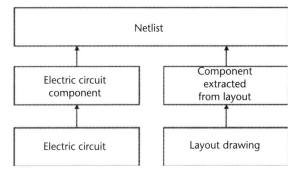

Figure 5.8   Simplified diagram of relationship between the circuit, the layout, the netlist, and list of IC components.

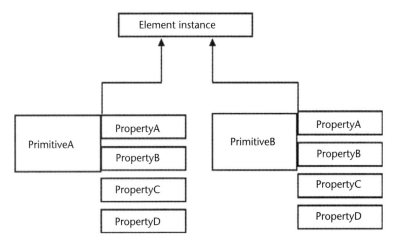

Figure 5.9   Description of the device instance and its corresponding features and primitives in the layout design flow.

Figure 5.10 shows a general approach to the abstraction or creation of a generalized semiconductor device model.

5.6   Determining the Basic PDK Components and Standard Elements List

A process design kit is a representation of a specific IC manufacturing process in the appropriate software format. The list of elements presented in Table. 5.5 is the basis of any modern PDK. The developer should have full access to all the components of such a PDK, starting with a process-specific parameters description and ending with the PDK implementation procedures within the framework of a selected IC design software package.

Table 5.6 lists the process-specific PDK components, the parameters of which should be standardized.

5.7   Features of the Digital Libraries Development for Designing Custom ICs with Submicron Design Rules

Cells of any application-specific integrated circuit (ASIC) can be divided into three groups. The first group includes IP-blocks. Such elements are designed beforehand and often represent integrated blocks that in most cases are purchased from external IP-blocks suppliers. Examples of such elements are analog (PLL, DAC), interface (USB, I2C), processors (ARM, PowerPc), and memory compilers (RAM, ROM).

The second group consists of standard cells, which are still the basic building blocks in the on-chip systems (OCS). They are used as coupling logic between multiple IPs on the same IC, as well as to create integrated end systems.

The last group of blocks components are input/output elements, which form the interface between the IC and the casing it is placed in.

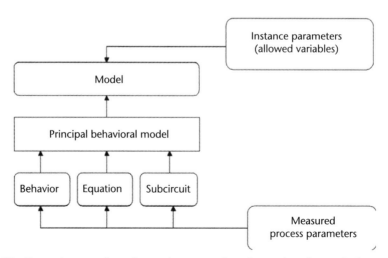

Figure 5.10   General approach to abstraction or creation of a semiconductor device model.

Table 5.5   Elements Forming the Basis of Any PDK and Source of Their Origination (Search)

Element	Description	Foundry	Software	Documentation
Process documentation				
Device specifications	Supported device list, including specifications	+		
Layoutrules	Standards for coding, layout design rules, device design features, sample projects (layouts)	+		+
Design guidelines	Guidelines describing the optimal procedure for IC design, taking into account the features of process implemented in a PDK	+		+
Device library symbols	Graphic (schematic) view that includes a list of device library element properties		+	+
Parameterized cell, p-cell	Examples of the correct layout design projects and automatic p-cells generation	+		+
Placement calculations	Indirect calculations for dependent parameters (callbacks)	+		+
Technology file	Definition of type and order of layout layers, as well as the layout description rules	+		
SPICE models	Device models, subcircuits, and behavioral patterns required to simulate the characteristics of devices included in the PDK	+	+	+
Set of rules for DRC	Verification of conformity of the designed layout to the rules set in the specification			
Set of rules for LVS	Verification of correctness and conformity of layoutrepresentation of the designed circuit to the results obtained during general-circuit simulation			
Set of rules for LPE	Extraction of netlist as well as capacitance and resistance nominal for verification of their conformity to the SOW requirements			
Set for rules RCX	Extraction (search) of new nodes and parasitic elements according ot the layout results for the designed IC			

Not so long ago, the choice of library was, in fact, the choice of technology, based on the required circuit performance, area, and cost (e.g., 0.35 or 0.25).

Within the chosen technology, there used to be only one logic cell library, and probably two I/O cell libraries. Selection of the I/O cells was performed by developer based on a tradeoff between I/O requirements and circuit solution with account being taken of the logic cells limitations: small basic element meant a lot of I/O cells; large basic element meant fewer I/O cells (Figure 5.11).

Historically, the standard cells were characterized by a very limited number of processes, voltages, and temperatures. There were only a few options available for timing models: for example, worst-case model (SS low speed, low voltage, high temperature), best-case model (FF high speed, high voltage, low temperature), and a typical case—TT. Timing parameters were used in the worst case to verify the time settings (setup) and in the best case to verify the hold time.

Table 5.6   Standard Components of a Basic PDK

Standard	Description
Devices/process types list	A set of presets covering a standards region for determinination of class (transistor, resistor, capacitor, and so on) and type (MOS transistor, poly-resistor) of components, which can be represented within the PDK standard. For example, it is not necessary to include LDMOS-transistor in PDK for digital design.
Symbolic/schematic representation and permissible dialects	The schematic symbol is used for the graphic imaging of a device. It is usually a symbol (alphabetic) indicating the device, contact location, and relative positioning of contacts for a particular class and type of device, as well as a simplified graphical representation and its permissible variations.
Device instance properties and names	These are properties and parameters relevant to the device instance included in the project and are the circuit design and layout properties. Here is a simple example: L, W for MOS devices. The standard may be expanded to include additional necessary calculations for the parameter values associated with the device location on a diagram (layout).
Layout preview (Technology file/ names and numbers of layers and their functions)	This must include a unified methodology, agreed upon by several manufacturers (foundries) for coding of layers and structures being part of the device.
Presentation of standard devices and LVS verification methods	See above.
Presentation of data type used in modeling and implementation, including subcircuits	The most complete representation of the device netlist depending on the selected modeling software. Inclusion of internal (optional) device parameters is possible, including parasitic elements and representation of the circuit in the form of device instances or a set of subcircuits.
Organization and methods of design rules structure development	Involvement of as many foundries as possible in the process of creating a standard and unified set of parameters, as well as requirements for the description of the device models, layout design rules, and so on.
Means of automatic layout binding (nodes, boundaries)	Development of standardized area of device bindings location used in the physical representation of the device, standardized for all EDA tools
Basic cells layout (templates for p-cells and parameter sets)	Standardization of a generalized set of parameterized cells (p-cell) layout for the basic types and classes of devices
PDK directories and files naming convention	Setting the list and file naming rules, as well as a list of basic and additional directories used for PDK data storage.
PDK quality control methodology	Suggest the nomenclature and test methods used to determine the degree of conformity (quality) of the tested PDK to the requirements of the standard for its development.

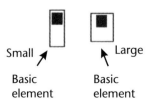

Figure 5.11   I/O cell choice principle based on core size.

Figure 5.12 shows five most commonly used typical boundary conditions: TT (NMOS—typical, PMOS—typical); SS (NMOS—slow, PMOS—slow), FF (NMOS—fast, PMOS—fast), FS (NMOS—fast, PMOS—slow), and SF (NMOS—slow, PMOS—fast).

Table 5.7 contains the main characteristics of the processes in question.

In the transition to technologies with 90 nm or less element size, and using advanced methods for supply voltage control, additional libraries have been developed that were available to any designer, making it possible to make the right choice depending on criteria such as performance, dynamic power, leakage current, occupied space, and cost. In addition, the libraries have appeared that are tailored to the specific technology parameters (such as gate insulator thickness T_{ox}, threshold voltage V_t, and so on).

In Figure 5.13 we can see a conditional positioning map in the coordinates performance-power of the various commercial radio-electronic devices. From this image we can see how the scope of the device actually determines the choice of technology. A similar map may be composed also for the electronic devices of the space vehicles. A number of libraries have become available for battery-operated products requiring power-optimized libraries. At the other end of the spectrum, we can see a number of performance-optimized libraries, such as graphics accelerators, but such devices consume maximum power. In the middle of the spectrum there are libraries that represent a tradeoff between performance and power consumption. Each kit includes cell libraries with different voltage values. This allows optimizing some IC elements in terms of power consumption and other elements in terms of speed—all within one project. Simultaneous use of both general-purpose and

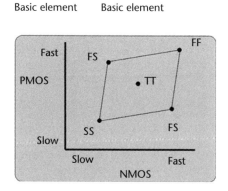

Figure 5.12   Boundary conditions of standard libraries in old technologies.

Table 5.7   Minimum Set of Boundary Conditions Parameters

Process	TT	SS	FF
Supply voltage, V	3.3	3.0	3.6
Temperature, °C	25	125	−40

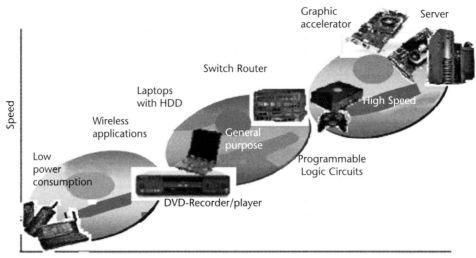

Figure 5.13   Selection of a specific library depending on the scope of application of the designed product.

low-power elements is inadmissible, since in this case the other process parameters and the level of elements will be different.

Table 5.8 shows that there is always a tradeoff between performance and leakage in the same library.

For example, the highest saturation current value (which determines the IC speed) is typical for LVt cells in the high speed cell libraries. The lowest leakage current can be obtained by selecting a cell from HVt low power consumption library. These libraries are overlapping in some way: SVt cell speed in a general-purpose library corresponds to LVt cell in a low power consumption library. In addition, from leakage data we can see a certain overlap between the cells of the general-purpose and high-speed libraries.

Table 5.8   Some Cell Library Selection Criteria for nMOS Transistors

	Cell Type	U/M	Low Power	General Purpose Cells	High Performance
v_{DD}		V	12 | 0.84	1.0	1.2
V_t	HVt	V	0.6	0.45	0.4
	SVt	V	0.5	0.35	0.35
	LVt	m	0.4	0.30	0.35
I_{dsat}	HVt	$\mu A/\mu m$	400	500	850
	SVt	$\mu A/\mu m$	500	650	950
	LVt	$\mu A/\mu m$	600	750	1000
I_{off}	HVt	$nA/\mu m$	0.1	1	10
	SVt	$nA/\mu m$	0.2	10	40
	LVt	$nA/\mu m$	0.4	80	90

It should be noted that cell libraries with low power consumption are character-ized by lower voltages in order to further reduce power consumption [1].

Table. 5.9 shows a list of some 90 nm elements from the well-known Taiwan Semiconductor Manufacturing Company (TSMC), very frequently used by domestic fabless companies to place their orders with foundries (see Section 3.6).

A number of important points should be noted for Table 5.9:

- The general purpose library is also characterized by voltage of 1.2V, provid-ing an opportunity to increase performance;
- Another library with ultra-high V_t value was added to a low-power cell library with a view to further leakage reduce.

Let us briefly review the change in number of boundary conditions required for a developer while changing to the 90-nm technology.

Usually additional boundary conditions, which are described for modern librar-ies, are present not only because of various voltages in the developing project, but due to the account being taken of the elements functioning at lower voltages.

Thus, looking at the fragment of UMC company products catalog (Table 5.10), with 90 nm design rules, we can see that there are various options available within the same library. It has some added elements with low leakage values.

Since for technologies with lower supply voltage requirements and values it is difficult to definitively specify the temperatures at which the slowest/fastest elements can be obtained, two corners (variants of boundary conditions) with ultra-low tem-peratures are added in the library.

Thus, an increasing number of cell libraries are obtained as a result of micro-circuit design, and an increasing number of such basic elements are obtained that the developers often call *corners* (from the English "core").

Table 5.9   Example of 90-nm TSMC Libraries

Process	Library Name	Supply Voltage (V)
CLN90GT	TCBN90GTHP	1.2
	TCBN90GTHPHVT	
	TCBN90GTHPLVT	
CLN90G	TCBN90GHP	1.0
	TCBN90GHPHVT	
	TCBN90GHPLVT	
	TCBN90GHPOD	1.2
	TCBN90GHPODHVT	
	TCBN90GHPODLVT	
CLN90LP	TCBN90LPHP	
	TCBN90LPHPHVT	
	TCBN90LPHPLVT	
	TCBN90LPHPUHYT	

Table 5.10   Minimum Set of Boundary Conditions for 90-nm UMC Libraries

Library Name	Process (nMOS—pMOS)	Temperature (°C)	Supply voltage (V)	Notes
TTNT1p20v	Typical—typical	25	1.2	Standard corner
SSHT1p08v	Slow—slow (SS)	125	1.08	Slow corner
FFLT1p32v	Fast—fast (FF)	−40	1.32	Fast corner
FFHT1p32v	Fast—fast (FF)	125	1.32	High loss corners
SSLT1p32v	Slow—slow (SS)	−40	1.32	Low temperature corners
SSLT1p08v	Slow—slow (SS)	−40	1.08	
Low-voltage operating conditions: same library for low voltages				
TTNT0p80v	Typical—typical	25	0.80	Standard corner
SSHT0p70v	Slow—slow (SS)	125	0.70	Slow corner
FFLT0p90v	Fast—fast (FF)	-40	0.90	Fast corner
FFHT0p90v	Fast—fast (FF)	125	0.90	High leakage corners
SSLT0p90v	Slow—slow (SS)	−40	0.90	Low temperature corners
SSLT0p70v	Slow—slow (SS)	−40	0.70	

Thanks to the latest technologies the designer is aware that an increasing number of cells becomes available to her, and this makes it possible for the design tools to select the cell with the most relevant drive signal that makes sense in terms of power consumption and performance (see Table 5.11).

The technology widely used in recent years (e.g., those ensuring 65, 45 nm design rules) provide access not only to basic logic cells and built-in memory, but also to the RF modules, nonvolatile memory cells, and so on.

Let us consider the elements that make up the basic digital libraries of the modern ICs. First of all, they are:

- Classic logic gates—AND, OR, flip-flops, drivers with different power, and so on;
- Special low-power logic elements:
  - Clock controls;
  - Multi-Vt elements;
  - Level shifters;
  - Isolators;
  - Combined level shifters and isolators;
  - Hold registers;
  - Keys;
  - Power controllers and so on.

It should be noted that modern libraries contain classic logic gates (AND, OR, flip-flops) necessary to implement the functionality of any designed product.

In addition, there are elements necessary to support low-power circuit designs:

- Clock controls are used to reduce the dynamic power in the clock circuit;
- Cells functioning at several threshold voltage values (multi-Vt) are used to trade off between increasing performance and reducing leakages;
- Level shifters are used in projects with multiple supply voltages.

Other elements are necessary for the safe cutoff of separate power domains.

The last on the list are elements with design and process variations (ECO cells)—cells without hardcoded functions that are sometimes added to the project. Typically this allows reducing the cost, if the function errors are detected after manufacturing pilot samples.

The digital circuit layout begins with routing structure. The routing grid is determined by the metal-interlayer contact principle. The width of power rails and height of elements are determined on the basis of power rails performance requirements (Figure 5.14, Table 5.11).

The cells height is measured in tracks (levels), the formation of which represents the first metallization level (M1). An element consisting of eight tracks is high enough to be able to hold eight Ml horizontal conductors. Cell libraries are developed for a certain number of tracks in height, which affects the timing and routing of the library that should be considered in the project.

Libraries with high tracks support more complex routing and transistors with higher buffer capacity and are usually configured for high performance. However, they may have higher leakage values. A library of 11 or 12 tracks is already considered a high-track library.

Libraries with low tracks are optimized for efficient use of space and are usually designed using transistors with less powerful buffers, so they are less suitable for high-speed designs.

Libraries with standard track heights are designed with a reasonable tradeoff between space efficiency and speed. These libraries are used in the majority of designs. A library of 9 or 10 levels is considered a standard track height library. In turn, the complex cells may be of double or triple height.

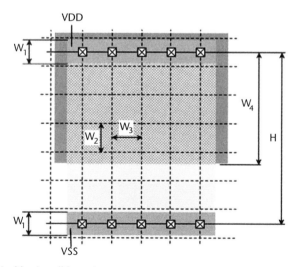

Figure 5.14   Classical logic cell layout.

Table 5.11   Typical Cell Parameters

Parameter	Symbol
Cell height (number of tracks)	H
Power rail width	W_1
Vertical grid	W_2
Horizontal grid	W_3
N-tab height	W_4

Clock controls (Figure 5.15) are widely used to reduce the clock circuit dynamic power. Synthesis tools can automatically replace the feedback multiplexer circuits with integrated clock gating (ICG). The typically used cells are multi-Vt cells, allowing us to choose between performance increase in reduction of leakage. However, it should be borne in mind that the use of multi-Vt cells requires additional masks during manufacture. This means that an increase in number of possible threshold voltages (i.e., the use of multi-Vt cells) will lead to an increase in the final cost of the chip, and this should not be disregarded. In practice, more than two V_t values are very rarely used, since the advantage in the gain obtained due to leaks and performance will grow smaller as the number of different threshold voltages increases [7].

Figure 5.16 illustrates the use of IC semiconductor areas with different supply voltages (multi-VDD). This methodology is based on moving away from the traditional use of a single fixed power rail for all parts of the device.

The main feature of this approach is division of the internal chip logic into several areas having the same voltage/power, each of these areas having its own

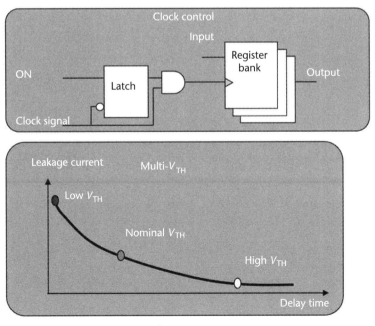

Figure 5.15   Typical use of IC clock controls.

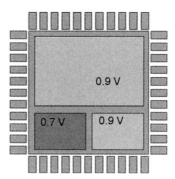

Figure 5.16   Use of areas with different voltages.

power supply. For example, the CPU speed requirements may be as high as the semiconductor technology permits. In this case, a relatively high supply voltage is required. On the other hand, a USB unit can operate at a fixed and fairly low frequency determined by the protocol used, and not by technology. In this case, the use of power rails with lower supply voltage may be sufficient to ensure that USB unit satisfies the performance requirements. The use of a rail with lower supply voltage means that its dynamic and static power will be lower.

5.8   Structural and Circuit-Related Features of Submicron ICs Library Basic Cells Design

Synthesis of even the most simple circuit solution using two different voltages presents certain difficulties for the developers, the essence of which is as follows [1].

- The synthesis of a voltage shifter device for transmission of signals between the blocks, where the rails have different power values, often requires a built-in shifter circuit (i.e., buffers transmitting the signal from one device to another and using different voltages).
- Statistical timing analysis (STA): when one power source is present, it is possible that such analysis is performed in one functional point. And this is the point the libraries are characterized for, and the standard design tools perform the analysis in normal mode. If some blocks work with different voltages and use the libraries that cannot be characterized with the precise values of the voltages used, then the STA becomes much more complicated.
- The design of the chip general layout, power grid routing. The use of methodology of several voltage domains requires more detailed design of the chip general layout, with the discretization power grid becoming more complex (Figure 5.16).

5.8.1   Voltage Level Shifters

When the problem of passing the signal between regions with different voltages is dealt with, it is necessary to use voltage level shifters. It may be particularly difficult to pass the signal from 1V to 5V region, since it is very likely that with the difference in 1V, the threshold for 5V may not be reached. However, in modern microcircuits, the internal voltages are firmly tied to the value 1V. Why are voltage level shifters required when passing the signals from 0.9V domain to 1.2V domain? The main reason is that a 0.9V signal supplied to the 1.2V gate will simultaneously activate the circuits with n-channel and p-channel transistors, resulting in the unnecessary crowbar currents.

The best solution to this problem is enabling permissible voltage ranges (as well as rise and fall times) to be transmitted to each domain. This is usually done by setting special level shifters between any domains that use different voltages. This approach reduces the problems of voltage swing and voltage domain timing delimitation, leaving the internal clock of each individual domain untouched.

At first glance, a simple switching from the output buffer to a higher voltage rail will not cause any problems. In this case there are no problems associated with short-circuits or breakdowns, or the short rise time in comparison with the topmost CMOS logic header and footer switching levels. However, certain special-purpose step-down components are necessary for secure timing closure.

HL shifters [1, 3] may be quite simple. They are basically two series-connected inverters. The HL shifter introduces only a buffer delay, so its influence on timing is insignificant.

Here, the key issue is the passing of signals from cells connected to the low voltage power rail to a cell with high supply voltage. Several design methods are known, where the buffered and inverted form of a lower voltage signal during the direct process is used to control the cross-feedback transistor structure operating at a higher voltage.

The simplest HL shifter is shown in Figure 5.17. It requires only one VDDH voltage.

The LH shifter is shown in Figure 5.18 and requires two voltages and, most frequently, a two-level system for construction of basic cells.

Additional library cells needed for the low-power design:

- Isolation cells;
- Power circuits;
- Retention flip-flops;
- Always-on buffers;
- Special pad cells.

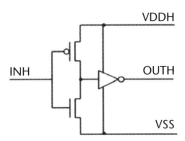

Figure 5.17   HL shifter.

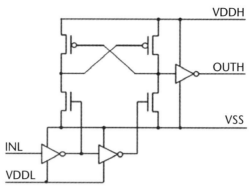

Figure 5.18   LH shifter.

5.8.2 Power Gating

In order to reduce the overall leakage current on the chip, it is highly desirable to implement mechanisms for powering down the unused blocks. This method is known as power gating. It is designed to ensure the two power modes: power-up and power-down. The task consists of switching between the two modes in appropriate time and in appropriate manner in order to maximize energy savings with minimum impact on performance.

The difficulty in the design of such power gating cells is that the voltages at the outputs of a power gated block may fall very slowly. The result of this slow process may be that the voltage at these outputs will be unstable most of the time, generating considerable crowbar currents in the always-on block. To prevent such crowbar currents from being generated, special isolators are placed between the output of a power gated block and the inputs of all circuits in the block. These isolators must be designed so that they quench the crowbar currents when the voltage on one of the inputs exceeds the threshold with the control input powered down. The power gating controller provides the isolation signal.

For some of these power gated blocks it is highly desirable to save the internal state of the block when powered down and to restore this state when powered up. This method can save a significant amount of time and power during powering-up. One way to implement this saving method is the use of save registers instead of usual flip-flops. Save registers usually have an auxiliary or shadow register, which is slower that the main register but has significantly lower leakage currents. This shadow register is always enabled and saves the contents of main register during power gating. Save registers must be notified using special signals alerting when to save the current contents of the main register in the shadow register and when to return the data back to the main register. This process is controlled by the power gated controller.

Always-on buffers in on/off blocks are used for signal routing from the active block through the inactive block to another active block. For connection of n-tubs and p-tubs with global power and ground rails, the special filler cells are used.

5.8.3 Isolator Cells (Isolators) in Submicron Microcircuits

These elements are used to isolate the switching power pads having identical voltages. Every interface of a power gated region needs to be managed. We need to be sure that powering down the region will result in crowbar current in any inputs of the powered up blocks. Also we need to be sure that none of the floating outputs of the power-down block will result in spurious behavior in the power-up blocks.

The outputs of the power gated block are the primary concern, since they can cause electrical or functional problems in other blocks. The inputs to the power gated blocks usually are not an issue—they can be driven to valid logic values by powered up blocks, without creating electrical (or functional) problems in the powered down block.

The basic approach to controlling the outputs of powered down blocks is to use isolation cells to clamp the output to a specific, legal value. There are three basic types of isolation cells, those that clamp signal to "0," those that clamp it to "1," and those that latch it to the most recent value. In most cases, it is sufficient to clamp the output to an inactive state. When using active high logic, the most common approach is to clamp the value to "0." AND-gate function accomplishes this. With active low logic, OR-gate function parks the output at logic "1." Clamp library cells are designed to avoid crowbar currents and leakage paths when the signal input floats, as long as the control input is in the appropriate (isolate) state. In addition, their synthesis models typically have extra attributes to ensure these cells never get optimized away, buffered incorrectly or inverted as part of logic optimization.

Figure 5.19 shows an example of AND-style isolation clamp-low. When the low isolation control signal ISOLN is in "1" state of logic, the transmitted signal reaches output; otherwise (the signal is in "0" state of logic), the output is clamped low.

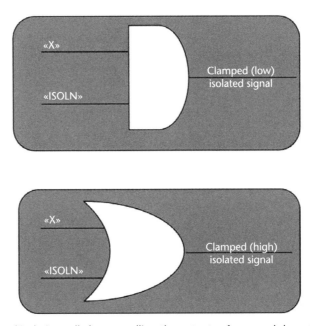

Figure 5.19   Use of isolation cells for controlling the outputs of powered down blocks.

Figure 5.19 shows an example of an OR-style isolation clamp-high. When the active high logic isolation control signal ISOL is high, the output is clamped high; when low, the signal passes to the output. These clamp gates add delay to the signals they are isolating. For some critical paths, this added delay may not be acceptable (e.g., on cache memory interfaces).

Power control library cells (sleep transistors) are used as switches to power up parts of the circuit that are in standby mode. A sleep transistor is a p- or n-channel transistor with high V_{th} (threshold voltage), which connects the constant power source with the circuit power supply, commonly referred to as virtual power supply. The p-channel sleep transistor is used for supply of V_{DD} and is called a header switch; n-channel sleep transistor controls the VSS ground rail and is called footer switch [7, 8] (Figure 5.20).

An example of footer switch layout (height as that of standard elements or double) is shown in Figure 5.21.

5.8.4 Always-On Buffers

In some cases there is some need for buffering of signals in powered-down areas. For these purposes, always-on buffers are used. In these always-on cells, the switchable VDD and/or VSS rails may have a variable value [8].

Always on VDD/VSS rails in such cells may be presented as extra inputs. During digital circuit routing, these inputs are connected to the nonswitchable power supply/ground.

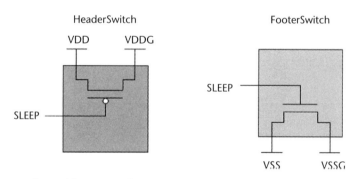

Figure 5.20 Header and footer switches.

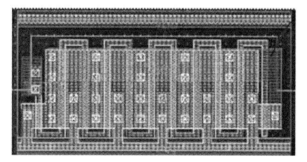

Figure 5.21 Footer switch layout implementation.

The buffering of signals in powered down blocks is used for:

- Signals passing between the active blocks, which require buffering in the powered down block;
- Power gating signals.

Always-on VDD and VSS outputs are typically characterized by the following differences:

- They are not connected directly to the power rails;
- During routing they are connected to nonswitchable power supply/ground.

Figure 5.22 shows such a conventional inverter (buffer) layout and always-on buffer. As we can see, always-on inverter is not connected to VDD_local/VSS_local.

In special filler cells, if compared to standard fillers, n-tub and/or p-tub pins are not connected to VDD/VSS supply rails. The voltage on well pins determines when the cells are in forward-bias or back-bias mode. This bias voltage is usually traced as signal output or as a special supply grid.

Special filler cells play an important role in power gating (Figure 5.23). If using header a filler cell, VNW output is connected to a VDD global rail; if using a footer switch, then the VPW output is connected to a VSS global rail. This circuit keeps the wells powered up when the region is powered down. When a header switch is used as a sleep transistor in the power gating system, the VNW output is connected to constant power supply in order to avoid the floating n-tub. Conversely, if a footer cell is used, then the VPW output is connected to always-on ground in order to prevent the floating p-tub.

It is known that the floating tubs can create a lot of problems for chip designer, such as parasitic transistors, leakage currents, or latches.

The so-called engineering change order cells (ECO) are often used as part of PDK libraries. Their main feature is that they are cells without functionality that have been added during the design (filler cells) and are used only if any problems arise after the chip is manufactured.

Their connection requires new metal masks and vias, and only in this case will such cells have the desired functionality.

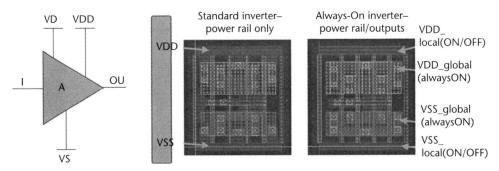

Figure 5.22 Always-on buffer implementation.

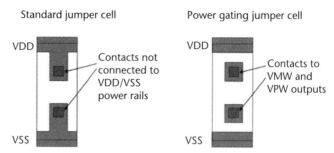

Figure 5.23   Implementation of special filler cells.

ECO cells (or sets of such cells) may implement the more complex functions AND, NAND, NOR, XOR, flip-flop, multiplexer, inverter. The principle of their use is identical to the use of reserve cells.

The authors know from their experience that ECO elements can perform some other functions associated with specific intelligence activities. They are sometimes used to perform a Trojan function when you want to deliver a chip with a back-door to the customer. But this is subject of another study, which is unrelated to the purpose of this book.

In order for the design to be completed at a low cost only by changing the metal layer, some dummy cells are added to the chip. When a functional problem is identified immediately after the manufacture of a chip, these dummy cells are converted to the functional cells by changing the metal layers. Of course, the performance of such cells will be lower than normal.

In addition, during the layout design before the first production, some other dummy cells are sometimes added, These cells are called reserve cells. They are added at the routing phase so the designers could make changes to the design at a later stage. This provides an opportunity to more accurately save the previous cell planning and routing using the spare cells, which can significantly reduce the time of release of the design for implementation.

The same library cells class includes I/O libraries. The majority of low-power I/O devices may be included in a standard I/O library (Figure 5.24). The standard I/O devices are those that do not have specific requirements for the casing, connections,

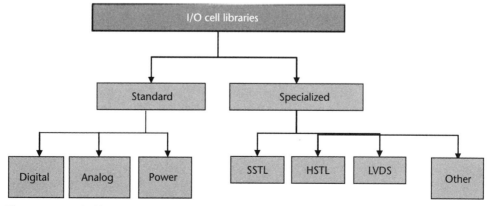

Figure 5.24   I/O libraries classification.

or signal forms, compared to special I/O devices. Standard I/O devices are divided into three main groups—digital, analog, and power. There may be multiple versions of these I/O devices (Figure 5.24).

The main features of these I/O devices are:

- The ability to manage heavy loads, as the pin regions must be able to manage several pF, in contrast to several fF loads inside the IC;
- Interface designed to operate at a different voltage, due to the use of various onboard and in-chip supply voltages;
- Low switching noise, as due to the inductance of the casing and circuit board tracks, an excessive deviation of voltage from nominal value may occur at a certain current:
- Protection against electrostatic discharge, as a person or a machine performing the assembly may accumulate a charge up to 2 kV and 500V, respectively. Such voltages will cause damage to an IC unless adequate ESD protection is ensured.

5.9   Standard PDK Data Files

A standard cell library contains multiple files. Thus, TSMC 90 nm low-power cell library contains more than 50,000 files, the main of which are:

- Physical data (LEF), used for chip planning and routing;
- Information on temporary parameters, power consumption, and functionality: LIB files are used by synthesis and layout design tools;
- Cell description at RTL register transfer level: verilog/VHDL for modeling.

Let as look at a specific example of a library exchange format (LEF) library file.

LEF includes almost all information of a typical cell as of a black box, such as layout layers, interlayer contacts, spacing, node type, and cell macros. A LEF file is, in fact, an ASCII representation of a library.

Almost all the information about the library can be found in one LEF file. However, such a file will be large and difficult to use. Instead, the information can be divided into two files: a process LEF file and a library LEF file. A process LEF file contains all the information about LEF technology for the design, such as planning and routing rules, as well as layout layers process data: the contents of LEF-file include cell geometry, output geometry, gaps, antenna effect data, and so on.

A fragment of such file looks like this:

CLASS BLOCK;
FOREIGN single_port_bbb;
ORIGIN 0 0;
SIZE 774 BY 547;
SYMMET MACRO single_port_bbb

RYXYR90;
PLNOUT
DLRECTLON INPUT;
USE SIGNAL;
PORT
LAYER M3;
RECT 420.180 625.650 420.960 625.810;
END
ANTENNAPARTIALMETALAREA 1.929 LAYER M1 ;
ANTENNAGATEAREA 0.377 LAYER M1 ;
END OUT
OBS
LAYER M1 ;
RECT 0.000 0.000 774.000 547.000;
END
END single_port_bbb

The second group of files—LIB-files—use the synthesis and layout design tools. Let us consider in more detail a specific example of a library file with .*lib* extension.

This file format is intended for modeling, synthesis, and testing. It is generated during the description of library parameters and contains data on all timing parameters and energy consumption of cells. In addition, this file contains information about the cell logical functions; propagation delays; duration of their rising and falling edges; setting, hold, deletion, and recovery times; and values of minimum pulse time, leakage power, switching power, cell space, output directions, terminals capacitance, and more.

A sample .*lib* file listing is as follows:

library (Digital_Std_Lib) {
technology (cmos);
delay_model :table_lookup;
capacitive_load_unit (l,pf);
lu_table_template(cap_tr_table) {
variable_1 :input_net_transition;
variable_2: total_output_net_capacitance;
index_1 («0.12, 0.24»);
index_2(«0.01, 0.04»);}
cell (inv) {
area: 3;
cell_leakage_power: 0.0013;

pin(OUT) {

direction: output;

function: «!IN»;

timing() {

related_pin : «IN» ;

timing_type: «combinational»;

timing_sense : «positive unate» ;

*cell_rise(cap_tr_table) { values(«l.0020, 1.1280,” *

«1.0570, 1.1660»);}

*rise_transition(cap_tr_table) { values(«0.2069, 0.3315,” *

«0.1682, 0.3062»);}

*cell_fall(cap_tr_table) {values(« 1.0720, 1.2060,” *

«1.3230, 1.4420»);}

*fall_transition(cap_tr_table) { values(«0.2187, 0.3333,” *

«0.1870,0.3117»);}}}

The microcircuit designers regret to say that the current trend is the desire to make more and more libraries because of the increased number of boundary conditions and stresses by which the libraries should be characterized. Furthermore, since there are more threshold voltages V_t, the number of available libraries has also increased. As a result, theselibraries store an increasing number of files.

Another important aspect of the libraries is the fact that the timing, noise, and power data of the designed microcircuits should be as accurate as possible, which is especially important for submicron technologies when it is not enough just to have a similar .lib format containing nonlinear delay models and power consumption models. More accurate models are required, and this can be achieved, for example, by using CCS models instead of NLDMs. Since a lot of software developers are present on the market, we can recommend that microcircuit designers use only two models: effective current source model (ECSM) by Cadence and composite current source (CCS) by Synopsys.

5.10   PDK Standard Current Source Models (CCS)

Upper level metallization has a greater resistance with reduced width of metal, resulting in interconnection resistance that is significantly higher than cell resistance. For 90 nm and less, the cell capacitance varies considerably between linear and nonlinear signal areas. In addition, the input capacitance has become a kind of function of the steepness of edge of the transmitted signal. Due to this problem, the NLDM (nonlinear) method that has been widely used by developers before is not now suitable for modeling the input port capacitance. The use of a CCS model improves the output driver and receiver modeling due to more accurate timing (Figure 5.25).

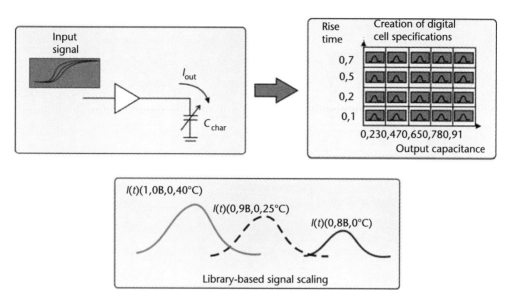

Figure 5.25 CCS model.

It should be noted that there are two similar current source model formats: ECSM and CCS. They have the following features [8]:

- They provide full description of the outputs instead of previously used rates of rise/delay.
- They provide more advanced receiver models indicating the terminal capacitances.
- They are made available by CAD vendors as extensions to existing lib-models.

The first group—CCS models—is built in the following manner. At the time of drawing-up the digital cell specification by changing various parameters (rise time, output capacitance, and so on) the output current and waveform (Figure 5.26) are measured, and all that is automatically stored in the library. Then, knowing the specific cell operating modes (load, rate of rise, process, supply voltage, temperature, and so on), the CCS models determine the necessary signals, based on the scaling, calculated by adjacent signals characterized for different conditions. What is important for IC designers is that the CCS models are scalable, which means that as boundary conditions increase, the calculations accuracy also increases and the number of the given boundary conditions is reduced considerably.

Figure 5.26 shows an example of a PDK library based on variables for nominal voltage and ±20% deviation, as well as low and high values of operating temperatures.

CCS scaling enables us to analyze circuit design for certain ranges of voltages and temperatures with fewer libraries. The analysis of solutions in the range of 1.20 to 0.80V with 5 mV increment at temperatures −40, 25, and 125°C requires only six libraries compared to 27 libraries necessary for the analysis of nonlinear delay models (NLDM) without scaling. In fact, 6 voltage-temperature (VT) combinations will be required for a complete representation of a library with several supply voltages.

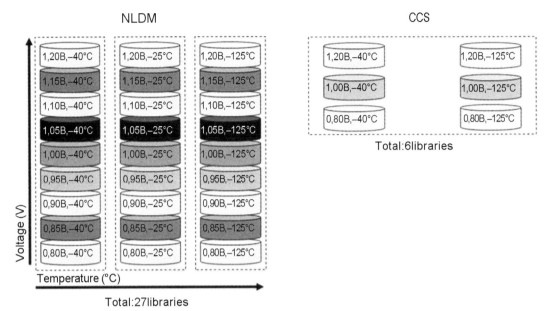

Figure 5.26 An example of CCS model use.

Of course, these boundary conditions will increase to at least 18, when taking into account the scatters of the process used by manufacturer. For this case, the number of conditions can be reduced to six, using statistical models for different process variations.

5.11 Methods and Examples of Adaptation of Standard IC Design Tools for Designing Microcircuits with 90-, 65-, 45-nm Design Rules

5.11.1 Teaching (Educational) Design Kit by Synopsys: Possibilities, Applications, Prospects

Here one will find a description of the free access educational design kit (EDK) by Synopsys, which supports the design process with 90 nm design rule and includes all the necessary components, including design rules, models, process files, verification and team extraction, scripts, symbol libraries, and parameterized cells (p-Cells). EDK also includes digital standard cell library (DSCL), which supports all modern low power device designing methods; I/O standard cell library (IOSCL); set of memories (SOM) with different volume and data width, and phase-locked loop (PLL). EDK components cover any type of project for educational and research purposes. Although EDK contains information about a particular semiconductor foundry, it allows you to implement a project for 90-nm technology with high accuracy and efficiency.

As known from the periodical press, in the era of nanometer technology universities tend to conduct the most advanced research of the highest quality in the field of IC design. In addition to EDA tools of the leading software developers, EDKs for various IC manufacturing technologies are also in demand. However, creating such

EDKs is associated with numerous difficulties, including both time-consuming development and high-complexity check of project results. However, the most important of these problems is the limitations of intellectual property (IP), imposed by the IC manufacturers, which do not allow universities to copy their technology into EDKs. That is why, in order to actually implement its marketing policy, Synopsys needed to create one open EDK, which, on the one hand, would not contain confidential information of manufacturers and, on the other hand, would have characteristics that are sufficiently close to the actual design kits provided by IC foundries.

5.11.2 Brief Overview of EDK by Synopsys

Synopsys has developed an open EDK, which contains no intellectual property restrictions and is intended to be used for research and educational purposes. This EDK is focused on programs designed to train highly qualified specialists in the field of microelectronics in various universities, educational institutions, and research centers. The EDK is also intended to support students so that they can better master modern advanced design methodology and possibilities of modern IC design tools of this company (Synopsys). The design kit from Synopsys even allows students developing various ICs using 90-nm technology and Synopsys' design tools. This design kit also allows the use of existing methods of devices designing with low power consumption, which is especially important for space applications.

Synopsys EDK includes the following components: technology kit (TK), digital standard cell library (DSCL), I/O standard cell library (IOSCL), set of memories (SOM), and phase-locked loop (PLL).

Some abstract 90-nm technology was used to develop EDK. EDK, which is described herein, does not contain the actual confidential information from any semiconductor foundry. Nevertheless, it is quite close in its characteristics to the real 90-nm technology. Using the abstract 90-nm technology has allowed Synopsys to create this open EDK, which provides a good opportunity to use it for study and research of the real characteristics of microelectronic devices with 90-nm design rule.

Let us consider a brief description of the EDK components.

The EDK is based on TK, a set of technical files necessary for the implementation of the physical representations of the project (e.g., layout). Standard technology design kit contains:

1. *Design rules*. These EDK components were created through the use of scalable CMOS process design rules of MOSIS company. They provide better portability of projects than newly developed rules for the 90-nm process, because the sizes in the rules for 90 nm can be greater by 5–20% in comparison with real foundry regulations. Figure 5.27 shows an example of the basic rules used in chip design.
2. *The rules for the topology of the microcircuit*. This part of the TK contains a description of the available basic elements and design rules of their topology. This kit contains all the components offered by the standard 90-nm technology to any factory with settings 1.2V/2.5V. Figure 5.28 shows examples of such semiconductor structures forming.

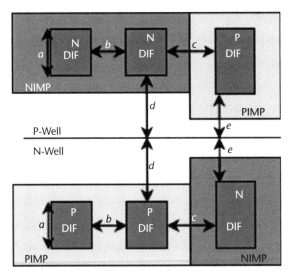

$a=0,12, b=0,14, c=0,18, d=0,24, e=0.2\mu m$

Figure 5.27   Example of the design rule in EDK.

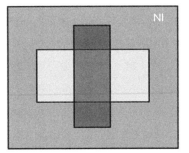

1,2V thin oxide standard n-MOS   2,5V thin oxide p-MOS

Figure 5.28   Examples of cells layout.

3. *Topological map of GDSII layers.* This part of the TK contains the names of layers and their numbers in GDSII format, used in 90-nm process. Some layers such as dummies, markers, and text, are added to the MOSIS layers map. To form a universal process, any numbers of layers can be chosen. Table 5.12 gives an example of a layers map.

4. *Technology description.* This part of TK contains approximate values of dielectric and metal layers thickness.

Table 5.12   Example of Initial Part of the Microchip Layout Layers Map

Layer #	Data Type	Tape Out Layer	Drawing or Composite Layer	Layer Name TechMAp Fle	Layer Name in DRC	Layer Name in LVS	Layer Usage Description
1	0	YES	Drawing	NWELL	NWELLi	NWELLi	NWELL
2	0	YES	Drawing	DNW	DNWi	DNWi	Deep NWELL

5. *Universal library of Spice models.* These models are based on the so-called predictive technology model [1].

The SPICE model library contains the following transistors and diodes:

- Transistors:
  a.   Devices with supply voltage of 2.5V: MOS-transistors with thick oxide layer;
  b.   Devices with supply voltage of 1.2V: MOS-transistors with thin oxide layer and typical values (high and low) of threshold voltages. For each of these devices five models of boundary conditions (corners) are determined: TT—both devices are typical; FF—both devices are fast; SS—both devices are slow; SF—slow n-MOS/fast p-MOS; FS—fast n-MOS/slow p-MOS.
- Diode (P + polysilicon resistor without silicides).

To assess the accuracy of SPICE models, the parameters of the models were scaled to 0.25-μm process technology in order to compare them with the characteristics of known 0.25-μm patterns (Figure 5.29). The number of DC transfer characteristics was received and the average curve of the common set was chosen as a typical limit value for devices with a supply voltage of 2.5V, which is close to the actual semiconductor technology.

FF, SS, SF, and FS models with their boundary conditions were formed by changing the threshold voltages (V_{th0}) and oxide thickness (t_{ox}) within the range of $\pm 5\%$.

Figure 5.30 shows the transfer characteristics for TT, FF, and SS boundary conditions based on the model of n-MOS transistor with a thin oxide.

1. *Milky way technological file.* This file contains the rules used by design tools of Synopsys.

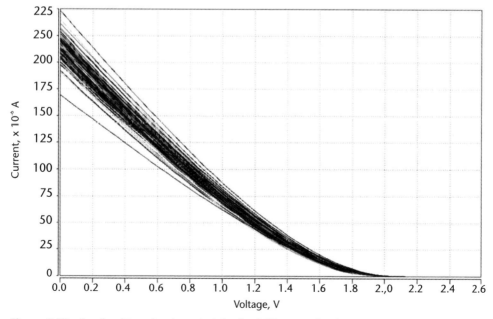

Figure 5.29   Family of transfer characteristics for 0.25-μm technology.

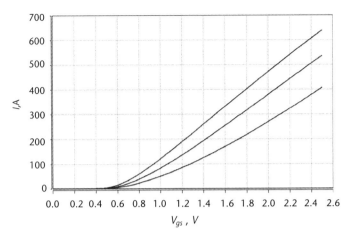

Figure 5.30   Transfer characteristics with TT, FF, and SS boundary conditions for 0.25-μm n-MOS transistor with thin oxide.

2. *The universal symbol library and parameterized cells (PCell).* Universal symbols and PCells are contained in the libraries of MOS transistors, resistors, diodes, and bipolar transistors. Parameterized cells are constructed using TCL scripting language to work in the environment of Synopsys Cosmos Schematic Editor.

3. *DRC and LVS rules.* These design rules are needed for Synopsys Hercules software to perform the design rule check and the circuitry layout comparison.

4. *Extracting files.* These files are used in Synopsys Star-RCXT software for extraction of parasitic components: ITF, TLU+, files for conversion, and batch files.

5. *Support scripts.* To accompany the design process many additional scenarios are necessary. For example, a script to convert the SPICEnetlist for a particular technology and the installation script for PCell.

5.11.3 Synopsys Digital Standard Cell Library

The digital standard cell library (DSCL) is used for designing different ICs according to 90-nm design rule using Synopsys CAD software. The DSCL was created on the basis of 1P9M 1.2V/2.5V design rules and is focused on the optimization of the main characteristics of IC design.

DSCL contains a total of 251 cells and includes typical combinational logic cells with different current carrying capacity.

DSCL library also contains all of the cells that are required for any type of low-power design [5]. These cells allow designing ICs with different internal supply voltages to minimize the dynamic consumption and leakage current (clock signal control box, noninverting delay line—0.5–2.0 ns; pass transistors; bidirectional switches, isolation cells, LH and HL shifters; retention flip-flops, power off and ground cells; always open noninverting buffers, etc.).

DSCL libraries also contain mixed cells, which complement the library. The composite current source (CCS) technology is used, which is a simulation technique for the characterization of cells to meet the requirements of modern product design methods with low power consumption. CCS technology provides an analysis of timing, noise and power consumption analysis tests for devices manufactured by nanometer technology.

To fully meet the requirements of the product design methods with low power consumption, DSCL library was characterized for 16 conditions of the process/voltage/temperature, shown in Table 5.13.

As the DSCL developers say, its functionality has also been tested with many other modeling environments. As a result, it was shown that DSCL meets all necessary requirements.

Selection of the physical structure of digital cells was performed to ensure maximum cell density in the digital designs, as well as to meet the requirements of product design methods with low power consumption. That is why there are single (Figure 5.31) and double (Figure 5.32) height structures, the parameters of which are shown in Table 5.14.

5.11.4 I/O Standard Cell Library

I/O standard cell library (IOSCL) is used to develop a variety of integrated circuits on 90-nm technology with the use of the designs by Synopsys. This library has been generated using the design rules with 90-nm 1P9M 1.2 V/2.5V, to develop

Table 5.13 Boundary Conditions of Devices Operation

Block Number	Process (NMOS/Characterized for 16 Conditions of the Process/Voltage/Temperature)		
FFHT1p32v	Typical–typical	25	1.2
TTHT1p20v	Typical–typical	125	1.2
TTNT1p20v	Typical–typical	−40	1.2
FFLT1p32v	Slow–slow	25	1.08
SSHT0p07v	Slow–slow	125	1.08
TTLT1p20v	Slow–slow	−40	1.08
SSLT0p07v	Fast–Fast	25	1.32
FFNT1p32v	Fast–fast	125	1.32
SSNT0p07v	Fast–fast	−40	1.32
SSLT1p08v	Typical–typical	25	0.8
SSNT1p08v	Typical–typical	125	0.8
SSHT1p08v	Typical–typical	−40	0.8
TTHT0p08v	Slow–slow	25	0.7
TTNT0p08v	Slow–slow	125	0.7
TTLT0p08v	Slow–slow	125	0.7
FFHT0p90v	Slow–slow	125	0.7
FFNT0p90v	Fast–fast	125	0.9
FFLT0p90v	Fast–fast	−40	0.9

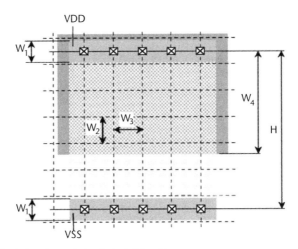

Figure 5.31   Physical structure of single-height cell.

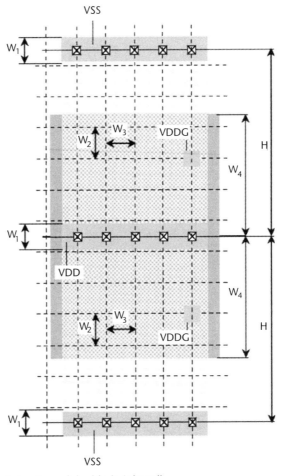

Figure 5.32   Physical structure of double-height cell.

Table 5.14   Sizes of Physical Structures

Parameter	Symbol	Value
Cell height	H	2.88 μm
Rail width	W_1	0.16 μm
Grid pitch (vertical)	W_2	0.32 μm
Grid pitch (horizontal)	W_3	0.32 μm
Height of N-tub	W_4	1.68 μm

the educational center of the Synopsys-Armenia company Synopsys-Armenia Educational Department (SAED).

IOSCL, providing a full range of standard features, contains 36 cells (including CMOS noninverting input buffer; CMOS noninverting bidirectional cell; 2/4/8/12/16 output driver tri-state, noninverting analog bidirectional no-resistor pads with ESD electrostatic discharge protection; basic power; input-output power; input-output ground pad; diode crosstalk; IOVSS to VSS; decoupling capacitors VDD and VSS, IOVDD, and IOVSS; damage cell, filler cell; and pad). CCS modeling technology has been used for the characterization of IOSCL. Moreover, all cells have dimensions of 65×300 μm or less.

5.11.5 Standard Set of PDK Memory Modules

The set of memories (SOM) was developed by Synopsys using technological process SAED 90 nm 1P9M 1.2 V/2.5V. It contains a number of static RAM (SRAM) with a small number of words (word depth—m) and bits per word (data width—n). All SRAM included in SOM are synchronous dual-port SRAM which are write enable, output enable, and chip enable on each port. In addition, SOM includes 16 SRAM blocks with the same architecture but different nxm size ratios (width × depth), where n = 4, 8, 16, 32, and m = 16, 32, 64, 118. Under these synchronous dual-port nxm-SRAMs have two ports (primary and double) for the same memory cell. From both ports, one can independently gain access to read or write operations.

5.11.6 Phase-Locked Loop

Phase-locked loop (PLL) is actually a clock multiplier that should generate a stable, high-speed clock from the slower clock. It was designed using 90-nm process SAED 1P9M 1.2V/2.5V. PLL has three operating modes: normal, with external feedback, and in bypass mode. The external feedback mode reverse input clock is aligned in phase with the input clock signal. These aligned clocks allow you to remove clock delay and phase shift of the timing between the devices. In bypass mode, the reference clock signal is shunted on the output.

5.11.7 The Geography of Application and Prospects of EDK

Currently Synopsys EDK is used for educational and research purposes. EDK is used in almost all institutes and universities around the world, which offer courses in chip design, including Syracuse University (New York), University of California Extension (Santa Cruz), Purdue University (Indiana), Oregon State University (Corvallis), California State University, Northridge (Los Angeles), Silicon Valley Technical Institute (San Jose), University of California (San Diego), San Francisco State University (San Francisco), University of Tennessee (Knoxville), Kate Gleason College of Engineering (New York), Rochester Institute of Technology (New York), Indian Institute of Technology Kanpur (Kanpur, India), Armenian State University of Technology (Yerevan, Armenia), Yerevan State University (Yerevan, Armenia), State Armenian-Russian Slavic University (Yerevan, Armenia), and Moscow Institute of Electronic Technology (Moscow, Russia).

EDK is also used in a number of famous centers of learning, including Synopsys' Customer Education Services, Synopsys' Corporate Application Engineering team, and Sun Microsystems.

There are many examples of how EDK promotes education in the field of microelectronics. All universities are included in the programs of the Synopsys company on cooperation in the field of microelectronics design education [6], using the new learning system developed by Synopsys [7]. This learning system is used in laboratory work, course and diploma projects, and master's and Ph.D. theses [8].

To keep pace with the proposals from the industry, the new versions of EDK are designed for 65- and 45-nm design rules. These versions are developed using the same methods, and they have roughly the same functionality as 90-nm EDK. Currently, these are new cells package SOM, particularly for memory cells that have already been used in the prior known projects (such as OpenSPARC project).

5.12   Contents of Educational Design Kits Provided by IMEC

In order to teach students within the framework of cooperation programs (EURO-PRACTICE and others), Interuniversity Microelectronics Center (IMEC), Eindhoven, Belgium, and Microelectronics Training Center (MTC) may provide the following PDKs:

- 130-nm CMOS mixed-signal RF general purpose;
- 130-nm CMOS logic general purpose;
- 90-nm CMOS mixed-signal low power RF;
- 90-nm CMOS mixed-signal general purpose RF;
- 90-nm CMOS logic low power;
- 90-nm CMOS logic general purpose.

Table 5.15 shows the basic characteristics of TSMC 90-nm digital libraries intended for design of low-power ICs. Such libraries (PDK 90-nm CMOS logic low power), provided by EROPRACTICE, are available for educational and research purposes.

It is recommended to use these libraries while learning digital ICs design flow assisted by Cadence software within academic subjects such as basics of CAD use in microelectronics, IC layout design, and basics of IC schematic design.

Table 5.15   Components of Libraries with IMEC Design Kits

Library	Cell Type	Name	Description
TCBN90LPHDBWP	Standard cell	90-nm low-power 1.2V/2.5V standard cell library, high-density, characterized for 1V, release 200f	90-nm logic 1.2V/2.5V low power process (1P9M, core 1.2 V), low Vt, 7-track library. 0.28-μm x-pitch, total 645 cells (include 620 base cells, 9 level shifter cells, and 1 tapcell), raw gate density = 560 Kgate/mm^2, support multi-Vdd design, low-voltage range is 1.0V +/– 10%
TCBN90LPHDBWPHVT	Standard cell	90-nm low-power 1.2V/2.5V standard cell library, high-density, high Vt, characterized for 1V, release 200f	90-nm logic 1.2V/2.5V low power process (1P9M, core 1.2V), high Vt, 7-track library. 0.28-μm x-pitch, total 645 cells (include 620 base cells, 9 level shifter cells, and 1 tapcell), raw gate density = 560 Kgate/mm^2, support multi-Vdd design, low-voltage range is 1.0V +/– 10%
TCBN90LPHDBWPLVT	Standard cell	90-nm low-power 1.2V/2.5V standard cell library, high-density, low Vt, characterized for 1V, release 200f	90-nm logic 1.2V/2.5V low power process (1P9M, core 1.2V), low Vt, 7-track library. 0.28-μm x-pitch, total 645 cells (include 620 base cells, 9 level shifter cells, and 1 tapcell), raw gate density = 560 Kgate/mm^2, Support multi-Vdd design, low-voltage range is 1.0V +/– 10%
TCBN90LPHP	Standard cell	90-nm low-power 1.2V/2.5V standard cell library, high-performance, release 150j	90-nm logic low-power process 1.2V/2.5V (1P9M,core 1.2V), 0.28-μm x-pitch, nominal-Vt, total 867 cells (include filler cells), 9-track library, raw gate density = 436 KGate/mm^2, support multi-Vdd design (include level shifter cell and isolation cell inside)
TCBN90LPHPCG	Standard cell	Coarse-grain 90-nm low-power 1.2V standard cell library, high-performance, release 150e	Coarse-grain MTCMOS library, standard-Vt, total 20 cells (include special cells): power switch header cell; front-sync trigger cell; always on cell
TCBN90LPHPHVT	Standard cell	90-nm low-power 1.2V/2.5V standard cell library, high-performance, high Vt, release 150j	90-nm low-power process (1P9M,1.2V/2.5V), 0.28-μm x-pitch, high-Vt, total 867 cells (include filler cells), 9-track library, raw gate density = 436 KGate/mm^2, support multi-Vdd design (include level shifter cell and isolation cell inside)
TCBN90LPHPHVTCG	Standard cell	Coarse-grain 90-nm low-power 1.2V standard cell library, high-performance, high Vt, release 150d	90-nm low-power process (1P9M,core 1.2V), coarse-grain MTCMOS library, high-Vt, 20 cells (include special cells): power switch header cell; front-sync trigger cell; always on cell

(continues)

Table 5.15 Components of Libraries with IMEC Design Kits *(Cont.)*

Library	Cell Type	Name	Description
TCBN90LPHPHVTWB	Standard cell	90-nm low-power 1.2V/2.5V standard cell library, high-performance, high Vtwith bias, release 150d	90-nm low-power process (1P9M,1.2V/2.5V), 0.28-μm x-pitch, high-Vt, total 845 cells (base cell: 805 cells, ECO cell: 32 cells, 7 filler cells + 1 tap cell), 9-track library, bias voltage = 0.6V, raw gate density = 451 Kgate/mm^2
TCBN90LPHPLVT	Standard cell	90-nm low-power 1.2V/2.5V standard cell library, high-performance, low Vt, release 150j	90-nm low-power process (1P9M,1.2V/2.5V), 0.28-μm x-pitch, low-Vt, total 867 cells(include filler cells), 9-track library, raw gate density = 436 KGate/mm^2, support multi-Vdd design (include level shifter cell and isolation cell inside)
TCBN90LPHPLVTCG	Standard cell	90-nm coarse-grain low-power 1.2V standard cell library, high-performance, low Vt, release 150d	90-nm Low-power process (1P9M, core 1.2v), coarse-grain MTCMOS library, low-Vt, 20 cells (include special cells): power switch header cell; front-sync trigger cell; always on cell
TCBN90LPHPLVTWB	Standard cell	90-nm low-power 1.2V/2.5V standard cell library, high-performance, low Vt with bias, release 150c	90-nm low-power process (1P9M,1.2V/2.5V), 0.28-μm x-pitch, low-Vt, total 845 cells (base cell: 805 cells, ECO cell: 32 cells, 7 filler cells + 1 tap cell), 9-track library, bias voltage = 0.6V, raw gate density = 451 Kgate/mm^2
TCBN90LPHPUD	Standard cell	90-nm low-power 1.2V/2.5V standard cell library, high-performance, standard Vt, release 200a	90-nm low-power process (1P9M,1.2V/2.5V), 0.28-μm x-pitch, standard-Vt, total 837 cells (+ 7 filler cells), 9-track library, raw gate density = 436Kgate/mm^2. 1.0V (1.0V ± 10%)
TCBN90LPHPUDHVT	Standard cell	90-nm low-power 1.2V/2.5V standard cell library, high-performance, high-Vt, release 200a	90-nm low-power process (1P9M,1.2V/2.5V), 0.28-μm x-pitch, high-Vt, total 837 cells (+ 7 filler cells), 9-track library. Raw gate density = 436Kgate/mm^2. 1.0V (1.0V ± 10%)
TCBN90LPHPUDLVT	Standard cell	90-nm low-power 1.2V/2.5V standard cell library, high-performance, low-Vt, release 200b	TSMC 90-nm low-power process (1P9M,1.2V/2.5V), 0.28-μm x-pitch, standard-Vt, total 837 cells (+ 7 filler cells), 9-track library, raw gate density = 436 Kgate/mm^2. 1.0V (1.0V ± 10%)
TCBN90LPHPULVT	Standard cell	90-nm low-power 1.2V/2.5V standard cell library, high-performance, ultra lowVt, release 200a	90-nm low-power process (1P9M,1.2V/2.5V), 0.28-μm x-pitch, ultra low Vt, total 867 cells (include filler cells), 9-track library, raw gate density = 436 Kgate/mm^2, supports multi-vdd design (include level shifter cell and isolation cell inside)

Table 5.15 Components of Libraries with IMEC Design Kits *(Cont.)*

Library	Cell Type	Name	Description
TCBN90LPHPWB	Standard cell	90-nm low-power 1.2V/2.5V standard cell library, high-performance, with bias, release 150c	90-nm low-power process (1P9M,1.2V/2.5V), 0.28-μm x-pitch, nominal Vt, total 845 cells (base cell: 805 cells, ECO cell: 32 cells, 7 filler cells + 1 tap cell), 9-track library, bias voltage = 0.6V, raw gate density = 451 Kgate/mm^2
tpan90lpnv2	Analog standard I/O	90nm Low-power 1.2V/2.5V standard I/O library, analog I/O, release 200a	N90LP, 1.2V/2.5V, universal analog standard I/O
tpan90lpnv3	Analog standard I/O	90-nm low-power 1.2V/3.3V universal analog I/O compatible with linear universal standard I/O, release 210a	1.2V/3.3V, universal analog I/O compatible with linear universal standard I/O
tpdn90lpnv2	Digital standard I/O	1.2V/2.5V, regular, linear universal standard I/O library, release 200c	1.2V/2.5V, regular, linear universal standard I/O
tpdn90lpnv3	Digital standard I/O	1.2V/3.3V, regular, linear universal standard I/O library, release 210b	1.2V/3.3V, regular, linear universal standard I/O
tpbn90gv	Standard I/O	Standard I/O bond pad library, release 140a	Standard I/O bond pad library, release 140a

References

[1]   Belous, A. I., V. A. Emelyanov, and A. S. Turtsevich, *Fundamentals of Circuit Design of Microelectronic Devices*, M.: Technosfera, 2012.

[2]   OK Technical Committee, "Design Objectives Document," Wolfgang Roethig, Ed., October 2003.

[3]   Belous A. I., A. S. Turtsevich, and S. A. Efimenko, "Fundamentals of Design and Application of Microelectronic Devices for Power Electronics," Gomel F. Skaryna State University, 2013.

[4]   Belous, A. I., and V. A. Solodukha, "Fabless Business Model in a Foundry: Myths and Reality," *Components and Technologies*, No. 8, 2012, pp. 14–18.

[5]   Keating, M., et al., *Low Power Methodology Manual for System on Chip Design*, Synopsys, Inc. &v ARM Limited, New York: Springer, 2007.

[6]   CCS Timing Technical White Paper Version 2.0, 12/20/06.

[7]   CCS Power Technical White Paper Version 3.0, 24/08/06.

[8]   CCS Noise Technical White Paper Version 1.2, 12/01/06.

List of Acronyms and Abbreviations

AFT   algorithmic functional tests

ALU   arithmetic logic unit

BE   basic elements

BiCMOS LSIC   bipolar complementary metal-oxide-semiconductor large scale integrated circuit

BIT   burn-in test

BLE   basic logical element

CCD   charge-coupled devices

CE   coordination (matching) elements

CF   catastrophic failure

CMOS IC   complementary metal-oxide semiconductor integrated circuit

DC   design center

DM   dielectric-metal

DS   dielectric-semiconductor

EAS   Earth artificial satellite

ECB   electronic component base

ECL   emitter-coupled logic

ECS   electrostatic charge

EEPROM   electrically erasable programmable read only memory

EMR   electromagnetic radiation

ENPO   experimental scientific and production enterprise

EPROM   erasable programmable read only memory

ERS   Earth remote sensing

FC   functional control

FFL   functional failure level

FFT   fast Fourier transform

FOL   faultless operation level

FSS   fast surface states

HUP   heavy uncharged particles

I2L   integrated injection logic

IC   integrated circuit

IIL   information integrity level

IIR   impulse ionizing radiation

IOA   integrated operational amplifier

IR   ionizing radiation

ISS   International Space Station

IVC   integrated voltage comparator

LE   logical element

LSIC   large-scale integrated circuit

LV   launch vehicle

MC   memory cell

MCC   minority charge carrier

MDS   metal-dielectric-semiconductor

ME   memory elements

MNOS   metal-nitride-oxide-semiconductor

MOS   metal-oxide-semiconductor

MOST   MOS-transistor

OA   operating amplifiers

ODC   onboard digital computer

OT   outage time

PE   protection elements

PROM   programmable read only memory

RAM   random access memory

RC   reentry capsule

RD   radiation defects

REE   radio and electronic equipment

ROM   read-only memory

RT   radiation tolerance

SC   Spacecraft

SCOA   spacecraft onboard equipment

SDP   superdeep penetration

SEF   special exposure factors

SIR   steady ionizing radiation

SOI   silicon-on-insulator

SOS   silicon-on-sapphire

SPA   Scientific-Production Association

SRAM   static random-access memory

SS   surface states

STC   scientific and technical center

STTL   Schottky transistor-transistor logic

SXR   superhard X-ray

SV   space vehicle

TE   thyristor effect

TR   technical requirements

TTL   transistor-transistor logic

V-C curve   capacitance-voltage curve

VAC   volt-ampere characteristic

VC   voltage comparators

VLSIC   very large-scale integrated circuit

VS   voltage stabilizer

VTC   voltage temperature coefficient

About the Authors

Anatoly Belous is a member of the National Academy of Sciences of the Republic of Belarus, and an academician, doctor of technical sciences, professor, laureate of the State Award of the Republic of Belarus, and Meritorious Inventor of the Republic of Belarus.

In 1973, he graduated from the Minsk Radio-Technical University in the specialty of electronic engineering. He was a chief design engineer of more than 70 projects in the development of microelectronic devices for the space rocket industry, including for the carrier rockets *Energy*, *Angara*, and *Proton*, the ballistic missile *Satan* (SS-18), space vehicles in the Quantum series, the orbital station Mir, the shuttle *Buran*, and Earth remote scanning satellites of the Kanopus series for the international orbital station.

He was awarded the USSR's Gold Medal for Labor Valor. He has more than 300 scientific publications, and more than 150 patents, 18 monographies, and 5 tutorials. He is a member of the editorial boards of four science journals, chairman of the state experts council of the Republic of Belarus on microwave electronics, photonics, microelectronics, and nanoelectronics.

He is deputy chairman of the program-drafting committee of two International Annual Conferences on Microelectronics.

For a number of years, he conducted special courses of lectures on space electronics in the technical universities of Russia, Belarus, China, India, Bulgaria, Vietnam, Poland, and Ukraine. He delivers regular reports on space electronics at conferences, seminars, and working meetings in the Russian state space corporation Roskosmos.

He is the editor in chief of this tome.

Vitali Saladukha is general director of INTEGRAL holding (Minsk), the largest semiconductor manufacturer in the territory of the former USSR, which includes design centers, semiconductor factories, and assembly lines. In 1980, he graduated Belarussian State University (Minsk) with a bachelors degree in radiophysics and electronics. Currently, he is a Ph.D. Laureate of the State Award of the Republic of Belarus.

He specializes in the area of design development and industrial applications of microelectronic technologies. He is the author of more than 100 science publications, among them seven books and two tutorials. He is a scientific research manager for a number of major science programs involving microelectronics development for space applications. His new technological solutions in the area of microelectronics are protected by 50 patents. His best-known books include *Base Technological Processes for Fabrication of the Semiconductor Devices and Integrated Circuits on Silicon* in three volumes and *Fundamentals in Technology of Packaging the Integrated Circuits and Semiconductor Devices*.

Siarhei Shvedau is director of the Design Center of the holding Integral (Minsk). In 1980, he graduated from the Minsk Radio-Technical University in the specialty of physics of semiconductors. He is a Laureate of the State Award of the Republic of Belarus.

He specializes in the area of design and production implementation of radiation-resistant element base for the nuclear and space industries. He was the chief design engineer on more than 30 projects, is the author of more than 100 science publications, and has more than 20 patents in the area of creation of radiation-resistant integrated circuits.

Index